THE HANDBOOK OF

OPTICAL COMMUNICATION NETWORKS

The Electrical Engineering Handbook Series

Series Editor
Richard C. Dorf
University of California, Davis

Titles Included in the Series

The Handbook of Ad Hoc Wireless Networks, Mohammad Ilyas
The Avionics Handbook, Cary R. Spitzer
The Biomedical Engineering Handbook, Second Edition, Joseph D. Bronzino
The Circuits and Filters Handbook, Second Edition, Wai-Kai Chen
The Communications Handbook, Second Edition, Jerry Gibson
The Computer Engineering Handbook, Vojin G. Oklobdzija
The Control Handbook, William S. Levine
The Digital Signal Processing Handbook, Vijay K. Madisetti and Douglas Williams
The Electrical Engineering Handbook, Second Edition, Richard C. Dorf
The Electric Power Engineering Handbook, Leo L. Grigsby
The Electronics Handbook, Jerry C. Whitaker
The Engineering Handbook, Richard C. Dorf
The Handbook of Formulas and Tables for Signal Processing, Alexander D. Poularikas
The Handbook of Nanoscience, Engineering, and Technology, William A. Goddard, III,
   Donald W. Brenner, Sergey E. Lyshevski, and Gerald J. Iafrate
The Handbook of Optical Communication Networks, Mohammad Ilyas and
   Hussein T. Mouftah
The Industrial Electronics Handbook, J. David Irwin
The Measurement, Instrumentation, and Sensors Handbook, John G. Webster
The Mechanical Systems Design Handbook, Osita D.I. Nwokah and Yidirim Hurmuzlu
The Mechatronics Handbook, Robert H. Bishop
The Mobile Communications Handbook, Second Edition, Jerry D. Gibson
The Ocean Engineering Handbook, Ferial El-Hawary
The RF and Microwave Handbook, Mike Golio
The Technology Management Handbook, Richard C. Dorf
The Transforms and Applications Handbook, Second Edition, Alexander D. Poularikas
The VLSI Handbook, Wai-Kai Chen

Forthcoming Titles

The CRC Handbook of Engineering Tables, Richard C. Dorf
The Engineering Handbook, Second Edition, Richard C. Dorf

THE HANDBOOK OF

OPTICAL COMMUNICATION NETWORKS

EDITED BY

Mohammad Ilyas
Hussein T. Mouftah

CRC Press
Taylor & Francis Group
Boca Raton London New York

CRC Press is an imprint of the
Taylor & Francis Group, an **informa** business

CRC Press
Taylor & Francis Group
6000 Broken Sound Parkway NW, Suite 300
Boca Raton, FL 33487-2742

First issued in paperback 2019

© 2003 by Taylor & Francis Group, LLC
CRC Press is an imprint of Taylor & Francis Group, an Informa business

No claim to original U.S. Government works

ISBN-13: 978-0-8493-1333-2 (hbk)
ISBN-13: 978-0-367-39526-1 (pbk)

Library of Congress Card Number 2002041927

Library of Congress Cataloging-in-Publication Data

The handbook of optical communication networks / Mohammad Ilyas, Hussein T. Mouftah [editors].
    p.   cm. -- (Electrical engineering handbook series ; 30)
    Includes bibliographical references and index.
    ISBN 0-8493-1333-3 (alk. paper)
    1. Optical communications. I. Ilyas, Mohammad, 1953 - II. Mouftah, Hussein T. III.
Series.

    TK5103.59 .H365 2003
    621.382'7—dc21
                                  2002041927

Visit the CRC Press Web site at www.crcpress.com

Visit the Taylor & Francis Web site at
http://www.taylorandfrancis.com

and the CRC Press Web site at
http://www.crcpress.com

Preface

During the last 3 decades, the field of telecommunications has witnessed tremendous growth. Proliferation of the Internet has started a true revolution that is expected to continue through the foreseeable future. Three factors have played major roles in the unprecedented growth of this field:

- Users' incessant demand for high-speed communication facilities for heavy-duty applications such as rich-content video
- Availability of high-speed transmission media such as optical fibers
- Availability of high-speed hardware such as high-resolution video cameras and high-speed processors

These factors are leading towards an integrated high-speed (and high-bandwidth) communication environment where all communication needs will be supported by a single communication network. The latest trends indicate that bandwidth needs double every 100 days. The volume of data traffic has surpassed the volume of voice traffic. Such a monumental demand for bandwidth can only be met by using optical fiber as transmission media. Other bottlenecks such as bringing fiber to the desktop, or to the home, still exist. However, eventually these obstacles will be overcome. Emerging optical communication networks represent a step in that direction.

The Handbook of Optical Communication Networks is a source of comprehensive reference material for such networks. The material presented here is intended for professionals in the communications industry who are designers and/or planners for emerging telecommunication networks, researchers (faculty members and graduate students), and those who would like to learn about this field.

The handbook is organized in the following seven parts:

- Introduction and optical networks architectures
- Protocols for optical network architectures
- Resource management in optical networks
- Routing and wavelength assignment in WDM networks
- Connection management in optical networks
- Survivability in optical networks
- Enabling technologies for optical networks

Each part consists of 2 to 5 chapters dealing with the topic, and the handbook contains a total of 21 chapters. Although this is not precisely a textbook, it can certainly be used as one for graduate and research-oriented courses that deal with optical communication networks. Any comments from readers will be highly appreciated.

Many people have contributed to this handbook in their unique ways. The first and the foremost group that deserves immense gratitude are the highly talented and skilled researchers who have contributed the 21 chapters to this handbook. All have been extremely cooperative and professional. It has also been a pleasure to work with Nora Konopka, Helena Redshaw, and Amy Rodriguez of CRC Press, and we are extremely grateful for their support and professionalism. Our families have extended their unconditional love and strong support throughout this project, and they all deserve very special thanks.

Mohammad Ilyas
Boca Raton, Florida

Hussein T. Mouftah
Ottawa, Ontario, Canada

About the Editors

Dr. Mohammad Ilyas earned his B.Sc. degree in electrical engineering from the University of Engineering and Technology, Lahore, Pakistan, in 1976. From March 1977 to September 1978, he worked for the Water and Power Development Authority, Pakistan. In 1978, Dr. Ilyas was awarded a scholarship for his graduate studies, and he completed his M.S. degree in electrical and electronic engineering in June 1980 at Shiraz University, Shiraz, Iran. In September 1980, he joined the doctoral program at Queen's University in Kingston, Ontario, Canada. He earned his Ph.D. degree in 1983. His doctoral research was about switching and flow control techniques in computer communication networks. Since September 1983, Dr. Ilyas has been with the College of Engineering at Florida Atlantic University, Boca Raton, where he is currently Associate Dean for Graduate Studies and Research. From 1994 to 2000, he was Chair of the department. During the 1993–94 academic year, he was on his sabbatical leave with the Department of Computer Engineering, King Saud University, Riyadh, Saudi Arabia.

Dr. Ilyas has conducted successful research in various areas including traffic management and congestion control in broadband/high-speed communication networks, traffic characterization, wireless communication networks, performance modeling, and simulation. He has published one book and over 130 research articles. He has supervised 10 Ph.D. dissertations and 32 M.S. theses to completion and has been a consultant to several national and international organizations.

Dr. Ilyas is a senior member of IEEE and an active participant in several IEEE technical committees and activities.

Hussein Mouftah joined the School of Information Technology and Engineering (SITE) of the University of Ottawa in September 2002 as a Canada research chair (Tier 1) professor in optical networks. He was previously full professor and department associate head of the Department of Electrical and Computer Engineering at Queen's University (1979–2002). He has three years of industrial experience, mainly at Bell Northern Research of Ottawa (now Nortel Networks) (1977–1979). He also spent three sabbatical years at Nortel Networks (1986–1987, 1993–1994, and 2000–2001), conducting research in the areas of broadband packet switching networks, mobile wireless networks, and quality of service over the optical Internet. He served as editor-in-chief of the *IEEE Communications Magazine* (1995–1997) *IEEE Communications Society* director of magazines (1998–1999), and chair of the Awards Committee (2002–2003). Dr. Mouftah is the author or co-author of 4 books, 17 book chapters, and more than 650 technical papers and 8 patents in this area. He is the recipient of the Association of Professional Engineers of Ontario (PEO) 1989 Engineering Medal for Research and Development, and the Ontario Distinguished Researcher Award of the Ontario Innovation Trust. He is the joint holder of the Best Paper Award for a paper presented at SPECTS'2002, and the Outstanding Paper Award for papers presented at the IEEE HPSR'2002 and the IEEE ISMVL'1985. He is also a joint holder of an honorable mention for the Frederick W. Ellersick Price Paper Award for best paper in the *IEEE Communications Magazine* in 1993. He is the recipient of the IEEE Canada (Region 7) Outstanding Service Award (1995). Dr. Mouftah is a Fellow of the IEEE (1990).

Contributors

Mohammed A. Alhaider
Electrical Engineering Department
King Saud University
Riyadh, Saudi Arabia

Mohamed A. Ali
Department of Electrical
  Engineering
City College of the City University
  of New York
New York, New York

Toshit Antani
Department of Electrical and
  Computer Engineering
University of California, Davis
Davis, California

Chadi Assi
Department of Electrical
  Engineering
City College of the City University
  of New York
New York, New York

David Benjamin
Nortel Networks
St. Laurent
Quebec, Canada

Imrich Chlamtac
Department of Electrical
  Engineering
University of Texas at Dallas
Dallas, Texas

Shirshanka Das
Department of Computer Science
University of California,
  Los Angeles
Los Angeles, California

W.R. Franta
GATX Capital
San Francisco, California

Aysegül Gençata
Department of Computer
  Engineering
Istanbul Technical University
Istanbul, Turkey

Mario Gerla
Department of Computer Science
University of California,
  Los Angeles
Los Angeles, California

Peter Green
Nortel Networks
Ottawa, Ontario, Canada

Mounir Hamdi
Department of Computer Science
Hong Kong University of Science
and Technology
Kowloon, Hong Kong

Pin-Han Ho
Department of Electrical
and Computer Engineering
University of Waterloo
Waterloo, Ontario, Canada

Mohammad Ilyas
Department of Computer
Science and Engineering
Florida Atlantic University
Boca Raton, Florida

Tariq Iqbal
City of Riviera Beach
West Palm Beach, Florida

Jason P. Jue
Department of Computer Science
University of Texas at Dallas
Dallas, Texas

Hussein T. Mouftah
School of Information
Technology and Engineering
University of Ottawa
Ottawa, Ontario, Canada

Biswanath Mukherjee
Department of Computer Science
University of California, Davis
Davis, California

C. Siva Ram Murthy
Department of Coputer Science and
Engineering
India Institute of Technology
Madras, India

Kanna Potharlanka
Department of Electrical
and Computer Engineering
University of California, Davis
Davis, California

M. Yasin Akhtar Raja
Physics and Optical
Science Department
University of North
Carolina, Charlotte
Charlotte, North Carolina

Byrav Ramamurthy
Department of Computer
Science and Engineering
University of Nebraska, Lincoln
Lincoln, Nebraska

Matthew N.O. Sadiku
Department of Electrical
Engineering
Prairie View A&M University
Prairie View, Texas

Laxman Sahasrabuddhe
Department of Computer Science
University of California, Davis
Davis, California

Chava Vijaya Saradhi
Department of Computer
Science and Engineering
Indian Institute of Technology
Madras, India

Abdallah Shami
Department of Electrical
Engineering
Lakehead University
Thunder Bay, Ontario, Canada

Narendra Singhal
Department of Computer Science
University of California, Davis
Davis, California

Emmanuel A. Varvarigos
Department of Computer
  Engineering and Informatics
University of Patras
Patras, Greece

Theodora Varvarigou
Department of Electrical
  and Computer Engineering
National Technical University
  of Athens
Athens, Greece

Evangelos Verentziotis
Department of Electrical
  and Computer Engineering
National Technical University
  of Athens
Athens, Greece

Hooman Yousefizadeh
Department of Electrical
  Engineering
Florida Atlantic University
Boca Raton, Florida

Xiaohong Yuan
Computer Science Department
North Carolina Agricultural
  and Technical State University
Greensboro, North Carolina

Ding Zhemin
Department of Computer Science
Hong Kong University of Science
  and Technology
Kowloon, Hong Kong

Jun Zheng
School of Information Technology
  and Engineering
University of Ottawa
Ottawa, Ontario, Canada

Bin Zhou
Department of Electrical
  and Computer Engineering
Queen's University
Kingston, Ontario, Canada

Ali Zilouchian
Department of Electrical
  Engineering
Florida Atlantic University
Boca Raton, Florida

Contents

chapter one

Overview of optical communication networks: Current and future trends*

Aysegül Gençata
Istanbul Technical University
Narendra Singhal
University of California, Davis
Biswanath Mukherjee
University of California, Davis

Contents

* This work has been supported in part by the U.S. National Science Foundation Grant No. ANI-98-05285. Aysegül Gençata was a visiting scholar at U.C. Davis when this work was performed.

1.1   Introduction

The focus of this chapter is to present technological advances, promising architectures, and exciting research issues in designing and operating next-generation
optical wavelength-division multiplexing (WDM) networks, which are scalable
and flexible. We discuss important building blocks of optical WDM networks
and overview access, metropolitan, and long-haul networks separately. Special
attention has been paid to the long-haul network because there is a tremendous
need to develop new intelligent algorithms and approaches to efficiently design
and operate these wide-area-optical-mesh networks built on new emerging technologies. We present several research topics including routing and wavelength
assignment, fault management, multicasting, traffic grooming, optical packet
switching, and various connection-management problems. The Internet is developing rapidly with the ultimate goal being to provide us with easy and fast
access to any desired information from any corner of the world. Information
exchange (or telecommunications) technology, which has been evolving continuously since the telephone was invented, is still striving to meet the users'
demands for higher bandwidth. This demand is attributed to the growing popularity of bandwidth-intensive networking applications, such as data browsing
on the World Wide Web (WWW), java applications, video conferencing, interactive distance learning, on-line games, etc. Figure 1.1 plots the past and projected growth of data and voice traffic as reported by most telecom carriers.[1] It
shows that, while voice traffic continues to experience a healthy growth of
approximately 7% per year, data traffic has been growing much faster. To support this exponential growth in the user data traffic, there is a strong need for
high-bandwidth network facilities, whose capabilities are much beyond those
of current high-speed networks such as asynchronous transfer mode (ATM),
SONET/SDH* etc.[2]

  Fiber-optic technology can meet the previously mentioned need because
of its potentially limitless capabilities:[3] huge bandwidth (nearly 50 terabits
per second [Tbps] for single-mode fiber), low signal attenuation (as low as
0.2 dB/km), low signal distortion, low power requirement, low material
usage, small space requirement, and low cost. Given that a single-mode

---

* SONET and SDH are a set of related standards for synchronous data transmission over fiber
optic networks. SONET is short for Synchronous Optical NETwork and SDH is an acronym for
Synchronous Digital Hierarchy.

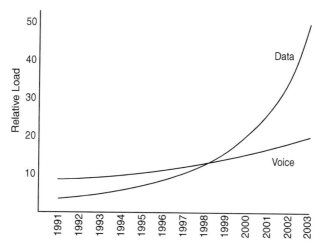

Figure 1.1 Past and projected future growth of data and voice traffic.

fiber's potential bandwidth is nearly three orders of magnitude higher than electronic data rates of a few tens of gigabits per second (Gbps), we need to tap into this huge optic-electronic bandwidth mismatch. Because the maximum rate at which an end user — which can be a workstation or a gateway that interfaces with lower-speed sub-networks — can access the network is limited by electronic speed, concurrency among multiple user transmissions should be introduced to exploit the fiber's huge bandwidth.

WDM is a favorite multiplexing technology in optical communication networks because it supports a cost-effective method to provide concurrency among multiple transmissions in the wavelength domain. Several communication channels, each carried by a different wavelength, are multiplexed into a single fiber strand at one end and demultiplexed at the other end, thereby enabling multiple simultaneous transmissions. Each communication channel (wavelength) can operate at any electronic processing speed (e.g., OC-192 or OC-768).* For example, a fiber strand that supports 160 communication channels (i.e., 160 wavelengths, each operating at 40 Gbps) would yield an aggregate capacity of 6.4 Tbps.

Today's telecom network can be considered to consist of three sub-networks: access (spanning about 1 to 10 km), metropolitan (covering about 10 to 100 km), and long haul (extending to 100s or 1000s of km) (see Figure 1.2); and fiber is being extensively deployed in all three sub-networks. Typically, the network topology for access can be a star, a bus, or a ring; for metro a ring; and for long haul a mesh. Each of these sub-networks has a different set of functions to perform; hence, each has a different set of challenges, technological requirements, and research problems. For example, for the long-haul network, carriers are more concerned with capacity, protection,

---

* OC-n stands for an "optical channel" with data rate of n x 51.84 Mbps approximately.

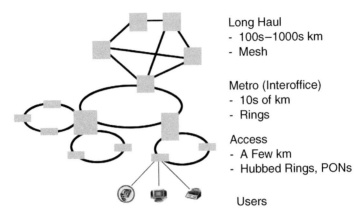

Long Haul
- 100s–1000s km
- Mesh

Metro (Interoffice)
- 10s of km
- Rings

Access
- A Few km
- Hubbed Rings, PONs

Users

Figure 1.2 Telecom network overview.

and restoration, while for the metro or access network, carriers are more concerned with service provisioning/monitoring, flexibility, etc.

The focus of this chapter is to present technological advances, promising architectures, and exciting research issues in designing and operating next-generation optical WDM networks that are scalable and flexible. The next section provides a brief discussion on the important building blocks of optical WDM networks. This is followed by an overview of access, metropolitan, and long-haul networks separately. Special attention has been paid to the long-haul network because there is a tremendous need to develop new intelligent algorithms and approaches to efficiently design and operate these wide-area-optical-mesh networks built on new emerging technologies. We present several research topics including routing and wavelength assignment (RWA), fault management, multicasting, traffic grooming, optical packet switching, and various connection-management problems.

1.2   Enabling WDM technologies

An important factor to consider in the design of a WDM network is the number of wavelengths to use. The maximum number of wavelengths is limited by optical device technology and is affected primarily by the total available bandwidth or spectral range of the components (including the fiber) and the spacing between channels. Conventional fibers have a low attenuation region between 1335 and 1625 nm with a "water-peak window" at 1385 nm. New "all-wave" fibers do not have this water peak and hence can use a larger spectrum (see Figure 1.3). Channel spacing itself is affected by several factors such as the channel bit rates, optical power budget, non-linearities in the fiber, and the resolution of transmitters and receivers. In *dense* wavelength-division multiplexing (DWDM), a large number of wavelengths (>160) is packed densely into the fiber with small channel spacing. An alternative WDM technology with a smaller number of wavelengths (<

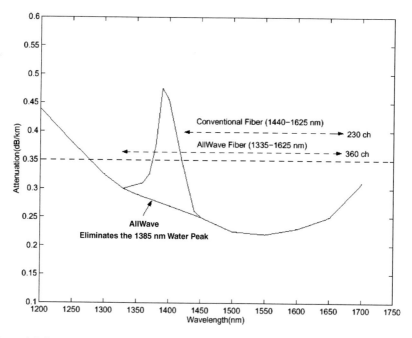

Figure 1.3 Low-attenuation region of all-wave fiber vs. conventional fiber.

10), larger channel width, larger channel spacing, and much lower cost is termed as *coarse* WDM (CDWM).

Although new approaches and technologies are constantly under development, this section highlights some of the emerging and novel technologies that can revolutionize the design and effectiveness of WDM networks. Optical components employed in building a typical point-to-point optical WDM transmission system are depicted in Figure 1.4. Several optical signals sent by transmitters (lasers) are coupled together using a (wavelength) *multiplexer* into a fiber. Signals are amplified, when necessary, using *amplifiers* such as erbium-doped fiber amplifiers (EDFAs) to compensate for signal attenuation. At intermediate nodes, these signals can be dropped and new signals can be added using optical add drop multiplexers (OADMs). At the receiving end, a (wavelength) *de-multiplexer* is used to segregate the individual wavelengths arriving on the fiber, which are then fed into the receivers (filters).

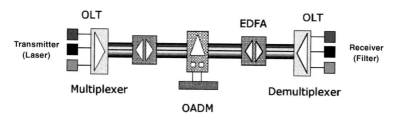

Figure 1.4 A typical point-to-point optical fiber communication link.

The EDFAs with gain spectrum of 30 to 40 nm each (typically in the 1530 to 1560 nm range; see Figure 1.4) can be interconnected to broaden their gain bandwidth. This "amplifier circuit" is referred to as an ultrawide-band EDFA, which can fully exploit the expanded low-attenuation region of the new "all-wave fiber" (see Figure 1.4).

OADMs — also referred to as wavelength add-drop multiplexers (WADMs) — are employed to take in (add) and take out (drop) individual wavelengths from an optical fiber completely in the optical domain (i.e., without any conversion of the optical signal into electronic domain). OADMs, which can add and drop a specific predefined channel (or a group of channels), are said to be fixed-tuned (or static) and the technology for manufacturing them is mature. Reconfigurable OADMs (ROADMs), with add-drop wavelengths that can be controlled by an external stimulus (e.g., by software) are said to be tunable (or dynamic). ROADMs are more powerful because they can adapt to the fluctuating traffic demand but the technology for building ROADMs is still in nascent stage.

When the network topology is a mesh, where nodes are interconnected by fibers to form an arbitrary graph, an additional fiber interconnection device is needed to route the signals from an input port to the desired output port. These devices are called optical crossconnects (OXCs). They can either be *transparent* (to bit rates and signal formats) in which signals are switched all-optically or be *opaque* in which incoming signals are converted from optical to electronic domain and switched electronically. A possible architecture for a transparent OXC is presented in Figure 1.5, where all signals on a particular wavelength (e.g., λ_1) arriving on M input fibers are switched separately by a wavelength-specific MxM switch. As more and more wavelengths are packed into a fiber, the size of the OXCs is expected to increase. Among several technologies (e.g., bubble, liquid crystal, thermo-optic, holographic, electro-optic, $LiNbO_3$, etc.) used for building all-optical OXCs, MEMS (micro-electro-mechanical-systems) based OXCs are becoming popular because of their compact design, low power consumption, and promise

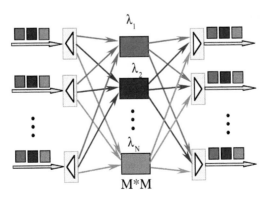

Figure 1.5 An optical crossconnect (OXC) of size NM x NM (N is the number of wavelengths; M is the number of incoming/outgoing fibers).

for high port count.[4] For details on emerging MEMS-based OXC architectures, please refer to Reference 5.

1.3   Access networks

The access network connects the subscribers (home or business) to the service providers; in other words, it serves as the "last mile" (as well as the "first mile") of the information flow. To meet the growing traffic demand, service providers expend most of their effort on increasing the bandwidth on their backbone network. But little has changed in the access network. It is now the general opinion that the last mile has become a bottleneck in today's network infrastructure.[6] Optical technology is a promising candidate for solving the bandwidth problem in access networks because it can provide at least 10 to 100 times more bandwidth over a larger coverage area. The next wave in access network deployment will bring the fiber to the building (FTTB) or to the home (FTTH), enabling Gbps speeds at costs comparable to other technologies such as digital subscriber line (DSL) and hybrid fiber coax (HFC).[6]

Three optical technologies are promising candidates for the next-generation access networks: point-to-point topologies, passive optical networks, and free-space optics.

1.3.1   Point-to-point topologies

Point-to-point dedicated fiber links can connect each subscriber to the telecom central office (CO), as illustrated in Figure 1.6a. This architecture is simple but expensive due to the extensive fiber deployment. An alternative approach is to use an active star topology, where a curb switch is placed close to the subscribers to multiplex/de-multiplex signals between the subscribers and the CO. This alternative in Figure 1.6b is more cost effective in terms of deployed fiber. A disadvantage of this approach is that the curb switch is an active component that requires electrical power as well as backup power at the curb-unit location.

1.3.2   Passive optical networks

Passive optical networks (PONs) replace the curb switch with a passive optical component such as an optical splitter (Figure 1.6c). This is one of the several possible topologies suitable for PONs including tree-and-branch, ring, and bus. PON minimizes the amount of fiber deployed, total number of optical transceivers in the system, and electrical power consumption. Currently, two PON technologies are being investigated: ATM PONs (APON) and Ethernet PONs (EPON). APON uses ATM as its layer-2 protocol; thus, it can provide quality-of-service features. EPON carries all data encapsulated in Ethernet frames, and can provide a relatively inexpensive solution compared to APON. EPON is gaining popularity and is being

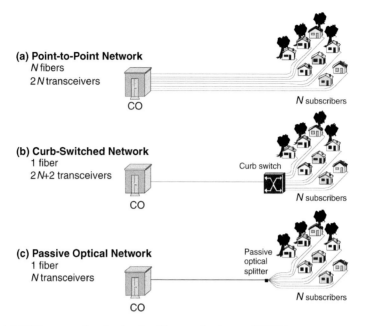

Figure 1.6 Different technologies for fiber-to-the-home (FTTH).

standardized as a solution for access networks in the IEEE 802.3ah group. In Kramer and Pesavento (2000), design issues are discussed and a new protocol (IPACT) for EPON is proposed.[7]

1.3.3   Optical wireless technology (free space optics)

Low-power infrared lasers can be used to transmit high-speed data via point-to-point (up to 10 Gbps) or meshed (up to 622 Mbps) topologies.[8] An optical data connection can be established through the air via lasers sitting on rooftops aimed at a receiver. Under ideal atmospheric conditions, this technology can provide a transmission range of up to 4 km.[8] Several challenges need to be addressed for optical wireless technology, including weather conditions, movement of buildings, flying objects, and safety considerations.

1.4   Metropolitan networks

Metropolitan-area (or metro) networks serve geographic regions spanning several hundred kilometers, typically covering large metropolitan areas. They interconnect access networks to long-haul backbone service providers. Currently, SONET/SDH-based rings form the physical-layer infrastructure in metro networks. SONET rings utilize a single channel (at 1310 nm wavelength) with a TDM (time-division multiplexing) technique. With TDM, a high-bandwidth channel (e.g., OC-192) can be divided into several low-bandwidth sub-channels (e.g., OC-1, OC-3, etc.), and each sub-channel can carry a different low-rate traffic stream. A TDM sub-channel is carried physically in a time slot traveling through the ring. A data stream can be added to a time slot at the source node and travel to the destination where it is extracted by SONET add/drop multiplexers (ADMs).

With the emergence of WDM technology, a logical step is to upgrade the one-channel SONET ring to a multiple-channel WDM/SONET ring. In a WDM/SONET ring network, each wavelength can operate similar to a SONET TDM channel. However, bandwidth upgrade comes with a price: in a simple-minded solution, a SONET ADM is needed for each wavelength at each node, increasing the total number of ADMs in a network W times, where W is the number of wavelengths. Fortunately, it may be possible to have some nodes on some wavelengths where no add/drop operation is needed on any time slot (see Figure 1.7 for an example). The total number of ADMs can be reduced by carefully packing the low-bandwidth connections into wavelengths. Packing low-speed traffic streams into high-speed traffic streams to minimize the resource usage is called *traffic grooming*, and it is a research subject that has received a lot of attention.[9–14]

To realize an architecture with grooming, a new optical component should be used at each node: a wavelength ADM (WADM) that can selectively bypass some of the wavelengths and extract the others from a fiber (see Section 1.2). In Chiu and Modiano (2000), a unidirectional WDM ring network is considered where the number of SONET ADMs is minimized.[9] In Gerstel et al. (2000), the authors address the problem of designing WADM rings for cost-effective traffic grooming.[10] They propose and analyze a collection of WADM ring networks considering that the network cost includes the number of wavelengths, transceiver cost, and the maximum number of hops. Another work on cost-effective design of WDM/SONET rings[11] minimizes the number of wavelengths and the total number of ADMs for a given static traffic; this work is applicable to both unidirectional and bidirectional

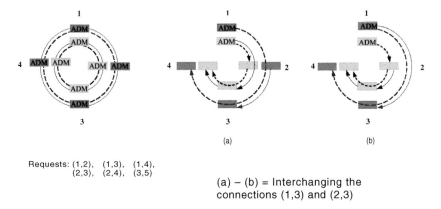

Requests: (1,2), (1,3), (1,4),
         (2,3), (2,4), (3,5)

(a) – (b) = Interchanging the
connections (1,3) and (2,3)

Figure 1.7 Reducing the number of ADMs by traffic grooming. The ADM at node 2 on the outer wavelength channel is not needed if the connections (1,3) and (2,3) in (a) are interchanged.

rings. In Berry and Modiano (2000), the dynamic traffic case is considered, where the traffic is given as a set of matrices.[12] The authors formulate the problem as a bipartite graph-matching problem and they develop algorithms to minimize the number of wavelengths that must be processed at each node. A formal definition of the problem is given through integer linear program (ILP) formulations in Wang et al. (2001), and a simulated-annealing-based heuristic is proposed to solve the problem.[13] Dutta and Rouskas (2002) present a framework for computing bounds for traffic grooming in ring networks, which can be used to evaluate the performance of heuristic algorithms.[14] For a survey of traffic grooming in ring topologies, please refer to Modiano (2001).[15]

The WDM/SONET ring architecture may be the next step to provide a higher-bandwidth solution in the metro network, but the TDM-based infrastructure poses challenges toward a more flexible, data-driven metro network. New questions arise as the data traffic grows to be the main component of the overall demand, and consequently brings more "burstiness" and unpredictability. The choice of the future would be a metropolitan optical network architecture that is scalable, flexible, capable of providing just-in-time connection provisioning, and exploiting the full advantages of a WDM system. WDM mesh topologies are the logical candidates to achieve these goals.[8]

1.5   Long-haul networks

The long-haul network (spanning hundreds to thousands of km) typically has OXCs at its nodes interconnected by a mesh of fibers (see Figure 1.2). Traffic from the end users (which could be an aggregate activity from a collection of terminals) is collected by the access networks and fed into the long-haul networks through metro networks. This high-bandwidth traffic is carried on a long-haul WDM network from one end to the other by the wavelength channels available on fibers. In this section, we present significant research issues concerning provisioning and maintenance of wavelength-channel-based connections.

1.5.1   Routing and wavelength assignment

In a wavelength-routed WDM network, end users communicate with one another via end-to-end (possibly all-optical) WDM channels, which are referred to as *lightpaths*.[16] These lightpaths are used for supporting a *connection* in a wavelength-routed WDM network, and may span multiple fiber links. Figure 1.8 shows several nodes in a network communicating among themselves through lightpaths (e.g., a lightpath connection from CO to NJ spans across the physical links CO–TX, TX–GA, GA–PA, and PA–NJ). In the absence of wavelength converters, a lightpath must occupy the same wavelength on the fiber links through which it traverses; this property is known as the *wavelength-continuity constraint*. Given a set of connections, the problem of

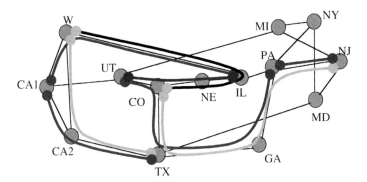

Figure 1.8 Lightpath connections in a WDM mesh network. (Solid circles mark the end points of a lightpath.)

setting up lightpaths by routing and assigning a wavelength to each connection is called the *routing and wavelength assignment* (RWA) problem.[17,18] Note that fiber links in a mesh network (e.g., the one in Figure 1.8) are bidirectional, while the lightpaths may be unidirectional or bidirectional.

Typically, connection requests may be of three types: static, incremental, and dynamic. With static traffic, the entire set of connections is known in advance, and the problem is then to set up lightpaths for these connections in a global fashion while minimizing network resources such as the number of wavelengths in the network. Alternatively, one may attempt to set up as many of these connections as possible for a given fixed number of wavelengths per fiber link (typically, all fibers are assumed to have the same number of wavelengths). The RWA problem for static traffic is known as the *static lightpath establishment* (SLE) problem.[18] In the incremental-traffic case, connection requests arrive sequentially, a lightpath is established for each connection, and the lightpath remains in the network indefinitely. For the case of dynamic traffic, a lightpath is set up for each connection request as it arrives, and the lightpath is released after some finite amount of time. The *dynamic lightpath establishment* (DLE) problem[18] involves setting up the lightpaths and assigning wavelengths to them while minimizing the connection blocking probability or maximizing the number of connections that can be established in the network over a period of time.

In a wavelength-routed WDM network, the wavelength-continuity constraint can be eliminated if we can use *wavelength converters* to convert the data arriving on one wavelength on a fiber link into another wavelength at an intermediate node before forwarding it on the next fiber link. Such a technique is feasible and is referred to as *wavelength conversion*.[19] Wavelength-routed networks with this capability are referred to as *wavelength-convertible networks*. A wavelength converter that can convert from any wavelength to any other wavelength is said to have full-range capacity. If there is one wavelength converter for each fiber link in every node of the network, the network is said to have *full wavelength-conversion* capability. A wavelength-convertible network with full wavelength-conversion capability at each node is equivalent to a circuit-switched telephone network; thus, only the routing problem needs to be addressed, and wavelength assignment is not an issue.[17]

Three basic approaches are used for the routing subproblem: *fixed routing, fixed-alternate routing, and adaptive routing.*[20] Fixed routing is a straightforward approach in which same fixed route is always chosen to route a connection for a given source-destination pair. One example of such an approach is *fixed shortest-path routing.* In fixed-alternate routing, each node in the network maintains a routing table containing an ordered list of a number of fixed routes to each destination node. For example, these routes may include the shortest-path route, the second shortest-path route, the third shortest-path route, etc. When a connection request arrives, the source node attempts to establish the connection on each of the routes from the routing table in sequence, until a route with a valid wavelength assignment is found. If no available route is found from the list of alternate routes, then the connection request is blocked. In adaptive routing, the route from a source node to a destination node is chosen dynamically, depending on the network state. This approach has lower connection blocking than fixed and fixed-alternate routing, but it is more computationally intensive.

Once a path has been chosen for a connection, a wavelength must be assigned to it such that any two lightpaths that are sharing the same physical link are assigned different wavelengths. Assigning wavelengths to different lightpaths that minimizes the number of wavelengths used under the wavelength-continuity constraint reduces to the *graph-coloring* problem.[18,21,22] This problem has been demonstrated to be NP-complete, and the minimum number of colors needed to color a graph G (called the chromatic number $\chi[G]$ of the graph) is difficult to determine. However, there are efficient *sequential graph-coloring* algorithms, which are optimal in the number of colors used. Other RWA heuristics such as First-Fit, Least-Used, Most-Used, etc. can be found in Reference 22.

1.5.2   Fault management

In a wavelength-routed WDM network (as well in other networks), the failure of a network element (e.g., fiber link, crossconnect, etc.) may result in the failure of several optical channels, thereby leading to large data and revenue losses. Several approaches are used to ensure fiber-network survivability against fiber-link failures.[23] Survivable network architectures are based either on reserving backup resources in advance (called "protection"),[24] or on discovering spare backup resources in an online manner (called "restoration").[25] In protection, which includes automatic protection switching (APS) and self-healing rings,[26,27] the disrupted service is restored by utilizing the precomputed and reserved network resources. In dynamic restoration, the spare capacity, if any, available within the network is utilized for restoring services affected by a failure. Generally, dynamic restoration schemes are more efficient in utilizing network capacity due to the multiplexing of the spare-capacity

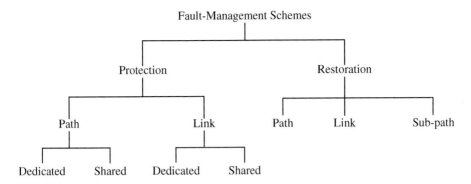

Figure 1.9 Fault-management schemes.

requirements, and they provide resilience against different kinds of failures, while protection schemes have a faster restoration time and guarantee recovery from disrupted services. Although protection schemes are suitable for the optical layer (with wavelength routing), dynamic restoration schemes are suitable for Layer 3 (IP packet switching). Below, we examine these two fault-management schemes for mesh networks (see Figure 1.9).

1.5.2.1 Protection

Existing connections in a network can be protected from fiber failures either on a link-by-link basis (which we call *link protection*) or on an end-to-end basis (which we call *path protection*).

In *link protection*, during connection setup, backup paths and wavelengths are reserved around each link on the primary path. In the event of a link failure, all the connections traversing the failed link will be rerouted around that link and the source and destination nodes of the connections traversing the failed link would be oblivious to the link failure.

In *path protection*, during connection setup, the source and destination nodes of each connection statically reserve a primary path and a backup path (which are link and/or node disjoint) on an end-to-end basis. When a link fails, the source node and the destination node of each connection that traverses the failed link are informed about the failure (possibly via messages from the nodes adjacent to the failed link) and backup resources are utilized. Although *path protection* leads to efficient utilization of backup resources, *link protection* provides faster protection-switching time.

The *link-* and *path-protection* schemes can either be *dedicated* or *shared*.

In *dedicated-link protection*, at the time of connection setup, for each link of the primary path, a backup path and wavelengths are reserved around that link and they are dedicated to that connection.

In *shared-link protection*, the backup wavelengths reserved on the links of the backup path are shared with other backup paths. As a result, backup channels are multiplexed among different failure scenarios (which are not expected to occur simultaneously). Therefore, *shared-link protection* is more capacity efficient when compared with *dedicated-link protection*.

In *dedicated-path protection*, at the time of connection setup for each primary path, a link-disjoint backup path and wavelength are reserved, and dedicated to that connection. The primary and the backup paths can carry identical traffic from the source to the destination simultaneously (referred to as 1+1 protection) or the backup path, although reserved for use in the event of a failure of the primary path, can carry lower-priority preemptive traffic (referred to as 1:1 protection).

In *shared-path protection*, the backup wavelengths reserved on the links of the backup path may be shared with other backup paths. In general, a scheme where M primary paths share N backup paths is known as M:N protection.

1.5.2.2   Restoration

Dynamic restoration schemes can be used to restore the failed link, or the failed paths, or sub-paths (see Figure 1.10).

In *link restoration*, the end nodes of the failed link dynamically discover a route around the link, for each connection (or "live" wavelength) that traverses the link. In the event of a failure, the end nodes of the failed link participate in a distributed algorithm to dynamically discover a new route around the link, for each active wavelength that traverses the link. When a new route is discovered around the failed link for a wavelength channel, the end nodes of the failed link reconfigure their OXCs to reroute that channel onto the new route. If no new route and associated wavelength can be discovered for a broken connection, that connection is dropped.

In *path restoration*, when a link fails, the source and the destination node of each connection that traverses the failed link are informed about the failure (possibly via messages from the nodes adjacent to the failed link). The source and the destination nodes of each connection independently discover a backup route on an end-to-end basis (such a backup path could be on a different wavelength channel). When a new route and wavelength channel is discovered for a connection, network elements such as OXCs are reconfigured appropriately, and the connection switches to the new path. If no new route (and associated wavelength) can be discovered for a broken connection, that connection is blocked.

In *sub-path* restoration, when a link fails, the upstream node of the failed link detects the failure and discovers a backup route from itself to the corresponding destination node for each disrupted connection.[28] Upon successful discovery of resources for the new backup route, intermediate OXCs are reconfigured appropriately and the connection switches to the new path. A connection is dropped in the absence of sufficient resource availability.

Link restoration is the fastest and path restoration is the slowest among the above three schemes. Sub-path restoration time lies in between those of link restoration and path restoration. For a comprehensive review of the

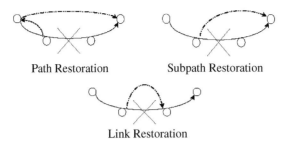

Path Restoration   Subpath Restoration

Link Restoration

Figure 1.10 Mechanisms for restoring connections after fiber failure.

literature on the design of survivable optical networks, please consult the literature.[29–33]

1.5.3. Multicasting

Multicasting is the ability of a communication network to accept a single message from an application and deliver copies of the message to multiple recipients at different locations.[34] One of the challenges is to minimize the amount of network resources that are employed by multicasting. To illustrate this point, let us assume that a video server wants to transmit a movie to 1000 recipients (Figure 1.11a). If the server were to employ 1000 separate point-to-point connections (e.g., TCP* connections), then 1000 copies of the movie would have to be sent over a single link, thus making poor use of the available bandwidth. A scalable and efficient implementation of multicasting permits a much better use of the available bandwidth by transmitting at

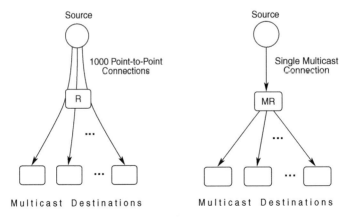

Figure 1.11 An example that illustrates the amount of network resources employed by (left) unicasting a movie to 1000 different users as opposed to the amount of network resources employed by (right) multicasting the movie. (R = standard router, MR = multicast router)

* TCP stands for "Transmission Control Protocol," which is widely used in today's Internet.

most one copy of the movie on each link in the network as illustrated in Figure 1.11b.

Today, many multicast applications exist, such as news feeds, file distribution, interactive games, video conferencing, interactive distance learning, etc. Many more applications are expected to emerge to exploit the enormous bandwidth promised by the rapidly growing WDM technology,[35] but the implementation of these applications is not necessarily efficient because today's long-haul networks were designed to mainly support point-to-point (unicast) communication. In the future, as multicast applications become more popular and bandwidth intensive, there emerges a pressing need to provide multicast support in the underlying communication network.

In an optical WDM network, a *lightpath* provides an end-to-end connection from a source node to a destination node. A *light-tree* is a *point-to-multipoint* generalization of a lightpath and provides "single-hop"* communication between a "source" node and a set of destination nodes, which makes it suitable for multicast applications.[36] A light-tree enables a transmitter at a node to have many more logical neighbors, thereby leading to a denser virtual interconnection diagram and a lower hop distance. A multicast-capable WDM long-haul network can not only support efficient routing for multicast traffic, but it may also enhance routing for unicast traffic by allowing more densely connected virtual topologies (refer to Section 1.5.5). To realize multicast-capable WDM long-haul networks, we need to develop multicast-capable switch architectures and design efficient RWA algorithms, as outlined below.

1.5.3.1 *Multicast-capable OXC architectures*

Two approaches are used to design switches capable of supporting multicasting. One approach is to use electronic crossconnects, which perform switching in the electronic domain and the other is to use "all-optical" switches for switching in the optical domain. Although switching in the latter is "transparent" to bit rate and bit-encoding schemes, switching in the former requires knowledge of bit rate and bit-encoding strategies, and hence is "opaque."[37]

Opaque switches. Figure 1.12 depicts an *opaque approach*, in which the incoming optical bit streams are converted to electronic data; the data is switched using an electronic crossconnect, and then the electronic bit streams are converted back to the optical domain. Observe that the signal in a channel arriving on the input fiber link *D* is replicated into three copies in the electronic domain. One copy is dropped locally at the node and the remaining two are switched to different channels on outgoing fiber links *1* and *2*. (Along with the light-trees, the switch can also be used to establish lightpaths from a source to a destination as presented in the figure by a unicast connection from input fiber link *2* to output fiber link *D*.) This "opaque" switch architecture is currently very popular due to

---

* A hop is an all-optical segment of a path and may span several physical links.

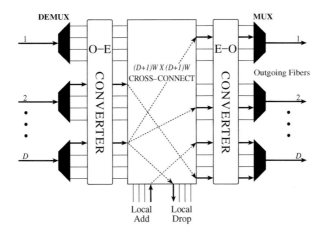

Figure 1.12 Opaque OXC architecture for supporting multicasting using electronic crossconnect.

the existence of mature technology to design high-bandwidth, multi-channel, non-blocking, electronic cross-connect fabrics at a low cost. Several companies are already shipping optical OXCs based on optical-electronic-optical (O-E-O) conversion, which can be used for building a multicast-capable OXC with O-E-O conversion.

Wavelength converters are not needed in a network where nodes are equipped with optical switches based on the *opaque approach* because, once an incoming bit stream in the optical domain is converted to electronic domain, it can be switched and converted back to the optical domain on any wavelength. In other words, full-range wavelength conversion[19] is an inherent property of such switches and the *wavelength-continuity constraint* need not be obeyed.

Transparent switches. Figure 1.13 illustrates a multicast-capable *all-optical switch* that crossconnects optical channels directly in the optical domain. Again, several companies are working toward building *all-optical* switches using various technologies, a popular one employing tiny mirrors based on micro-electro-mechanical-system (MEMS) technology. For multicasting in *all-optical* switches, "optical splitters" are needed to replicate an incoming bit stream to two or more copies as illustrated in Figure 1.13. A signal arriving on wavelength λ_b from input fiber link *D* is sent to the optical splitter **X** for splitting into three identical copies. One of the three replicas is dropped locally at the node while the other two are switched to output fiber links *1* and *2*. Observe that the signal arriving on wavelength λ_a from input fiber link *2* bypasses the node. In this architecture, amplifiers are required because the output signal power weakens when the input signal is split (e.g., a 3-dB attenuation in power occurs for a two-way, equal-power splitting of an optical signal). Wavelength converters are useful in such switches to reduce the probability of blocking of multicast sessions. In the

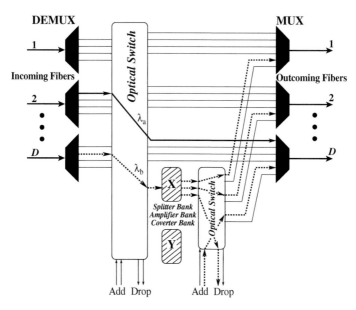

Figure 1.13 Transparent OXC architecture for supporting multicasting using optical splitters.

absence of wavelength converters, the light-tree-based multicast session would exhibit the *wavelength-continuity constraint*.

1.5.3.2 *Multicast routing and wavelength assignment*

The multicasting problem in communication networks is often modeled as a Steiner Minimum Tree (SMT), which is an NP-complete problem.[38] The problem complexity increases when several multicast sessions (which we expect to occur in the future) have to be established at a minimum aggregate cost. Heuristics are employed for routing and wavelength assignment of multicast sessions.[39,40]

A single fiber cut on a multicast tree can disrupt the transmission of information to several destination nodes on a light-tree. This loss would be large if several sessions were occupying the affected fiber link. In order to prevent the large loss of information, it is imperative to protect the multicast sessions through a protection scheme such as reserving resources along a backup tree. Protecting such multicast sessions using schemes (dedicated or shared) discussed in Section 1.5.2 is an important problem that needs to be studied.

1.5.4 *Traffic grooming in WDM mesh networks*

Today, each wavelength channel has the transmission rate of over a Gbps (e.g., OC-48 [2.5 Gbps], OC-192 [10 Gbps], or OC-768 [40 Gbps] in the near future). However, the capacity required by the traffic streams from client networks (IP, ATM, etc.) can be significantly lower, and they can vary in the range from OC-1 (51.84 Mbps) or lower, up to full wavelength capacity. In

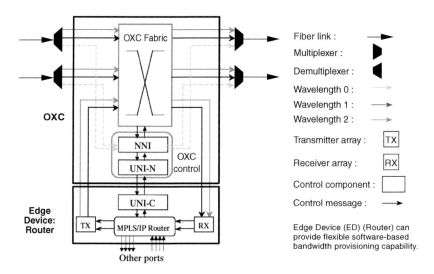

Figure 1.14 Node architecture for traffic grooming.

order to achieve the most efficient utilization of network resources, to reduce operating costs, and to maximize revenue from existing capacity, the low-speed traffic streams need to be efficiently "groomed" onto high-capacity optical channels (lightpaths).

The traffic-grooming problem has been well studied for WDM/SONET ring networks with the objective of minimizing the total network cost measured in terms of the number of SONET ADMs (see Section 1.4). As today's optical long-haul backbone networks are evolving from interconnected-rings topology to mesh topology, traffic grooming in WDM mesh networks has become a very important problem for both industry and academe.[41,42]

In order to support traffic grooming, each node in a WDM mesh network is equipped with an OXC that should be able to switch traffic at wavelength granularity as well as finer granularity. Figure 1.14 shows a simplified architecture of an OXC with grooming capability. In Figure 1.14, the grooming fabric (G-fabric) performs multiplexing, de-multiplexing, and switching of low-speed traffic streams. A transceiver array (T and R) is used to connect the G-Fabric to the W-Fabric (see Figure 1.14). The size of the transceiver array determines how many wavelength channels can be switched in and out of the G-fabric from the W-fabric. Hence, it determines the grooming capacity of an OXC. A lightpath is called a groomable lightpath if it is switched to the G-Fabric at its end nodes.

In a *static* grooming problem, all connection requests (of different bandwidth granularities) are known a priori. In a *dynamic* grooming problem, connection requests arrive randomly, hold for a finite duration, and require provisioning in real time and tearing down when they are over. The grooming of traffic can be either *single-hop* or *multi-hop*. While in the former, connections are allowed to traverse only a single lightpath hop, in *multi-hop* grooming a connection can be switched by the G-fabric at any intermediate node (i.e., it can traverse multiple lightpath hops).

When some low-speed connections require protection, they have to be prudently protected and groomed along with the other existing connections in order to maximize the network throughput. Again, these connections need not be just single-source, single-destination connections but they may be single-source, multiple-destination streams requiring multicasting of information to a group of nodes. Grooming of low-speed multicast connections, grooming along with protection, etc., are interesting and challenging areas that need more research.

1.5.5   IP over WDM

Rapid growth in data traffic and the predominance of Internet Protocol (IP) in data communication have led researchers to investigate the IP-over-WDM integration. In such architecture, network nodes employ OXCs and IP routers. Today's IP-over-ATM-over-SONET-over-optical approach reduces efficiency as well as the effective bandwidth provided by WDM technology. The trend is to converge the IP layer and the WDM layer by eliminating one or two layers of the protocol stack[43] and to offer a multi-protocol support (multi-protocol label switching or MPLS, see Section 1.5.7) for simplified network architecture.

An optical channel (i.e., a lightpath) can connect any two IP routers in an IP-over-WDM network. The set of lightpaths forms a virtual interconnection pattern called the virtual (logical) topology.[44] A lightpath is established by tuning the transmitter at the source node and the receiver at the destination node to an appropriate wavelength, and by configuring the OXCs along the path. The traffic between two nodes can be carried by the lightpath established between these nodes. Nodes that are not connected directly in the virtual topology can still communicate with one another using the multi-hop approach, namely, by using electronic packet switching at the intermediate nodes in the virtual topology. IP/MPLS routers, ATM switches, etc. can provide electronic packet switching. Interaction between the optical layer and the electronic layer (IP in this case) is a major issue including several functions, such as bandwidth provisioning, fault management, performance monitoring, etc. (see Section 1.5.2).

Bandwidth provisioning at the optical layer is related to the RWA problem. The latter is a hard (non-polynomial) problem, which includes minimizing the usage of network resources considering constraints on wavelength conversion, nodal-switching capabilities, and physical-layer connectivity (fiber layout).[3] The problem gets more complex when one considers the dynamism of the IP traffic. When traffic intensities between nodes change over time, the network may need to be re-optimized by online methods. This is a joint optimization problem involving IP routing, virtual-topology reconfiguration and therefore optical-layer routing and wavelength assignment.[45,47,49,51] To solve these problems, we need automated mechanisms that can interact with today's IP protocols (IPv4, IPv6, RSVP, etc.).

In such network architecture, the failure of a fiber link can lead to the failure of all of the lightpaths that traverse it. Considering that each lightpath operates at a rate of tens of Gbps, such a failure could cause a large amount of data and, consequently, revenue loss. Hence, network survivability is a crucial issue (see Section 1.5.2).

1.5.6   Call admission control based on physical impairments

Optical networking technology has many desirable features and, in general, offers better transmission-error characteristics compared to other physical-layer technologies, such as copper or radio. Its low error characteristics make it the best candidate to deploy for worldwide data-transmission backbones. However, even the optical-layer technology is far from being perfect, and at the scale of continent-wide or worldwide networks, physical-layer impairments may cause serious problems that we need to consider.

In a large-scale network, an optical signal may propagate through a number of nodes and long fiber spans (1000s of km) connecting the nodes. Throughout its propagation, the signal is subjected to degradation by several impairments: cross-talk from OXCs, amplified spontaneous emission (ASE) from EDFAs, four-wave mixing (FWM) from other signals propagating in the same fibers, laser phase noise at the transmitter, fiber dispersion and nonlinearities, etc.[3] As a result, the optical signal's bit-error rate (BER) may become too high to recover the original signal at the receiver. To exploit optical technology in long-haul mesh networks, and to make the future all-optical networks a reality, we need to develop intelligent approaches that can correct these undesirable effects.

To date, most of the studies on call admission and RWA problems assume an ideal physical layer that does not have any of the impairments cited above. The work in Ramamurthy et al. (1999) considers the physical-layer limitations by capturing the most significant impairments (ASE and cross-talk) before setting up a lightpath.[46] It estimates the on-line BER on candidate routes and wavelengths, and establishes a call on a lightpath only if the received BER is lower than a certain threshold (e.g., 10^{-12}).

Signal regeneration is another method to overcome signal degradation, and it may be performed in three forms:

1R-Regeneration: Re-amplifying. Signal is amplified using optical amplifiers, such as EDFA.

2R-Regeneration: Re-amplifying and re-shaping. The optical signal is converted to an electronic signal. It is both re-amplified and re-shaped. Re-shaping eliminates most of the noise and provides clear electrical 0s and 1s.

3R-Regeneration: Re-amplifying, re-shaping, and re-timing. The optical signal is converted to an electronic signal. Added to 2R-regeneration, it is also re-timed (or re-clocked). The time between bits at the receiver is not rigid, as it is at the source; re-timing adjusts the 1s and 0s so that they are equally spaced and match the bit rate of the system.

1.5.7   Network control and signaling

In an optical network, a control plane is needed to coordinate the necessary algorithms that provide the following functions:

1. A signaling protocol for setting up, maintaining, and tearing down the connections
2. A routing process for handling the topology and resource usage, and for calculating the routes
3. A naming and addressing scheme
4. A signaling protocol for providing communication between the entities requesting the services and those that provide the services.[48] Several initiatives are being developed to define and standardize such a control plane.

MPLS is a set of protocols for provisioning and managing core networks. It provides resource reservation and route set up to create data tunnels between ingress and egress nodes.* A label-switching mechanism ensures that all packets of the same data stream are routed through their predefined tunnel. Originally, MPLS was designed for packet-switching networks to overlay the Internet Protocol and to provide a standard interface that can communicate with several protocols (ATM, IP, frame relay, etc.). It has been generalized for optical networking, resulting in generalized MPLS or GMPLS.[50,52] GMPLS supports switching in time, wavelength, and space domains along with packet switching, by extending the signaling and routing protocols used in MPLS: Link management protocol (LMP), open-shortest-path-first/intermediate system to intermediate system (OSPF/ISIS) protocols, resource reservation protocol (RSVP), and constraint-based routing-label distribution protocol (CR-LDP).[53] GMPLS can provide traffic engineering** and fast rerouting mechanisms by the features of resource discovery, state information dissemination, path selection, and path management.[54]

Another effort, OIF-UNI (Optical Internetworking Forum-User Network Interface), defines the interoperation procedures for requesting and establishing dynamic connectivity between clients (e.g., IP, ATM, SONET devices, etc.) connected to an optical transport network (see Figure 1.15).[55] The UNI defines the set of services, signaling protocols used to invoke the services, the mechanisms used to transport signaling messages, and the auto-discovery procedures. Connection establishment, connection deletion, status exchange, auto-discovery, and information exchange (user data) are supported across the UNI.

---

* Nodes where traffic enters (ingress node) or leaves (egress node) a network.
** Traffic engineering is the process of controlling traffic flows in a network so as to optimize resource utilization and network throughput.

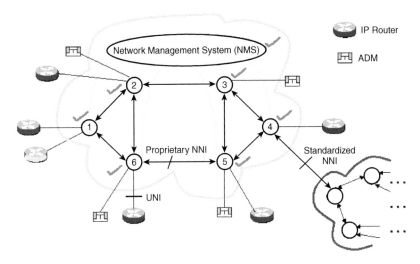

Figure 1.15 Network management system for IP-over-WDM networks.

1.5.8 *Optical packet switching*

WDM and optical-switching technology can provide the necessary band-width for the growing traffic demand. As data traffic starts to dominate the communication networks, the traffic even on the long-haul network becomes more data oriented (i.e., less predictable). In the long term, optical packet switching (OPS) could become a viable candidate because of its high-speed, fine-granularity switching, flexibility, and its ability to use the resources economically. The technology is still in a very early stage, and several issues need to be solved, including switch architectures, synchronization, contention-resolution schemes, etc. (see Yao et al. [2000] for a tutorial).[56]

An optical packet switch includes packet-synchronization stages, a switch fabric, and a control unit that extracts and reads the packet header to route the packet through the switch fabric to the proper output port. One main property here is to decide whether the network should operate synchronously. A globally synchronized (slotted) network can use aligned time slots as the holders of fixed-size packets. In such a network, the switch fabric at a node receives the incoming packets aligned, minimizing the packet contention; however, this switch architecture is more complex because of synchronization/packet-alignment stages. The other alternative is to build an asynchronus (unslotted) network where packets may have variable lengths. The switch architecture is simpler in this case, though packet-con-tention probability is higher. For a survey of different switch fabrics, please refer to Reference 57.

Contention of packets in a switch fabric is a major problem and has important impact on the performance of the network such as packet delay, packet-loss ratio, throughput, and average hop distance. Contention occurs when two or more packets try to leave the switch from the same output port at the same time. To resolve such a conflict, one of the packets is routed through the output port while others are routed elsewhere, depending on the contention-resolution scheme. In electronic switches, contention can be resolved by the store-and-forward technique where packets in contention are stored in a buffer and sent out when the port becomes available. Unfortunately, optical buffers similar to electronic counterparts (i.e., random-access memory [RAM]) do not exist; the only way to store an optical packet is to use optical delay lines (fixed length fibers). These "sequential access" buffers are less flexible than an electronic "random access" buffer because a packet entering a delay line will emerge from the other end of the line after a fixed amount of time. Several switch architectures that use optical buffers were proposed in Reference 56.

Another method for contention resolution is deflection routing, which is also called hot potato routing. In case of contention, one packet is routed along the desired link while others are forwarded on some other links, which may lead to longer paths. Deflection routing can be used along with optical buffers.

The unique advantage of WDM networks (i.e., several wavelength channels on the same link) can be used to create a third method for contention resolution, namely wavelength conversion. This is an attractive solution because it can achieve the same propagation delay and hop distance as the optimal case. All three or any two solutions can be combined to provide better performance.

Today, OPS still seems like a dream because of several technical obstacles. Our current vision for network planning is to implement a dynamically reconfigurable optical transport layer using fast OXCs, providing enough bandwidth to evolving data applications. If and when optical packet switching becomes available, we may choose to incorporate it into our existing optical circuit-switching architecture.[58] Packet switching is not limited only to wide-area networks; a tutorial on implementing OPS in metropolitan-area networks can be found in Reference 59.

1.5.8.1   Optical burst switching

As a midway solution between circuit switching and packet switching, optical burst switching (OBS) was proposed.[60] This approach is motivated by two problems: routing the IP traffic, which has a bursty characteristic, on a relatively-static circuit-switched network leads to poor usage of network resources; and OPS technology is not mature for the near future. In OBS, a control packet is sent first to reserve an appropriate amount of bandwidth and to preconfigure the switches along the path. The burst of data, which can consist of several packets forming a (possibly short) session, immediately follows the control packet, without waiting for an acknowledgment. OBS has a lower control overhead compared to OPS and it may lead to better resource usage compared to circuit switching because reserved resources are released after the completion of the burst. Several issues need to be addressed before OBS can be deployed, such as optical burst switch architecture, deciding on the time gap between the control packet and data burst (offset time), resolving resource conflicts without optical buffering, etc.[57,61]

1.6   Future directions

This chapter has provided an overview of several aspects of optical communication networks, specifically on WDM networks, and presented several challenging research advances involved therein. New technologies providing large bandwidth are being deployed in the long-haul and will migrate down to metro and access networks, moving the bottleneck closer to customers.

Optical Ethernet, a technology that extends Ethernet beyond the local area network (LAN) or access and into MANs and the long-haul networks is attracting attention. The transport style of optical Ethernet has the advantages of simplicity and ease of integration with long-haul DWDM systems, because a well-aggregated gigabit-Ethernet stream can either be mapped to a wavelength or aggregated further into a 10-gigabit Ethernet channel and then mapped to a wavelength for transport across the long-haul network.

As we feed more and more bandwidth to the insatiable customers, they are expected to create new "killer" applications that consume all the bandwidth — and will still be hungry for more — thereby creating the urgency for even more technological advancements. With every technological progress, new and exciting challenges and research problems are expected to sprout in the ever-expanding horizon.

References

1. http://www.rhk.com.
2. W.J. Goralski, *SONET*, 2nd Edition, McGraw-Hill, New York, May 2000.
3. B. Mukherjee, *Optical Commun. Networks*, McGraw-Hill, New York, July 1997.
4. D.J. Bishop, C.R. Giles, and G.P. Austin, The Lucent LambdaRouter: MEMS technology of the future here today, *IEEE Commun. Mag.*, vol. 40, no. 3, pp. 75–79, Mar. 2002.
5. P.B. Chu, S. Lee, and S. Park, MEMS: the path to large optical crossconnects, *IEEE Commun. Mag.*, vol. 40, no. 3, pp. 80–87, Mar. 2002.
6. G. Kramer and G. Pesavento, Ethernet passive optical networks (EPON): building a next-generation optical access network, *IEEE Commun. Mag.*, vol. 40, no. 2, pp. 66–73, Feb. 2002.
7. G. Kramer, B. Mukherjee, and G. Pesavento, IPACT: A dynamic protocol for an ethernet PON (EPON), *IEEE Commun. Mag.*, vol. 40, no. 2, pp. 74–80, Feb. 2002.
8. http://www.lightreading.com
9. A.L. Chiu and E.H. Modiano, Traffic grooming algorithms for reducing electronic multiplexing costs in WDM ring networks, *IEEE J. Lightwave Technol.*, vol. 18, no. 1, pp. 2–12, Jan. 2000.

10. O. Gerstel, R. Ramaswami, and G.H. Sasaki, Cost-effective traffic grooming in WDM rings, *IEEE/ACM Trans. Networking*, vol. 8, no. 5, pp. 618–630, Oct. 2000.

11. X. Zhang and C. Qiao, An effective and comprehensive approach for traffic grooming and wavelength assignment in SONET/WDM rings, *IEEE/ACM Trans. Networking*, vol. 8, no. 5, pp. 608–617, Oct. 2000.

12. R. Berry and E.H. Modiano, Reducing electronic multiplexing costs in SONET/WDM rings with dynamically changing traffic, *IEEE J. Selected Areas in Commun.*, vol. 18, no. 10, pp. 1961–1971, Oct. 2000.

13. J. Wang, W. Cho, V.R. Vemuri, and B. Mukherjee, Improved approaches for cost-effective traffic grooming in WDM ring networks: ILP formulations and single-hop and multihop connections, *IEEE J. Lightwave Technol.*, vol. 19, no. 11, pp. 1645–1653, Nov. 2001.

14. R. Dutta and G.N. Rouskas, On optimal traffic grooming in WDM rings, *IEEE J. Selected Areas in Commun.*, vol. 20, no. 1, pp. 110–121, Jan. 2002.

15. E.H. Modiano, Traffic grooming in WDM networks, *IEEE Commun. Mag.*, vol. 39, no. 7, pp. 124–129, Jul. 2001.

16. I. Chlamtac, A. Ganz, and G. Karmi, Lightpath communications: an approach to high bandwidth optical WANs, *IEEE Trans. Commun.*, vol. 40, no. 7, pp. 1171–1182, Jul. 1992.

17. R. Ramaswami and K. Sivarajan, Optical Routing and Wavelength Assignment in All-optical Networks, *IEEE/ACM Trans. Networking*, vol. 3, no. 5, pp. 489–500, Oct. 1995.

18. D. Banerjee and B. Mukherjee, A practical approach for routing and wavelength assignment in large wavelength-routed optical networks, *IEEE J. Selected Areas in Commun., Special Issue on Optical Networks*, vol. 14, no. 5, pp. 903–908, June 1996.

19. B. Ramamurthy and B. Mukherjee, Wavelength conversion in optical networks: progress and challenges, *IEEE J. Selected Areas in Commun.*, vol. 16, no. 7, pp. 1040–1050, Sept. 1998.

20. S. Ramamurthy and B. Mukherjee, Fixed-alternate routing and wavelength conversion in wavelength-routed optical networks, *IEEE Globecom*, Sydney, vol. 4, pp. 2295–2302, Nov. 1998.

21. D.W. Matula, G. Marble, and J.D. Issacson, Graph coloring algorithms, in R.C. Read, Ed., *Graph Theory and Computing*, Chapter 10, pp. 109–122, Academic Press, New York and London, 1972.

22. H. Zang, J.P. Jue, and B. Mukherjee, A review of routing and wavelength assignment and approaches for wavelength routed optical WDM networks, *Optical Networks Mag.*, vol. 1, no. 1, pp. 47–58, Jan. 2000.

23. H. Zang and B. Mukherjee, Connection management for survivable wavelength routed WDM mesh networks, *Optical Networks Mag., Special Issue on Protection and Survivability in Optical Networks*, vol. 2, no. 4, pp. 17–28, Jul. 2001.

24. S. Ramamurthy and B. Mukherjee, Survivable WDM mesh networks, Part I — Protection, *Proc., IEEE INFOCOM*, New York, pp. 744–751, March 1999.

25. S. Ramamurthy and B. Mukherjee, Survivable WDM mesh networks, Part II — Restoration, *Proc., IEEE Int. Conf. on Commun.*, (ICC '99) Vancouver, Canada, pp. 2023–2030, June 1999.

26. W.D. Grover, The self-healing network: a fast distributed restoration technique for networks using digital cross-connect machines, *Proc., IEEE Globecom*, pp. 28.2.1–28.2.6, Tokyo, Japan, Nov. 1987.

27. T. Wu, *Fiber Network Service Survivability,* Artech House, Norwood, MA, 1992.

28. J. Wang, L. Sahasrabuddhe, and B. Mukherjee, Path vs. sub-path vs. link restoration for fault management in IP-over-WDM networks: performance comparisons using GMPLS control signaling, *IEEE Commun. Mag.,* vol. 40, no. II, pp. 80–87, Nov. 2002.

29. S. Ramamurthy, Optical design of WDM network architectures, Ph.D. dissertation, Computer Science Department, University of California, Davis, Sept. 1998.

30. L.H. Sahasrabuddhe, Multicasting and fault tolerance in WDM optical networks, Ph.D. dissertation, Computer Science Department, University of California, Davis, Nov. 1999.

31. H. Zang, Design and analysis of WDM network architectures, Ph.D. dissertation, Computer Science Department, University of California, Davis, Apr. 2000.

32. O. Gerstel and R. Ramaswami, Optical layer survivability — a services perspective, *IEEE Commun. Mag.,* vol. 38, no. 3, pp. 104–113, Mar. 2000.

33. O. Gerstel and R. Ramaswami, Optical layer survivability — an implementation perspective, *IEEE J. on Selected Areas in Commun.,* vol. 18, no. 10, pp. 1885–1899, Oct. 2000.

34. L.H. Sahasrabuddhe and B. Mukherjee, Multicast routing algorithms and protocols: a tutorial, *IEEE Network,* vol. 14, no. 1, pp. 90–102, Jan./Feb. 2000.

35. R. Malli, X. Zhang, and C. Qiao, Benefit of multicasting in all-optical networks, *Proc., SPIE Conf. on All-Optical Networking,* vol. 2531, pp. 209–220, Nov. 1998.

36. L.H. Sahasrabuddhe and B. Mukherjee, Light-trees: optical multicasting for improved performance in wavelength-routed networks, *IEEE Commun. Mag.,* vol. 37, no. 2, pp. 67–73, Feb. 1999.

37. N.K. Singhal and B. Mukherjee, Architectures and algorithm for multicasting in WDM optical mesh networks using opaque and transparent optical cross-connects, Technical Digest, OFC, Anaheim, CA, paper TuG8, Mar. 2001.

38. S.L. Hakimi, Steiner's Problem in Graphs and its Implications, Networks, vol. 1, no. 2, pp. 113–133, 1971.

39. Y. Sun, J. Gu, and D.H.K. Tsang, Multicast routing in all-optical wavelength routed networks, *Optical Networks Mag.,* vol. 2, no. 4, pp. 101–109, Jul./Aug. 2001.

40. B. Chen and J. Wang, Efficient routing and wavelength assignment for multicast in WDM network, *IEEE J. on Selected Areas in Commun.,* vol. 20, no. 1, pp. 97–109, Jan. 2002.

41. K. Zhu and B. Mukherjee, Traffic grooming in an optical WDM mesh network, *IEEE J. on Selected Areas in Commun.,* vol. 20, no. 1, pp. 122–133, Jan. 2002.

42. K. Zhu and B. Mukherjee, On-line approaches for provisioning connections of different bandwidth granularities in WDM mesh networks, *Technical Digest,* OFC, Anaheim, CA, paper ThW5, pp. 549–551, Mar. 2002.

43. N. Ghani, S. Dixit, and T.-S. Wang, On IP-over-WDM integration, *IEEE Commun. Mag.,* vol. 38, no. 3, pp. 72–84, Mar. 2000.

44. B. Mukherjee, D. Banerjee, S. Ramamurthy, and A. Mukherjee, Some principles for designing a wide-area WDM optical network, *IEEE/ACM Trans. on Networking,* vol. 4, no. 5, pp. 684–696, Oct. 1996.

45. D. Banerjee and B. Mukherjee, Wavelength routed optical networks: linear formulation, resource budgeting tradeoffs, and a reconfiguration study, *IEEE/ACM Trans. on Networking,* vol. 8, no. 5, pp. 598–607, Oct. 2000.

46. B. Ramamurthy, D. Datta, H. Feng, J.P. Heritage, and B. Mukherjee, Impact of transmission impairments on the teletraffic performance of wavelength routed optical networks, *IEEE J. Lightwave Technol.*, vol. 17, no. 10, pp. 1713–1723, Oct. 1999.

47. J.-F.P. Labourdette and A.S. Acampora, Logically rearrangeable multihop lightwave networks, *IEEE Trans. on Commun.*, vol. 39, no. 8, pp. 1223–1230, Aug. 1991.

48. A. McGuire, S. Mirza, and D. Freeland, Application of control plane technology to dynamic configuration management, *IEEE Commun. Mag.*, vol. 39, no. 9, pp. 94–99, Sept. 2001.

49. A. Gencata and B. Mukherjee, Virtual-topology for WDM mesh networks under dynamic traffic, *Proc., IEEE INFOCOM*, New York, June 2002.

50. N. Jerram and A. Farrel, MPLS in optical networks, White Paper.

51. A. Gencata, L. Sahasrabuddhe, and B. Mukherjee, Virtual-topology adaptation with minimal lightpath change for dynamic traffic in WDM mesh networks, *Proc., Optical Fiber Commun. Conf. — OFC*, pp. ThGG119, Anaheim, CA, Mar. 2002.

52. Internet Engineering Task Force, http://www.ietf.org

53. A. Banerjee et al., Generalized multiprotocol label switching: an overview of routing and management enhancements, *IEEE Commun. Mag.*, vol. 39, no. 1, pp. 144–150, Jan. 2001.

54. D. Awduche and Y. Rekhter, Multiprotocol lambda switching: combining MPLS traffic engineering control with optical crossconnects, *IEEE Commun. Mag.*, vol. 39, no. 3, pp. 111–116, Mar. 2001.

55. Implementation Agreement OIF-UNI-01.0, http://www.oiforum.com, Sept. 2001.

56. S. Yao, B. Mukherjee, and S. Dixit, Advances in photonic packet switching: an overview, *IEEE Commun. Mag.*, vol. 38, no. 2, pp. 84–94, Feb. 2000.

57. L. Xu, H.G. Perros, and G. Rouskas, Techniques for optical packet switching and optical burst switching, *IEEE Commun. Mag.*, vol. 39, no. 1, pp. 136–142, Jan. 2001.

58. M.J. O'Mahony et al., The application of optical packet switching in future communication networks, *IEEE Commun. Mag.*, vol. 39, no. 3, pp. 128–135, Mar. 2001.

59. S. Yao, S.J.B. Yoo, B. Mukherjee, and S. Dixit, All-optical packet switching for metropolitan area networks: opportunities and challenges, *IEEE Commun. Mag.*, vol. 39, no. 3, pp. 142–148, Mar. 2001.

60. C. Qiao and M. Yoo, Optical burst switching (OBS) — a new paradigm for an optical internet, *J. High Speed Networks*, vol. 8, no. 1, pp. 69–84, 1999.

61. S. Verma, H. Chaskar, and R. Ravikanth, Optical burst switching: a viable solution for terabit IP backbone, *IEEE Network*, vol. 14, no. 6, pp. 48–53, Nov. 2000.

chapter two

Evolution of optical networks architecture

M. Yasin Akhtar Raja
University of North Carolina, Charlotte

Contents

2.1. Introduction

This chapter, together with Chapter 19, on optical transport networks, reviews the evolution of optical networks from the architectural and signal transport perspectives, respectively. The architectural delineation of optical networks can be considered from various geographical domains, multiplexing technologies, switching and routing functions, and transport capacity and technologies. Today, a revolution has occurred from earlier single-wavelength synchronous optical net (SONET)/synchronous digital heirarchy (SDH)-based point-to-point transport to various phases of multi-wavelength optical transmission networking and subsequent bandwidth explosion via advances in dense-wavelength division multiplexing (DWDM). Currently, an interconnection of various point-to-point optical links based on SONET/SDH rings, trees, and optical mesh topologies constitute the optical networks infrastructure. However, the evolution of the optical networks toward more flexible, survivable, scalable, and interoperable architectures is ongoing as we write these chapters. Despite all progress in the optical transport from high-speed TDM (OC-192) and DWDM, true all-optical networks have yet to be realized. In limited cases, long-haul and metro-area networks have achieved some wavelength-routing capabilities.

Meanwhile, although the optical-packet switching area has progressed with innovative protocols and optical labeling approach, it seems to be a technology of the future. The bottleneck in the optical networks obviously is the absence of all-optical switching and routing. Currently, several technologies are competing in a wide-open space. Namely, reconfigurable optical add/drop multiplexers (OADM), along with the rare-earth doped-fiber optical amplifiers (XDFA), are the enabling technologies of wavelength-routed DWDM optical networks using optical cross-connects (OXC) that are limited in size.

In this chapter, the discussion is confined to the geographical architecture only. Meanwhile, optical transport networks, which encompass all geographic domains, are considered in Chapter 19. Various multiplexing technologies are also included.

2.2   Background

The evolution of optical networks started from the SONET[1-4] with the definition of hierarchical electric time-domain multiplexing (ETDM) as synchronous transport signal (*STS-n*) for high bit rate in optical domain as optical channel of order "n" (*OC-n*). SONET is an interface (not a network) that aggregates all traffic at the electrical multiplexers using ETDM (i.e., all low-bit-rate streams add up to a high bit-rate using a common synchronized clock). The high bit stream *STS-n* is then converted (by the optical transmitter) to an optical signal *OC-n* that travels in fiber. At the physical layer synchronous time domain multiplexing (SONET/SDH) defines a frame format (125 μs duration) and TDM hierarchy as OC-n/STM-m with n = 3m, and the bit-rate compatibility first establishes between OC-3 and STM-1 (i.e., OC-1 has no equivalent bit-rate) and STM-m hierarchy involves 4× multiplexing.[1,4] A finer granularity exists at the OC-1 and STS-1 levels in North America. An equivalent compatible global standard exists outside of North America that is commonly known as the synchronous digital hierarchy (SDH), and uses synchronous transport module (*STM-m*) standards. An STM-1 is equivalent to STS-3, and STM-4 is equivalent to STS-12 and so on (see Chapter 19).

SONET/SDH-based networks consist of nodes or network elements (NE) that are interconnected with fiber cable over which user and network management information is transmitted. Such point-to-point circuit-entities and NEs are the building blocks of the SONET- based optical networks that exist today in various topologies (e.g., rings, trees, and meshes). SONET NEs receive signals from various sources such as access multiplexers, asynchronous transfer mode (ATM), and other LAN/MAN/and WAN gears. SONET NEs and SDH circuit-elements must have a proper interface to convert (or emulate) the incoming data traffic into the SONET/SDH format.[4] Overall, SONET-based optical networks (SONET-rings) have dominated the long-haul and metro-space, and use optical-to-electrical-to-optical (OEO) switching and routing functions. Since its first deployment in the 1980s, SONET/SDH has almost replaced copper in the long haul, and every year millions of miles of new fiber have been laid down all over the globe. As the SONET/SDH-based optical links evolved from STS-3 (STM-1)/OC-3 to STS-48 (STM-4)/OC-48 bit-rates and recently to OC-192, with experimental deployment of OC-768.[5] Concurrently, a revolution has occurred in the wavelength domain by multiplexing several wavelengths on a single fiber strand.

In parallel to wavelength division multiplexing (WDM) and TDM technologies, researchers are pushing optical time domain multiplexing (OTDM) that can take tens of gigabit streams to several hundreds of gigabits streams.[3] Although the nominal wavelength domain multiplexing had been implemented for 1310 and 1550 nm in the mid 1980s, an explosive growth occurred in the 1990s when EDFAs[6] became available and multiple wavelength signals could be amplified without de-multiplexing. The so-called coarse WDM migrated to dense-WDM (DWDM) and ultra-dense WDM, and new standards are still evolving.[7] This added dimension to the OC-n hierarchy has resulted in a truly explosive growth in capacity from single-wavelength 2.5 and 10Gb/s to hundreds of gigabits and even terabits with a potential trend towards petabits.[8,9] Although ultra high-bit and ultra broadband became reality in the long-haul space, the switching and routing still remains a major obstacle with only limited deployment of optical switches at a wavelength granularity. In literature,[1–10] one finds an array of terminology to describe the various optical networks related to architecture, geographical coverage, multiplex-technology, management, switching- and routing-based technologies and so on. It is virtually impossible to fully cover all the aspects of optical networks in a single chapter or even in a single monograph.

In the forthcoming sections, we will confine to an overview of optical networks architectural topologies with respect to geographical coverage only and few multiplexing technologies in Chapter 19. Meanwhile, other aspects such as grooming, and switching and routing will be also briefly reviewed in Chapter 19, whereas functionality, optical-packet switching, optical multicasting, self-healing, virtual private networks, IP-over WDM, management, and other issues of optical networks are left for the future. Another classification pertains to first-, second-, and third-generation of networks, and interested readers can find more details in *Optical Networks* by Ramaswami and Sivarajan (1998), and *Optical Networks* by Black (2002).[3,10] From the hybrid and all-optical networks perspective, based on the hardware infrastructure of the optical/photonic layer as defined in the layered model of communication networks,[2,3,10] today's networks fall under the hybrid categories. Generally, the all-optical networks will become a reality when an all-optical switching fabric replaces the optoelectronic (OEO) switching and routing nodes and photonics reaches the access/metro-edge. In Chapter 19, however, a more focused discussion will be given on the "optical transport networks," also known as "photonic transport networks," because their functionality encompasses all the categories from the edge to the core.

2.3   Optical networks architecture

Aside from the other enabling component technologies (switches and routers), a point-to-point optical transport link is the basic building block of all type of optical networks.[11] Interconnections of several point-to-point optical links constitutes an optical network with certain logical topologies (e.g., rings, trees, or meshes and various combinations thereof). Several books and countless articles, which describe a single-wavelength (SONET-based)[1–4] and multi-wavelength (i.e., open WDM[12] point-to-point optical links), are available in the literature. For clarity, we illustrate the concept of "open WDM architecture" in the point-to-point optical transport link in Figure 2.1. An open architecture allows the 1310-nm based short-reach SONET interfaces to communicate with the transponders. A single or multiple wavelengths are then assigned, multiplexed, amplified, and transported over a dual fiber for up/down traffic streams. At the open WDM terminals, the transponders convert/groom the incoming SONET/SDH as well as non-SONET signals and assign the available wavelengths from the ITU-grid and then multiplex all channels (wavelengths). Subsequently an EDFA (optical amplifier) boosts the signals before launching into the transport fiber (if the distance limitations so stipulate). Typically, for longer fiber spans, several EDFAs amplify the signals almost at every 120 km sections. At the other end, the incoming multi-wavelength signals are pre-amplified prior to de-multiplexing and sending into the transponders. A similar process occurs for the upstream traffic in the other fiber strand.

Several point-to-point optical transport links (such as Figure 2.1) constitute an optical network infrastructure. Depending upon the stretch of the deployed fiber, multiplexers, and switching and routing hardware, the

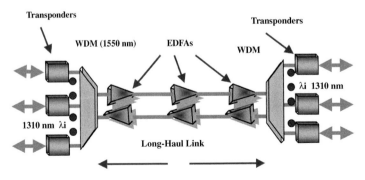

Figure 2.1 A schematic of a high-capacity open WDM point-to-point link, a building block of optical networks.

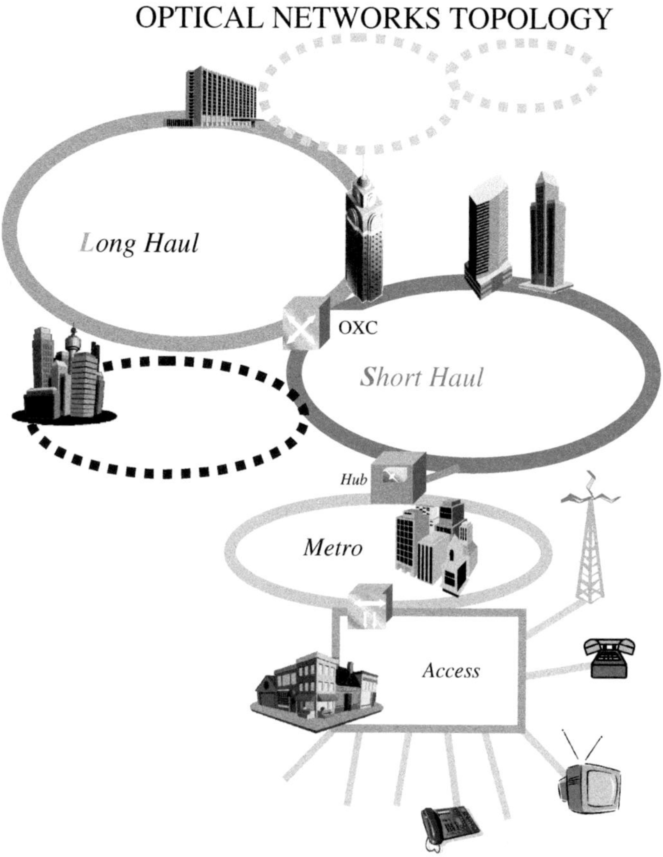

OPTICAL NETWORKS TOPOLOGY

Figure 2.2 A general view of optical networks' architecture (distribution, access/edge, metro-core, and long-haul subnets).

logical sub-network topology can be defined as rings, trees, and mesh and their combinations.

We shall only consider the geographical area definition of optical networks, which indeed follows the same approach as prevalent in the legacy communication networks.[13] Figure 2.2 illustrates the main concept that extends to various geographical areas, each covering certain sub-networks and interconnections. The overall network infrastructure is divided into three major zones: access/edge-optical, regional/metro-core optical, and long-haul/backbone optical networks. Access/edge-networks use a broad range of equipment technologies and protocols to provide the connectivity to user/client premises to the nearest node or carrier central office. These networks usually extend 1 to 10 km distances and are often referred to now as "the last/first-mile" of communication networks. Meanwhile, long-haul/core/backbone networks span greater regional and global distances, exceeding a 500 km or larger reach, and provide interconnectivity between various regional and metropolitan domains and subnets. Finally, metro (metropol-itan)/regional networks on the other hand bridge the gap between the access/edge and long-haul/backbone networks, and span the ranges of about 10 to 500 km. These networks can be delineated further into subdi-visions such as the "metro-edge" and "metro-core" domains of the optical networks.[14] Typically, these distance ranges can vary significantly and, therefore, have no sharp boundaries between these geographical domains. Usually, the density of traffic together with span/reach plays a role in categorizing various domains of optical networks.

Another perspective of optical networks pertains to various optical transport networks and distribution networks (in user/client domain). As one finds in literature,[1-4] the *optical transport networks* of course would include access/edge, regional/metro-core, and long-haul networks all integrated together. Meanwhile, distribution networks encompass more focused access/metro-edge domains.

2.4 Long-haul optical networks

For years, the main thrust in voice and data transmission has been high bit-rate transport over longer and longer distances. Fiber optics and related technologies have enabled such capabilities, and hence constitute the "back-bone optical networks" in the communication infrastructures. The first-gen-eration optical networks that provided high-speed and long-haul transport were based on SONET/SDH.[1-4] In such optical networks, the data packets are transported at high bit rate in the optical domain over long spans of fiber; however, circuit switching, traffic separation, routing, and protection functions are performed in electronic domain. This requires optical-to-elec-trical and electrical-to-optical (O-E-O) conversions, and thus can handle a single or at the most a few wavelengths. As the bit-rates increased to 2.5 and 10Gb/s in SONET/SDH and also DWDM emerged to increase the fiber-bandwidth utilization, traffic processing in long-haul transport and, at the intermediate nodes, became a complex, cumbersome, and expensive task. As a result, the optical networks then evolved into their second generation where several routing and switching functions are handled optically with electronic controls. A major breakthrough of course occurred due to the advent of EDFA,[15] which allowed optical amplification of multiwavelength signals simultaneously and eliminated the need for OEO conversion and regeneration in long-haul transport.

The other two equally important enabling component technologies in the second generation of optical networks are the optical add-drop multi-plexers (OADMs)[16,17] and the optical circulators.[3] These components provide crucial routing and switching functions at wavelength level. Second-gener-ation optical networks rely on such WDM technologies and constitute the optical-layer that provides services such as *lightpath, circuit-switching,* and *virtual-circuits.*[3] All-optical switching will take the networks to the next gen-eration of optical networks that is commonly recognized today as the third generation of optical transport systems.[10,14] The switching and routing functions of such third generation optical networks, of course, will be provided by the optical cross-connects. Optical cross-connects (OXC)[18,19] and photonic switches have been an area of active research over the past decade, and numerous types of devices have been demonstrated with attributes and limitations. Typical examples are MEMS[20] and bubble-jet (thermooptic)-based switches that provide switching and routing at wavelength level; but the time response and switching matrix size is limited, and lifetime and long-term reliability are concerns. Other developing optical switch technologies are based on liquid crystals, acoustooptic devices, electro-holographic, and various thermooptic methods. Today, there is no clear winner because the demand is on a whole list of factors (e.g., life, reliability, throughput, yield, size, speed, scalability, spectral-response, and cost issues), which makes the optical-switches quite a challenging technology.

Long-haul networks can further be divided into regional, national, and global domains with respect to their geographical spans. A major requirement on all the subnets is their throughput, interoperability, and transparency. The incoming data aggregation at the switching/routing nodes must have no latency and should not require extensive conversions and transformations. Only optical multiplexing and de-multiplexing with minimum grooming[21] would be needed.

2.5 Regional/metro optical networks

As mentioned previously, regional/metro area optical networks span geographical distances about 10 to 500 km and bridge the gap between user/client-based access and distribution-networks, and the high-capacity, high-speed, long-haul/backbone optical networks. Legacy metro-optical networks are primarily SONET/SDH-based networks that evolved from the need of standardization for multi-vendor interconnectivity and interoperability at the fiber level. However, over time, SONET/SDH high-capacity optical transport features have been augmented with other network functionalities, e.g., tributary cross-connection and routing/switching, add/drop multiplexing (ADM), regeneration and amplification, and protection, etc. In the emerging new technologies additional network functions such as wavelength domain multiplexing (WDM), optical ADM (OADM), and high-speed transport at OC-192/STM-64 (10Gb/s) are becoming more common. SONET/SDH-system-based metropolitan optical networks provide resiliency and exist today in various configurations such as linear chain (bus), tree, ring, and mesh.[1,4] Among all these logical configurations, however, ring topologies are by far the most dominant in metro/regional areas, although current trends indicate that mesh topologies may become ubiquitous in the future as optical switching technologies mature.

Regional/metro optical networks have received considerable attention recently because long-haul has matured and a natural next step is to bring the OC-192/STM-64, OC-768/STM-256,[5] and DWDM technologies into the

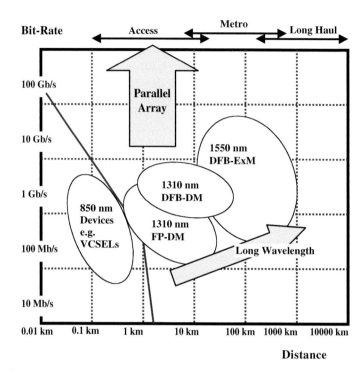

Figure 2.3 Nominal geographical spans of access, metro-core/regional, and long-haul networks as well as corresponding transmission rates with short- and long-wavelength transmitter devices.

metro domain. For a recent comprehensive review, the interested reader should see Reference 14. A finer delineation of metro optical networks can be metro-edge (closer to access) and metro-core (regional) optical networks. These geographical categories apparently are arbitrary but serve a purpose for equipment specifications and needs for restorations and protection (i.e., management functions).

Because metro-edge optical networks serve as interconnects and bridge the gap between the access networks and the high-speed metro-core and, subsequently, to long-haul/backbone optical networks, several equipment and protocol requirements need to be considered. Due to ubiquitous ring topology in the metro-space, we shall confine the discussion to metro-edge SONET/SDH rings and SONET over DWDM. Most of the metro-edge rings span an average length of 10 to 40 km and aggregate low-bit rate traffic from several central offices, usually at OC-3/STM-1 and OC-12/STM-4 rates and transport at higher speeds (e.g., OC-48/STM-16). Being at metro-edge, key network elements, such as add-drop multiplexers (ADM), must not only perform multiplexing functions, but also be able to groom the traffic from the client tributaries. From the access networks the incoming traffic may be comprised of various low-bit SONET rates and non-SONET digital signals up to hundreds of Mb/s. Metro-edge rings deal

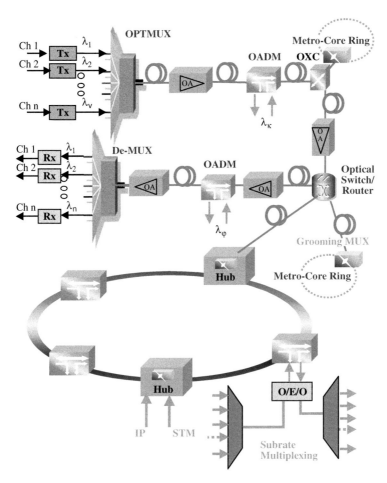

Figure 2.4 Typical long-haul DWDM optical networks architecture with opaque and transparent nodes and sub-rate-multiplexing at OADMs.

with hubbed-traffic patterns and are well suited for unidirectional path-switched rings (UPSR) approach. UPSR dual counter-rotating rings perform one-to-one receiver interconnections to provide protection signaling.[1] Subsequently the metro-edge traffic is routed on to larger metro-core rings at the inter-office level by connecting to key hub locations. In particular, this interconnection is performed via SONET/SDH digital cross-connects (DCS) nodes residing between metro-edge and metro-core (inter-office fiber) rings, e.g., commonly termed as wideband-DCS.[1]

In traditional metro-edge networks, an evolution of data networking protocols based on layering onto voice-centric infrastructure has occurred. Data protocols are mapped onto SONET/SDH frames via intermediate adaptation protocols and equipment such as ATM or frame relay (switches). Consequently, different layers are used to implement a complete service definition;[14] that is, SONET/SDH for transport and protection, ATM for bandwidth/traffic engineering, and IP for connectivity and routing. More recently, in advanced networks "direct packet over SONET" interfaces have been defined where IP packets are mapped onto the high-level data link control protocol and then framed into SONET/SDH. For example, many gigabit switches and routers provide such high-speed packets over SONET (POS) interfaces, i.e., OC-48c/STM-16c and Oc-192c/STM-64c, for connectivity to larger metro-core rings.

As discussed in the beginning of this section, regional/metro-core optical networks are comprised of a series of SONET/SDH rings that are categorized as metro-core rings. The metro-core rings infrastructure runs through several major central office hub locations, and constitutes the "core layer" as illustrated in Figure 2.5. These metro-core rings are then interconnected via DCS nodes as meshes and to backbone/long-haul infrastructures. These metro rings (Figure 2.5) are also known as "interoffice fiber" or collector rings and provide a higher level of aggregation and optical transport between metro-edge and backbone/long-haul networks. Legacy

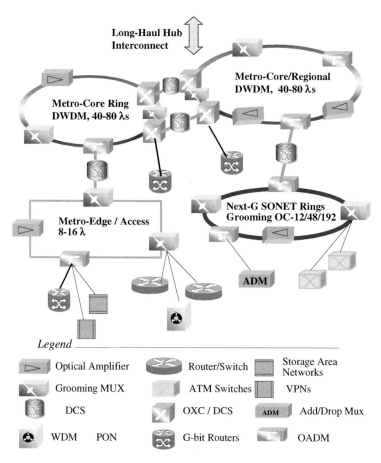

Figure 2.5 Metro optical networks architecture from edge-to-core.

metro-core setups commonly use OC-48/STM-16, and the new deployment is mainly OC-192/STM-64; thus, its demand is increasing rapidly.[22] Typical ring sizes range from 50 to 250 km, and for regional rings it can be approximately 500 km. Due to large geographic coverage and traffic demand in metro-core, rings are predominantly meshed between central offices and ADM locations. In this situation, bidirectional line-switched rings (BLSR) are suited for higher bandwidth efficiency. Two- and four-fiber BLSRs are quite common for performing loop-back span switching for logical (bi-fiber) and physical (quad-fiber) rings, respectively.[23] A recent detailed discussion on metro-core networks design and functionality can be found in Reference 14.

As will be discussed in the later sections of Chapter 19, dense wavelength division multiplexing (DWDM) has become dominant in the long-haul/backbone optical transport networks. It is now permeating rapidly into metro space, especially in the core first, and later will migrate all the way to the metro-edge domains. The metro-edge will play a vital role in grooming and mapping diverse traffic protocols onto the wavelengths and will enable end-to-end "lightpaths."[14] Eventually, it will spread into access networks in response to the growing bandwidth demand and will become enabling technology for the ultimate all-optical networks.

DWDM-over-SONET is a viable solution to provide scalable bandwidth reaching terabit per fiber per second while utilizing the existing fiber and optical amplifiers infrastructure. Currently, most of DWDM solutions are optimized as point-to-point transmission,[21,24] with fixed wavelength routing and optical add/drop multiplexing (OADM). The new metro DWDM solutions will include high-density transmission and reconfigurable wavelength routing over mesh and ring topologies. Clearly, dynamic or such reconfigurable OADMs will add more flexibility in DWDM application in metro space.

From the successful initial deployment of point-to-point DWDM systems in metro space, a natural next step was further interconnection of multiple wavelengths and spans of fibers into wavelength routing nodes that establish the *lightpaths*[2,3] with multihops for end-to-end connectivity over long-haul.[11] Hence, DWDM rings emerged from a natural progression and perform add/drop, pass-through, and protection functions. This will constitute "optical ring networks" (DWDM rings) that are very high capacity and offer scalability and transparency to variable high-bit-rate traffic. Several DWDM architectures such as static rings and reconfigurable/dynamic rings have been proposed in literature.[25] An elaborate account on such technology as applied to DWMD rings in metro area networks can be found in a recent reference.[14] The key component technologies include C- and L-band EDFAs[15,26] and static and reconfigurable OADM and OXC.[27] Using reconfigurable OADM and agile OXC various dynamic DWDM and self-healing rings architectures can be implemented. Optical rings and OXC would be deployed soon to support ring and mesh architectures in the metro area domains.[28] However, clearly fast "on-ramp" gateways will also be needed in the foreseeable future as the optical networks evolve to the next generation of all-optical networks. This concept has been tabled in the full-service access networks architectures (FSAN).[29]

2.6   Optical access networks (OAN)

Optical access networks[30] cover the "first-/last-mile" in the geographical topology and usually extend from 3 to 10 km. In the traditional telephone network infrastructure, the access networks constitute the "subscriber loops" where the copper twisted-wire pair connects the individual or institutional/ commercial customers to the local exchange carrier's network at the remote terminal or the central office level. Access networks are usually low-cost or support multiple cost structures to enable the evolution and services in an economically viable manner. Much like in the other segments of the optical networks, the optical hardware technologies are the enabler for generation, switching, transmission, and amplification of optical signals. The challenges pertain to physical layer issues and component technologies.[31] OAN must support the capacity requirements such as concentration and grooming for enterprise and residential users, and they must also support multi-bit-rate interfaces (e.g., constant and variable bit rates as well as synchronous and asynchronous data streams). Various architectures and one such example, based on the FSAN[29] concept, are illustrated in Figure 2.6.

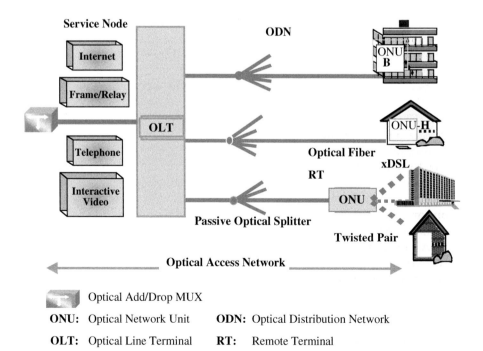

Figure 2.6 Typical optical access networks' architecture based on FSAN model.

Today, the bandwidth "bottleneck" has shifted to the first/last mile region, as growing end-user demands continue to drive traffic volumes. As growing bandwidth demand creates traffic jams on the "Internet highway," the worst congestions occur at enterprise sites in the suburbs or at university campuses. Hence, the problem does not concern the actual communication services; today's Internet service providers (ISP) and carriers already offer a multitude of products and services. Rather, the root of such problems lies in how these products and services are transported over the "optical highway." Clearly, a new breed of providers, called optical service providers (OSP), need a viable way to reach the suburban users and customers. One such solution involves DWDM technology. As mentioned earlier, this technology mixes different data signals using the same fiber, by modulating them on different independent wavelength channels. However, the issue with this technology is that it was geared and optimized for long-haul and metro-core networks; hence, it can be too expensive to allow services to be deployed from the metro-core/edge to the access networks. First, the deployment of optical access networks using optical add/drop multiplexers, EDFAs, and other optical components and subsystems are too costly for using in the last mile/subscriber loop. Second, traffic profiles and patterns change rapidly in the OAN arena, and hence the coexistence of optical and electrical infrastructures would not only be a desired feature but a necessity. Note that the "tree-and-branch" (ITU-T's G.983.1) or other switched digital architectures would probably be useful. Figure 2.6 shows an existing OAN architecture, which relies on the passive optical network (PON) concept, and constitutes a full-service access network (FSAN) topology.[29]

FSAN defines the requirements and solutions for low-cost-fiber-based access networks. It requires a very flexible, economical, and scalable service capability and components technologies. Typical examples are fiber to the curb (FTTC), fiber to the cabinet (FTTCab), and fiber to the home (FTTH). Important issues include cost-effective solutions for optical multiplexing and de-multiplexing, although the dominant costs are those associated with digging for conduits and terminal equipments. For FTTCab, low-cost VDSL chipsets are becoming available as well as are the low-cost optical network units (ONUs) for FTTH setups.[28] Currently, ATM-based PON (APON) can support up to 32 subscribers over a distance of about 20 km with bit rate exceeding 622Mb/s downstream and 155Mb/s upstream traffic.

Although today OAN is mostly supported with FTTCab in access space, few experimental/field trial type FTTH systems exist as well. In particular, some of the key requirements related to OAN deployments include:

1. Low-cost medium-range (over 20 km distance) and high-density (over 32) channel fan-out
2. High-density grooming and multiplexing to accommodate hundreds and thousands of users
3. Local dc power backup at customer's site

4. Remote service activation
5. Digging streets to add more fibers and/or investing in the DWDM infrastructure in the access domain

In addition, it needs to support integration of several existing and a variety of emerging access technologies. One promising solution would be to deploy new multiplexing techniques and exploit lower-cost components such as optical amplifiers, OADMs, and DWDM terminals. Many studies have shown steady price declines, and over time the economic viability of many of these technologies will be quite feasible.

2.7   All-optical networks — the wave of the future

Today, the emergence of an all-optical network (AON) is more likely than ever. Indeed, the exploding demand for bandwidth, driven by emerging technologies and applications, makes the development of an end-to-end fiber-rich network an inevitability. The AON era is approaching, and the telecom industry must decide how it is going to manage the transition in an efficient and cost-effective manner. The transition will be smooth if industry leaders choose the correct routes, architectures, and infrastructures along the way — both as short-term solutions in an optical-electrical-optical environment and as a framework for the long term. The optical industry must examine the different qualities and types of service enabled by different infrastructure configurations and then decide which services to offer over those infrastructures, knowing that those systems will be in a developmental flux for the next decade and standards will be still evolving.

Because companies in the optical industry must still be able to turn profits, such a fluid environment requires flexible, innovative approaches to mixed and hybrid network management. How successful industry is in maintaining revenues (i.e., backward-compatibility) while introducing new approaches to network development and management will determine the success of the transition to the future AON.

In order to help network operators take full advantage of optical networking technology, the International Engineering Consortium (IEC) and other standard bodies, e.g., ITU-T and IEEE, are playing an important role to identify the architectures and protocols. The very latest developments in optical technologies, networks architectures, and market trends driving the next-generation network will be enablers for future AON. With its range of perspectives on the promise of an AON, the legacy optical networks, their interoperability, and various architectural advances are critical resources for any communications company hoping to tackle today's optical networking challenges.

2.8   Acknowledgment

I wish to thank Dr. Nasir Ghani of Sorrento Networks, Inc. for his invaluable suggestions and a careful review of this chapter.

References

1. M. Chow, *Understanding SONET/SDH, Standards and Applications,* Andan Publishers, Holmdel, NJ, 1995.
2. B. Mukherjee, *Optical Communication Networks,* McGraw-Hill, New York, 1997.
3. R. Ramaswami and K.N. Sivarajan, *Optical Networks: A Practical Perspective,* Morgan Kaufmann Publishers, San Francisco, 1998.
4. S.V. Kartalopoulos, *Understanding SONET/SDH and ATM, Communications Networks for the Next Millennium,* IEEE Press, Piscataway, NJ, 2002.
5. Technical Digest Optical Fiber Communication OFC 2002, Anaheim CA, March 17–22, 2002; see, for example, ThEE sessions pp.600–613.
6. P.C. Becher, N.A. Olsson, and J.J.R. Simpson, *Erbium-Doped Fiber Amplifiers: Fundamentals and Technology,* Academic Press, San Diego, CA, 1999.
7. ITU-T, Draft Reccommendation, Extension of the nominal central frequencies grid included in Annex A of the present G92 draft rec for future version of G.692 Rec. http://www.ITU.Int/ITUDOC/ITU-T/COM15/DCONTR/DC-OCT98/306.html, 2002.
8. B. Mikkelson et al., High spectral efficiency (0.53 bit/s/Hz) WDM transmission of 160Gb/s per wavelength over 400 km of fiber, Changing at speed of light, *OSA Technical Digest, OFC 2001,* March 22, 2001, pp. ThF2–1–3.
9. U. Feiste et al., 160Gb/s transmission over 116 km field-installed fiber using 160Gb/s OTDM and 40Gb/s ETDM, Changing at speed of light, *OSA Technical Digest, OFC 2001,* March 22, 2001, pp. ThF3–1–3, and references therein.
10. U. Black, *Optical Networks: Third Generation Transport Systems,* Prentice Hall, Upper Saddle River, NJ, 2002.
11. K. Sato, Introduction strategy of photonic network technologies to create bandwidth abundant multimedia networks, *Trends in Optics and Photonics,* vol. 20, June 1998.
12. K. Bala, Multi-wavelength optical networks (WDM), *Handbook of Emerging Communications Technologies: The Next Decade,* Saba Zamir, Ed., CRC Press LLC, Boca Raton, FL, 2000.
13. J.J. Hecht, *Understanding Fiber Optics,* 4th ed., Prentice Hall, Upper Saddle River, NJ, 2002.
14. N. Ghani, J.-Y. Pan, and X. Cheng, Metropolitan optical networks, in *Optical Fiber Communications IVB: Systems and Impairments,* I. Kaminow and T. Li, Eds., Academic Press, San Diego, CA, 2002.
15. Y. Sun, A.K. Srivastava, J.J. Zhou, and J.W. Sulhoff, Optical fiber amplifiers for WDM optical networks, *Bell Labs Tech. J.,* vol. 4, no. 1, 1999, pp. 187–206.
16. J.J.E. Ford, V.A. Aksyuk, D.J.J. Bishop, and J.J.A. Walker, Wavelength add/drop switching using tilting micro-mirrors, *J. Lightwave Technol.,* vol. 17, no. 5, May 1999, pp. 904–911.
17. C.R. Giles and M. Spector, The wavelength add/drop multiplexer for lightwave communication networks, *Bell Labs Tech. J.,* vol. 4, no. 1, 1999, pp. 207–229.
18. C.R. Doerr, Proposed WDM cross-connect using a planar arrangement of waveguide grating routers and phase shifts, *IEEE Photon Technol. Letters,* vol. 10, April 1998, pp. 528–530.
19. Y.K. Chen and C.C. Lee, Fiber Bragg grating-based large nonblocking multiwavelength cross-connects, *J. Lightwave Technol.,* vol. 16, no. 10, 1998, pp. 1746–1756.

20. Special Feature Issue on "Optical Switching," *IEEE Commun. Mag.*, vol. 40, no. 3, March 2002, pp.72–101.
21. K. Bala, Grooming in the optical network, *SPIE Optical Network Mag.*, vol. 3, no. 1, Jan./Feb. 2002.
22. A. Shivji, SONET gear for the optical generation, *National Fiber Optic Engineers Conf. (NFOEC) 2000*, Denver, CO, August 2000.
23. T.H. Wu, *Fiber Network Service Survivability*, Artech House, Boston, 1992.
24. J.J.M.H. Elmirghani and H.T. Mouftah, Technologies and architectures for scalable dense WDM networks, *IEEE Commun. Mag.*, vol. 38, no. 2, Feb. 2000, pp.58–66.
25. P. Jaggi, H. Onaka, and S. Kuroyanagi, Optical ADM and cross connects: recent technical advancements and future architectures, *National Fiber Optic Engineers Conf. (NFOEC) 1998*, Orlando, FL, Sept. 1998.
26. A.M. Samara, M.Y.A. Raja, and S.K. Arabasi, High-gain short-length phosphate glass Er-Yb-doped fiber amplifier, *SPIE Symp. Optical Devices, Components and Systems, OPTO South East*, Clemson, SC, Oct. 4–5, 2001.
27. D. Stoll et al., Metropolitan DWDM: a dynamically configurable ring for the KomNet field trial in Berlin, *IEEE Commun. Mag.*, vol. 39, no. 2, Feb. 2001, pp. 106–113.
28. B. Khasnabish, Optical networking issues and opportunities: service providers' perspectives, *Optical Networks Mag.*, vol. 3, no. 1, Jan./Feb. 2002, pp.53–58.
29. ITU-T Recommendation G.983.1, Broadband optical systems based on passive optical networks (PON), 1998; Recommendation G.983.2, ONT management and control interface specifications for ATM PON, 2000; Recommendation G.983.3, A broadband optical access system with increased service capability by wavelength allocation, 2001.
30. Clavenna, Access Networks: DWDM's newly charted territory is metropolitan access, *Lightwave Magazine*, Special Reports, September 1999, pp. 49–59.
31. E. Harstead and P.H. van Heyningen, Optical access networks, in *Optical Fiber Telecommunications IVB: Systems and Impairments*, E. Kaminow and T. Li, Eds., Academic Press, San Diego, CA, 2002, pp. 438–513.

chapter three

Design aspects of optical communication networks

Kanna Potharlanka
University of California, Davis
Toshit Antani
University of California, Davis
Byrav Ramamurthy
University of Nebraska, Lincoln
Laxman Sahasrabuddhe
University of California, Davis
Biswanath Mukherjee
University of California, Davis

Contents

3.1   Introduction

Optical communication networks can support very high bandwidth because of the maturity of the wavelength-division multiplexing (WDM) technique. WDM transport systems with tera-bit-per-second (Tbps) capacity on a single strand of fiber are commercially available now; however, a number of technical challenges remain to be addressed before optical networks become the effective basis for a robust, high-capacity, and scalable next-generation information infrastructure. One such challenge is the development of standardized architectures and protocols for optical networks, appropriate at very high speeds, under different environments such as local, metro, and long-haul networks.

Although experimentation on a test-bed will reveal useful information about these issues, a simulation tool that can provide rapid and accurate analysis of various design choices is now essential for timely resolution of the engineering challenges posed by optical networks. In an optical network environment, simulation is often the dominant tool to analyze network behavior. For this purpose, we have designed and developed a tool called ARTHUR (A Routing And Wavelength Assignment Tool For Optical Networks). ARTHUR supports several useful features such as dynamic routing and wavelength assignment, wavelength conversion, and fault-management. ARTHUR can be used as a network design and planning tool and as a protocol analyzer. As a network designer/planner, ARTHUR can be used to create new network topologies and analyze their performance using various algorithms. The performance of various algorithms can be analyzed by testing them for different network topologies.

3.1.1   Optical networking and wavelength-division multiplexing

Wavelength-division multiplexing (WDM)[1-4] is a promising technology to exploit the enormous bandwidth of optical fibers. Multiple channels can be operated along a single fiber strand simultaneously, each on a different wavelength. Thus, WDM carves up the huge bandwidth of a single-mode optical fiber into channels whose bandwidths (for instance, 10 Gbps) are compatible with peak electronic processing speed. WDM-based optical networks are envisaged for spanning local, metropolitan, and wide geographical areas. Our focus here is on the design of wavelength-routed WDM networks that can provide wide geographical coverage.

The use of wavelength to route data is referred to as wavelength routing, and a network that employs this technique is known as a wavelength-routed network. Such a network consists of wavelength-routing switches (or routing nodes) interconnected using optical fibers. Some routing nodes are attached to access stations where data from several end users could be multiplexed on a single optical channel. The access station also provides optical-to-electronic conversion and vice versa to interface the optical network with conventional electronic equipment. A wavelength-routed network that carries data from one access station to another without any intermediate optical-to-electronic conversion is referred to as an all-optical wavelength-routed network.

In a wavelength-routed WDM network, end users communicate with one another via optical WDM channels, which are referred to as *lightpaths*. A lightpath may span multiple fiber links. When there is no direct lightpath on the same wavelength between the source and destination nodes, then in order to setup a lightpath, wavelength converters may be required. Networks that employ nodes equipped with wavelength converters are called wavelength-convertible optical WDM networks. In the absence of wavelength converters, a lightpath must occupy the same wavelengths on all fiber links through which it passes. Thus, given a set of lightpaths that need to be established on the network, and given a constraint on the number of wavelengths, we need to determine the routes over which these lightpaths should be set up and also determine the wavelengths that should be assigned to these lightpaths. This problem is known as *the routing and wavelength assignment* (RWA) problem.[3,5,6]

3.1.2   Control and management of WDM networks

Network control and management (NC&M) is important for any network. The cost of managing a large network in many cases dominates the cost of the equipment deployed in the network. Classically, network management consists of several functions:

- *Configuration Management* deals with the set of functions associated with managing orderly changes in a network. For example, this could include setting up and taking down connections in a network. (A connection between two nodes in the network consists of a path in the network, with a wavelength assigned on each link along the path.) Other functions include tracking the equipment in the network and managing the addition/removal of equipment, including any rerouting of traffic this may require.
- *Performance Management* deals with monitoring and managing various parameters that measure the performance of the network. Performance management is an essential function that enables network operators to provide quality-of-service guarantees to their clients and to ensure that clients comply with the requirements imposed by the operator.
- *Fault Management* is the function responsible for detecting failures when they happen, isolating the failed components, and restoring traffic that was interrupted due to the failure.
- *Security Management* involves protecting data belonging to network users from being tapped or corrupted by unauthorized entities. It includes administrative functions such as authentication of users as well.
- *Accounting Management* is the function responsible for billing and for developing lifetime histories of the network components.

3.1.3   Need for a design tool

In order to achieve high efficiency in network design and implementation, one needs to develop and analyze accurate models of the network. Due to the complexity of the network model, it may be very difficult or impossible to predict and analyze the behavior of the network. The RWA problem is also known to be NP-complete;[7] in other words, there is no algorithm that can complete computation in polynomial time. Hence, simulation is the dominant technique for the design and analysis of large, realistic networks.

For an optical network environment, we have designed and developed a tool called ARTHUR. ARTHUR supports several useful features such as dynamic routing and wavelength assignment, wavelength conversion, and fault management. Our work mainly concentrates on connection management in optical networks. Connection management can be defined as the process of managing the connection requests. Through connection management, connections can be added or dropped dynamically. Whenever an "add connection request" is made, depending on the availability of routes and wavelengths, a connection can be accepted or blocked. If a "delete connection request" is made, the design tool (ARTHUR) checks the status of the connection and drops it by freeing the network resources associated with the connection. The performance of various RWA algorithms can be determined by analyzing the statistics available after the simulation is completed. Our work mainly focuses on developing such a simulation tool to perform RWA for wavelength-convertible optical WDM networks, although it can be easily extended to support wavelength-continuous networks.

Section 3.2 discusses the background and briefly outlines the features and limitations of ARTHUR's competitors by introducing existing simulation packages and discussing their advantages and disadvantages. Section 3.3 provides the global architecture of ARTHUR. This section also discusses one of ARTHUR's distinguishing features, namely protection from faults. Also discussed is the communication-application program interface (API) used. Section 3.4 discusses the algorithms used to address the dynamic routing and wavelength assignment problem. Section 3.5 demonstrates various applications of ARTHUR. Section 3.6 concludes this chapter.

3.2   Background

To study the behavior of optical networks, simulation software packages are very useful. Only a few network simulation tools are in existence, however; simulation tools for optical networks are even more rare. The following subsections briefly discuss existing simulation tools.

3.2.1   NS — network simulator[8]

A very popular network simulation tool is network simulator (NS). NS is a discrete-event simulator targeted at networking research. NS provides substantial support for simulation of transmission control protocol (TCP), routing, and multicast protocols. The NS simulation description language is an extension of the tool command language (TCL). Using an NS command, a network topology is defined, traffic sources and sinks are configured, the simulation is invoked, and statistics are collected. By building upon a general-purpose language, actions can be programmed into the configuration; however, one of the limitations of NS is that it lacks a user-friendly interface. To know how to use NS, the users must first learn TCL, and write the script as an input file using TCL.

3.2.2   REALEDIT[8]

A more user-friendly network simulation tool is REALEDIT. REALEDIT is a visual editor for REAL 5.0 Network Simulator. It is written in Java to provide cross-platform portability. REALEDIT provides a user-friendly graphical user interface (GUI) to simplify both building a network for simulations and viewing the results. Just as NS, however, REALEDIT also oversimplifies the network architecture: it uses a circle to represent the network node and a line to represent the links between nodes. This design omits details that need to be represented in a network design.

Both NS and REALEDIT have been used to study nonoptical networks. They do not support features such as fiber length, number of wavelengths, etc. that are essential for the study of optical networks.

3.2.3   SIMON — simulator of optical networks[9]

SIMON was developed to provide functions that are required for optical networks. Some features of the simulation environment are as follows:

- A user-friendly GUI for the user to build up the network architecture. The user can easily draw the topology of an optical network.
- Detailed representation of network architecture relative to NS and REALEDIT. For example, the input and output ports in a network node can be shown unlike NS and REALEDIT, which do not provide this capability.
- Every call in the simulation can be traced while the simulation is in progress. Information on a particular call such as route, wavelengths used, and timestamp can be known.
- Tools to view statistics of a simulation such as number of calls offered, number of calls established, and number of calls blocked.

3.2.4   MERLiN — modeling evaluation and research of lightwave networks

Modeling Evaluation and Research of Lightwave Network (MERLiN) is a WDM network design and modeling environment that allows the development and evaluation of RWA algorithms. MERLiN was developed at the National Institute of Standards and Technology (NIST).

MERLiN has a client–server architecture where the client is a GUI, which allows a user to set up a network, make connection requests, and display results. The client code is divided into: Network Interface, Graphical Interface, and Components for the graphical interface, such as Node, Link, and Route. The server is a daemon process running in the background either on the same host machine as the client or on a remote machine.

MERLiN's limitations include no provision to address the dynamic RWA problem (see Section 3.3.3).[5] So, when a new set of connections need to be set up, all connections including existing ones are rerouted and then reassigned wavelengths. The network state (free and used wavelengths on all links of the network) is not maintained. MERLiN does not support wavelength conversion,[6] and hence, a lightpath must be set up on a single wavelength on all links on the route. This constraint is known as the wavelength-continuity constraint (see Section 3.4.3.1),[3,5] which could cause some new calls to be blocked. Also MERLiN lacks support for fault management.

3.3. ARTHUR

ARTHUR is based on a client–server architecture wherein several clients can connect to a single server. The client is a GUI, written in Java that connects to the server using UNIX sockets. The main purpose of the GUI is to provide a point-and-click facility to make connection requests (e.g., "add connection request" or "delete connection request") and display the results. The server is formed of the "core" and the "algorithms" modules. The algorithms module is where all RWA calculations are made. The core is an interface between client and algorithms. Both core and algorithms modules are written in C++.

A typical simulation would involve the following events: A set of connections is requested using the GUI. The GUI connects to the server. Once connected, the core parses the connection requests and invokes algorithms specified in the request. After computation, the algorithms module returns the results to core. Finally, the core sends back routes and other results including assigned wavelengths to the client, which are then displayed on the canvas of the GUI.

3.3.1   Design architecture

The architecture of ARTHUR consists of three major building blocks: core, plug-ins, and algorithms' library. Each block along with its components is a separate program that handles a specific aspect of the simulation process. This architecture provides a modular plug-and-play platform that lets users add new components such as algorithms and plug-ins and run them on different machines. The main reason for this distributed design is to facilitate the utilization of several computational resources (cluster of workstations) and distributed databases, as needed. Another advantage of this design is that it allows for language/system independent plug-ins and algorithms. A high-level view of the architecture is illustrated in Figure 3.1.

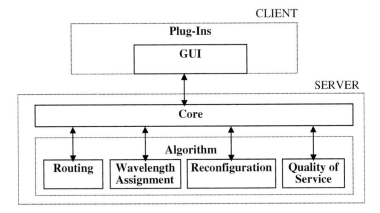

Figure 3.1 ARTHUR global architecture.

3.3.1.1   Core

The core of ARTHUR is a daemon process running in the background, which waits for connection requests. Given a script file as input that describes the topology, the routes, and the algorithms to be used, the core builds its own data representation of the network. Then, the core dispatches the topology to the requested algorithms. After the algorithms complete their computation, the core collects the results of the simulation and sends them back to the client.

3.3.1.2   Plug-ins

A plug-in is an interface module that allows easy access to core and to the algorithms database. It can be any stand-alone application that communicates with the core in ARTHUR's language format. A plug-in acts like a client in ARTHUR's client–server architecture. Its functionality is to send a request to the simulator and collect the results for the user.

The information exchanged between the plug-in and core (possibly over a network if the server is running on a remote machine) is in the form of a text-script file describing network components such as nodes and fibers, topology, a list of connection requests between pairs of source and destination nodes, and the RWA algorithms to be used.

The text-script file has to obey a number of syntax rules for the server/client to understand the requests/results. The different objects describing this file can be generated by the GUI. The GUI helps the user to build his/her own topology and make requests by simple mouse clicks. It then automatically

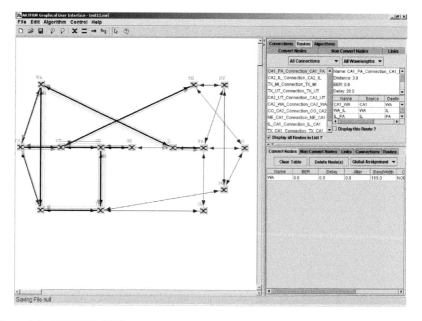

Figure 3.2 ARTHUR GUI.

generates the text file and sends it to the server when prompted. The plug-in interface allows the user to use the file format from other simulators and interface them to ARTHUR. Through dedicated plug-ins, it is possible to get information needed from a physical-layer simulator or higher-layer simulator (such as the NIST asynchronous transfer mode [ATM]/hybrid fiber coax [HFC] simulator) and translate the information into the format recognized by the core.

3.3.1.3   Algorithms

The algorithms module is a special kind of plug-in that communicates with ARTHUR's core. The main difference between an algorithm and a normal plug-in lies in the method used to exchange messages: the algorithms module uses ARTHUR's internal structures while a plug-in uses the script file representation. Another difference is that an algorithm is generally called by the core and not directly by the user. The algorithms library is divided into two parts: routing algorithms and wavelength-assignment algorithms.

3.3.1.4   Communication — application program interface (API)

Whenever communication is needed between a plug-in and core, or core and an algorithm, a communication pipe is opened. Then, messages are transferred both ways through a local or remote port depending on where the programs are running.

Communication between all the entities composing ARTHUR is conducted via a network interface using UNIX sockets. The packets generated have a specific format, as depicted in Figure 3.3.

The advantage of having such an interface is to make ARTHUR's plug-ins and algorithms language independent. The communication API of a plug-in only takes care of opening and closing the connection. Listed next is the description in pseudo-code of what a plug-in does:

- Connect to server communication port
- Call the pluginMain function
- Read an input file
- Send the file to server
- Wait for reply on the communication port
- If the answer is an error, return the error to the user
- If the answer is an acknowledgment message, read the parameters on the communication port
- Write the output file to disk
- Return from the pluginMain function
- Close the communication port

Size of the Packet	ARTHUR Command	Data

Figure 3.3 Packet format.

3.3.2   Implementation details

The server of ARTHUR is implemented in C++, and the client is implemented in Java. Object-oriented programming concepts have been widely used in both client and server designs. Advanced packages in Java such as Swing were used to implement the representation of the network topology on the GUI's canvas.

ARTHUR uses a framework of classes especially designed to simulate WDM networks. Listed next are the directories of the software that contain the source files:

- *Component* contains classes representing WDM components: link, cross-connect (see Section 3.3.3), wavelength, etc.
- *Math* contains the basic data types used in ARTHUR.
- *Arthur* contains source code of the server and some classes tightly bound to the ARTHUR internals.
- *Parser* contains code for the ARTHUR file parser and some classes tightly bound to the parser.
- *Tool* contains some useful classes such as list, array, matrix, and vector.
- *Network* contains classes used to manage communication between different entities of ARTHUR (server, plug-in, algorithms).

Apart from these classes, the parameter class represents the parameters list in and out of an algorithm. It manages several features that allow inserting newly developed parameters, accessing them, and removing them from lists. The communication process is hidden to the user. The server and client port managers are used in order to establish connections between the client(s) and server.

3.3.3   Dynamic routing and wavelength assignment[3,5,6]

When setting up a lightpath, we need to configure various switches in the network. The network control and management protocol is implemented in software and allows for connection *setup* and *teardown* via a user-friendly GUI.

The controller for a *wavelength-routing switch* (WRS) maintains the state of the connections flowing through the switch in a *connection switch table* (CST). A WRS uses an optical cross-connect (OXC), which switches signals arriving at different input ports to different output ports. In addition to the CST, the controllers maintain the topology of the network, and the availability of wavelengths on each link of the network. The network topology and wavelength information are maintained by a topology-update protocol. To set up a connection, a node needs to calculate the route for the connection, and a wavelength assignment for each link along the route. The following distributed protocol establishes a connection between two nodes:

- *Route and wavelength determination* — The originator of the connection finds a route to the destination, and an assignment of a wavelength to links along the route.
- *Reservation* — The originator of the connection requests all controllers along the route of the connection to reserve the selected wavelength on the links along the route.
- *Connection Setup/Release* — If reservation was successful at all controllers along the route, the connection is set up; otherwise, the reservation is released at all nodes along the route.

RWA schemes can be classified into static and dynamic categories depending on whether lightpath requests are known *a priori* or not. In dynamic RWA, lightpath requests between source-destination pairs arrive at arbitrary instants and each lightpath has an arbitrary holding time after which it is torn down. These lightpaths need to be set up dynamically by determining a route through the network connecting the source to the destination and assigning a free wavelength along this path. In contrast to the dynamic routing problem described previously, the static RWA problem assumes that all the connections that need to be set up in the network are known initially. The objective is to maximize the total throughput in the network (i.e., the total number of connections, which can be established simultaneously in the network).

3.3.4   Fault management in a WDM network

Providing resilience against failures is an important requirement for many high-speed networks. As networks carry more and more data, the amount of disruption caused by a network-related outage becomes more and more significant. Several techniques exist to ensure that networks can continue to provide reliable service even in the presence of failures. These fault-management techniques involve providing some redundant capacity within the network that is used to reroute traffic when a failure occurs. Moreover, it is desirable that the fault-management algorithms be implemented in a distributed manner without requiring coordination among all nodes in the network. This feature is necessary to ensure fast restoration of service after a failure.

We are mainly concerned with failures of network links. Links can fail because of a fiber cut. Our fault-management mechanisms are designed to protect against a single failure event. This assumes that the network is designed well enough that simultaneous multiple failures are very rare. In other words, it is unlikely that we will have another failure while we are trying to restore service from an earlier failure event.

3.3.4.1   Fault management in ARTHUR

A connection can be "protected" against link failure by calculating a backup path when the primary path for the connection is calculated.

When a link fails, the backup path kicks in. In ARTHUR, we support such protected connections.

We protect a connection using the k = 2 shortest-path routing algorithm. The algorithm calculates two paths for the connection (primary path and backup path). Wavelengths are assigned to both paths. Once the connection is established, only the primary path is shown on the GUI. A user can simulate a link failure by selecting a link and pressing the "Fail Link" button on the GUI. This causes all connections going through that link to switch over to their respective backup path, if any. At this point, the backup path of a connection is displayed on the GUI.[10]

Alternatively, one may choose not to protect a connection. This is done by using the shortest-path routing algorithm, which calculates only one path for the connection. The path is assigned wavelengths and shown on the GUI. If a user simulates a failure on one of the links used by the connection, the connection would cease to exist, as there is no backup path. The updated GUI would no longer show that connection.

3.4 Algorithms

ARTHUR has two kinds of algorithms:

1. Routing Algorithms (which perform route computation)
   a. Shortest-Path Routing Algorithm (k = 1)
   b. k = 2 Shortest-Path Routing Algorithm
2. Wavelength-Assignment Algorithms (WAA) (which perform wavelength assignment)[11]
   a. First-Fit WAA (perform wavelength conversion when necessary)

3.4.1 Generic events for any algorithm

An algorithm receiving a list of parameters from core performs some computation before returning an output list of parameters to core. Similar to the plug-ins, a communication API for the algorithms module exists. The communication API starts communication between an algorithm and core, and gets the parameters needed. Then, it calls a specific function named *algoMain*.

```
Void algoMain (ParameterList & param_in_list,
ParameterList &param_out_list);
```

The arguments are respectively the list of parameters needed as an input for the algorithm (filled by the communication API), and the output list of parameters for the algorithm. The output list of parameters contains the modified topology, accepted connections, and performance parameters (such as time to compute the routes) of ARTHUR.

3.4.2   Routing algorithms

3.4.2.1   Shortest-path routing algorithm (k = 1)[12]

The shortest-path routing algorithm computes a shortest path for a connection request between a node pair. We use Dijkstra's Algorithm[12] to compute the shortest path.

Input parameters.   Input Parameters are a description of the current topology that contains:

- A list of links
- A list of nodes
- A list of existing routes the algorithm considers in computing the new routes
- A list of connection requests

Output parameters.

- A list of new computed routes
- A list of all routes currently set in the topology — this includes previously existing routes in the given topology and newly computed routes

3.4.2.2   k = 2 shortest-path algorithm (edge-disjoint paths); Suurballe's algorithm[12]

In finding two edge-disjoint paths between a node pair, we use Suurballe's Disjoint-Pair Algorithm. The second path is edge disjoint with respect to the first path, which is useful for protection-related issues that are explained in Section 3.3.4.1. The algorithm calculates the optimal shortest-path pair instead of shortest path and next shortest path (which may not be link disjoint).

Input and output parameters.   The input and output parameters are the same as that of the shortest-path algorithm apart from the fact that, now, a connection will be associated with two routes.

3.4.3   Wavelength-assignment algorithm

3.4.3.1   Wavelength conversion

Consider the network in Figure 3.4. It depicts a wavelength-routed network containing two WDM cross-connects S1 and S2 and five nodes (A through E). Three lightpaths have been set up (from C to A on wavelength λ_1, from C to B on λ_2 and from D to E on λ_1). To establish any lightpath, we may require that the same wavelength be allocated on all of the links in the path. This requirement is known as the wavelength-continuity constraint[3] and wavelength-routed networks with this constraint are referred to as wave-

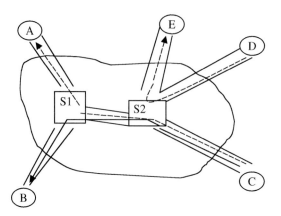

Figure 3.4 A wavelength-routed network.

length-continuous networks. The wavelength-continuity constraint distinguishes the wavelength-continuous networks from a circuit-switched network (or fully wavelength-convertible), which blocks calls only when no capacity exists along any of the links in the path assigned to the call.[13]

   Consider the portion of a network in Figure 3.5a. Two lightpaths have been established in the network: 1) between Node 1 and Node 2 on wavelength λ_1 and 2) between Node 2 and Node 3 on wavelength λ_2. Now, suppose a new lightpath between Node 1 and Node 3 needs to be set up. If only two wavelengths are available in the network, establishing such a lightpath from Node 1 to Node 3 is now impossible even though there is a free wavelength on each of the two links along the path from Node 1 to Node 3. This is because the available wavelengths on the two links are different. Thus, a wavelength-continuous network may suffer from higher blocking as compared to a wavelength-convertible network.

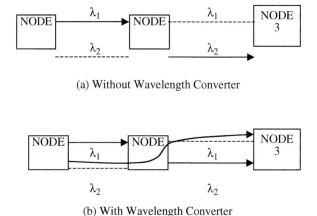

(a) Without Wavelength Converter

(b) With Wavelength Converter

Figure 3.5 Benefit of wavelength conversion.

It is easy to eliminate the wavelength-continuity constraint if we are able to convert data arriving on one wavelength along a link into another wavelength at an intermediate node and forward it along the next link. Such a technique is feasible and is referred to as wavelength conversion, and wavelength-routed networks with this capability are referred to as wavelength-convertible networks.[3,4,13] A wavelength-convertible network that supports complete conversion at all nodes is functionally equivalent to a circuit-switched network (i.e., lightpath requests are blocked only when no capacity is available on the path). In Figure 3.5(b), a wavelength converter at Node 2 is employed to convert data from wavelength λ_2 to λ_1. The new lightpath between Node 1 and Node 3 can now be established by using wavelength λ_2 on the link from Node 1 to Node 2 and then by using wavelength λ_1 to reach Node 3 from Node 2. In ARTHUR, we study networks equipped with wavelength-convertible nodes.

3.4.3.2 First-fit wavelength-assignment algorithm

The wavelength-assignment algorithm implemented is the first-fit algorithm that can perform wavelength conversion when necessary. This algorithm tries to allocate wavelength(s) to all the links on a route. In ARTHUR, other wavelength-assignment algorithms,[11] such as Random, Best-Fit, etc. can be implemented as well.

On invoking a simulation, core first calls the routing algorithm. The routing algorithm computes the route for the given connection and saves it in a global data structure, which includes the route, the links on the route, and the list of nodes on the route. The routing algorithm returns after performing route computation. Now, core invokes the wavelength-assignment algorithm with the following input parameters.

Input parameters. A description of the current topology that includes:

- Route computed for the connection
- A list of nodes
- A list of all the links on the routes
- Existing connections and their resource allocation

Output parameters.

- Wavelength list for the route

Algorithm. Let us consider the case of first-fit wavelength-assignment algorithm. Given the route and other input parameters, the algorithm performs as follows:

- For the given route, extract the first link on the route.
- Assign the first free wavelength on this link.
- Extract the second link.

- Search if the wavelength assigned on the first link is also available on the second link. (This may lead to two cases. If available, the algorithm assigns the same wavelength; otherwise, the algorithm searches for the first free wavelength that is available on all links and assigns that wavelength.)
- If any single wavelength is not available on all links, perform wavelength conversion at the node connected to the link where the wavelength being searched for was not available. In other words, stick to wavelength continuity and perform wavelength conversion only if necessary (COIN).
- If a link on the route has no wavelength available, the connection is blocked.

3.5   Applications of ARTHUR

3.5.1   A generic application

The generic events involved in setting up a connection are described next. It is assumed that the client (GUI) is connected to the server. A user first creates a topology on the GUI (client) by clicking the "create node" and "create link" buttons. The user may also edit link properties such as cost, etc.

By selecting the connection mode button, one may create a connection request between two nodes. To do so, the user first clicks on the source node and then the destination node. This event creates a new connection request, and it is displayed on the connection tab in the information panel.

Now, the user chooses a routing algorithm followed by a wavelength-assignment algorithm (WAA). By choosing the shortest-path algorithm, we instruct the simulator to calculate just one path for the connection. In other words, we are not requesting any protection for this connection. If one chooses Suurballe's routing algorithm, the simulator will calculate two paths (primary and backup) for the connection; hence, we request protection for this particular connection. The WAA is then chosen (say first fit).

Note that the first-fit WAA assigns wavelengths to both primary and backup, if any. Now, the user presses the "add connection" button, which sends all required data to the server. The server calculates the paths, assigns wavelengths, and returns the results to the client, which are then displayed on the canvas of the GUI.

3.5.2   Setup of a protected connection and link failure

This section explains how a protected connection is set up and the results displayed on the canvas on the GUI. Consider a network with the following elements that a user creates:

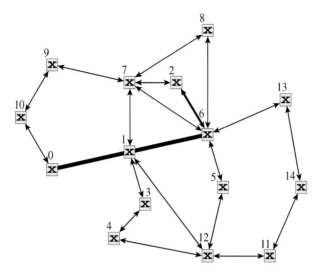

Figure 3.6 Primary path between Node 0 and Node 2.

Nodes: 0, 1, 2, …, 14.
Links: All links are bidirectional and have unit cost; connectivity is illustrated in Figure 3.6.

The user now creates a connection request (Source = Node 0; Destination = Node 2) and selects Suurballe's routing algorithm, first-fit WAA, and finally clicks the "add connection" button. Figure 3.6 (extracted from ARTHUR's GUI) illustrates the network and the first path that Suurballe's routing algorithm calculates. The cost of the first route is 3.

When a user requests a connection between Node 0 and Node 2 and chooses Suurballe's routing algorithm to perform routing, we actually receive two routes with both being assigned wavelengths. The GUI shows just the first route, as it is the primary path. The other route is a backup path and it is activated when one of the links on the primary path fails. One such case could be that the link between Node 1 and Node 6 fails. This is when the backup path kicks in. The alternate path, which the connection takes once the previously mentioned link failure occurs, is 0 →10 →9 →7 →2. The user can select multiple links from a given list and simulate a failure on them, causing the backup path, if any, to kick in.

3.5.3   Deleting a connection

On an existing network, which has a set of connections already set up, the user has a choice to delete one or more existing connections. Choosing a connection from the set of connections displayed in the connection tab and pressing the "delete connection" button does this job. At this point, the

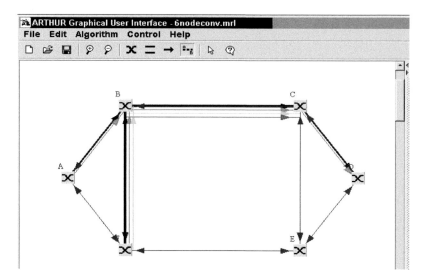

Figure 3.7 Illustration of wavelength conversion.

software first frees resources associated with this connection (e.g., wavelengths) and then displays the updated network.

3.5.4 Wavelength conversion (convert only if necessary)

This section demonstrates how ARTHUR assigns wavelengths. It follows the first-fit WAA as described in Section 3.4.3.2. Consider the network in Figure 3.7:

> Nodes: A, B, C, D, E, F
> Links: All links are bidirectional and have unit cost.

The number of wavelengths on each link is the number in the column "Lambda Ra ..." of Figure 3.8.

Name	Source	Destination	Lambda Ra...	Distance
Link_A_B	A	B	1	1.0
Link_A_F	A	F	4	1.0
Link_B_A	B	A	4	1.0
Link_F_A	F	A	4	1.0
Link_B_C	B	C	3	1.0
Link_B_F	B	F	4	1.0
Link_C_B	C	B	4	1.0
Link_F_B	F	B	2	1.0
Link_E_F	E	F	4	1.0
Link_F_E	F	E	4	100.0
Link_C_D	C	D	4	1.0
Link_C_E	C	E	4	1.0
Link_D_C	D	C	4	1.0
Link_E_C	E	C	4	1.0
Link_D_E	D	E	4	1.0
Link_E_D	E	D	4	1.0

Figure 3.8 Link details.

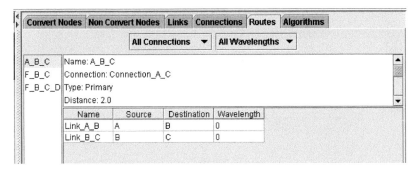

Figure 3.9 ARTHUR results for connection between Node A and Node C.

Consider the following connection requests:

Source = Node A; Destination = Node C; Route: A-B-C
Source = Node F; Destination = Node D; Route: F-B-C
Source = Node F; Destination = Node C; Route: F-B-C-D

Note that, first, the routing algorithm (which calculates the routes mentioned previously) is invoked, and then the WAA (which assigns wavelengths) is invoked. After the WAA is done, ARTHUR returns with the results in Figures 3.9, 3.10, and 3.11, for the preceding connection requests.

As we can see from Figure 3.9, the connection between A and C is wavelength-continuous (wavelength 0, route: A-B-C). After satisfying this connection, ARTHUR tries to allocate wavelengths for the connection from Node F to Node C. At this point, ARTHUR sees that it is possible to stick to wavelength continuity and hence assigns wavelength 1 (route: F-B-C). In order to satisfy the connection from Node F to Node D, ARTHUR finds that there is no single continuous wavelength on the route F-B-C-D. Hence, it performs wavelength conversion at Node B and assigns wavelength 2 on link B-C, wavelength 0 on link F-B and wavelength 0 on link C-D.

| Convert Nodes | Non Convert Nodes | Links | Connections | Routes | Algorithms |

All Connections ▼   All Wavelengths ▼

A_B_C	Name: F_B_C
F_B_C	Connection: Connection_F_C
F_B_C_D	Type: Primary
	Distance: 2.0

Name	Source	Destination	Wavelength
Link_F_B	F	B	1
Link_B_C	B	C	1

Figure 3.10 ARTHUR results for connection between Node F and Node C.

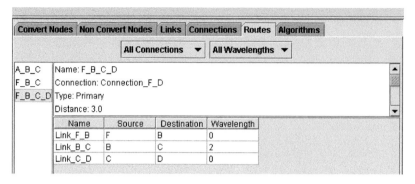

Figure 3.11 ARTHUR results for connection between Node F and Node D.

3.6   Conclusion

We presented an optical WDM mesh network design tool, ARTHUR, which supports dynamic RWA, fault management, and wavelength conversion. The tool provides features to design and configure the network topology, as well as dynamically add and delete a connection.

ARTHUR is most useful to network architects. Whenever a network topology is to be installed, it can be first instantiated in ARTHUR. The network topology can be created and the desired algorithms selected. A series of simulations can be run, and the results analyzed for the network's performance under different conditions. ARTHUR also provides an effective way to test the performance of various RWA algorithms under different network topologies.

Future extensions may include advanced features such as traffic grooming[14,15] (a term used to describe how traffic streams of different bandwidth granularities are packed and switched onto faster data streams), and changing the centralized architecture to a distributed approach. Another area that can be explored using ARTHUR is multicasting for light-trees.[16]

Wavelength-assignment algorithms such as best-fit, random, etc. can be implemented as well. New algorithms that can provide quality-of-service is another attractive enhancement.

Acknowledgments

This work has been supported in part by a research gift from Agilent Corp. We gratefully acknowledge Steve Joiner of Agilent for sponsoring this research, and Sree Rama Nomula and Mir Hussain Ali, graduate students at Univeristy of Nebraska-Lincoln, for their contributions to this project.

References

1. C.A. Brackett, Dense wavelength division multiplexing networks: Principles and applications, *IEEE Journal on Selected Areas in Commun.*, vol. 8, no. 6, pp. 948–964, Aug. 1990.
2. J.P. Laude, *Wavelength Division Multiplexing*, New York, Prentice Hall, 1993.
3. B. Mukherjee, *Optical Communication Networks*, New York, McGraw-Hill, 1997.
4. R. Ramaswami and K.N. Sivarajan, *Optical Networks: A Practical Perspective*, San Francisco, Morgan Kaufmann Publishers, 1998.
5. B. Ramamurthy, Efficient design of wavelength-division multiplexing (WDM)-based optical networks, Ph.D. dissertation, University of California, Davis, July 1998.
6. D. Banerjee and B. Mukherjee, A practical approach for routing and wavelength assignment in large wavelength-routed optical networks, *IEEE J. Selected Areas in Commun.*, vol. 14, no. 5, pp. 903–908, June 1996.
7. M.R. Garey and D.S. Johnson, *Computers and Intractability: A Guide to the Theory of NP-Completeness*, Freeman, 1979.
8. C. Yemparala, Integrated Simulation Environment for SIMON, Masters project report, University of Nebraska, Lincoln, Dec. 1999.
9. B. Ramamurthy, D. Datta, H. Feng, J.P. Heritage, and B. Mukherjee, SIMON: a simulator for optical networks, session on simulation tools and validation for optical devices and networks, in the *Fifth Ann. Conf. on All-Optical Networking* (part of the *SPIE Photonics East '99 Symp. on Voice Video and Data Communication*), Sept. 1999.
10. T. Antani, Dynamic Routing and Wavelength Assignment for Wavelength-Convertible WDM Optical Networks, Master of Science thesis, University of California, Davis, Jan., 2003
11. H. Zang, J. Jue, and B. Mukherjee, A review of routing and wavelength assignment approaches for wavelength-routed optical WDM networks, *Optical Networks Mag.*, vol. 1, no. 1, pp. 47–60, Jan. 2000.
12. R. Bhandari, *Survivable Networks: Algorithms for Diverse Routing*, San Francisco, Morgan Kauffman Publishers, 1999.
13. R. Ramaswami and K.N. Sivarajan, Routing and wavelength assignment in all-optical networks, *IEEE/ACM Trans. on Networking*, vol. 3, no. 5, pp. 489–500, Oct. 1995.
14. K. Zhu and B. Mukherjee, Traffic grooming in WDM optical mesh networks, *IEEE J. Selected Areas in Commun.*, vol. 20, no. 1, pp. 122–133, Jan., 2002.
15. E. Modiano and P. Lin, Traffic grooming in WDM networks, *IEEE Commun.*, vol. 39, no. 7, pp. 57–63, July 2001.
16. L.H. Sahasrabuddhe and B. Mukherjee, Light-trees: optical multicasting for improved performance in wavelength-routed networks, *IEEE Commun.*, vol. 37, no. 2, pp. 67–73, Feb. 1999.

chapter four

Evolution to an optical broadband services network

David Benjamin
Nortel Networks
Peter Green
Nortel Networks

Contents

0-8493-1333-3/03/$0.00+$1.50
© 2003 by CRC Press LLC

4.1 Introduction

New enterprise applications and explosive data traffic growth are now demanding line rates that previously were only available in the optical core of long-haul backbone networks. Coupled with these new line rates has been an equally dramatic increase in the number and variety of optical services. To meet this emerging market, service providers have started to deploy overlay networks, but at a large capital and operational cost. Service providers need to evolve their current networks in order to fully capitalize on these new service opportunities. This chapter discusses the new service drivers and proposes an architectural solution that transforms the current optical transport network into an *optical broadband services (OBS) network*. The OBS network's objective is the creation of an increasingly cost-effective infrastructure that allows carriers to better align their capital and operational expenses with their service revenues.

This chapter begins with a description of current transport networks and the changing market dynamics. Sections 4.3 and 4.4 detail emerging applications and their network requirements, from both a metro and a long-haul network perspective. Section 4.5 proposes a services architecture that addresses these new requirements. Section 4.6 then explores a key component of the architecture, *Service and Network Adaptation. Adaptation* maps all services in "standardized" network containers, creating a manageable, service-agnostic core. *Adaptation* also provides service transparency, ensuring that the customer's traffic, regardless of its protocol and rate, is delivered intact. Several new service opportunities enabled by this new network are presented Section 4.7.

4.2   Current optical transport networks

This section describes the current service and transport network architectures and the new services marketplace.

4.2.1   The evolution of the transport network

The role of the transport network is to provide point-to-point facilities between service flexibility points. At these service flexibility points, services are provisioned, managed, switched, and multiplexed into larger "containers" to be carried by the transport network. Twenty years ago, switched 64 kb/s voice traffic was the basic service unit, while DS1/DS3 circuits provided the transport layer. Over the years, the increasing demand for higher end user bandwidth services presented service providers with the opportunity to directly offer DS1 and DS3 service rates, previously only available in their transport networks. The evolution of network services and transport bandwidth is depicted in Figure 4.1.

Transforming the DS1/DS3 transport layer into a service layer required adding performance monitoring and limited switching capabilities at DS1/DS3 rates. Performance monitoring ensured that carriers could meet the service level agreements (SLAs) associated with these new services, while switching promoted faster service provisioning. In particular, digital cross-connects (DCS) were introduced for this purpose, although for the most part, provisioning remained a largely labor-intensive operation.

Coupled with the introduction of DS1/DS3 services was the need for a more cost-effective transport network technology, initially to target long-haul networks. The principal requirements were high reliability and cost points lower than existing electrical trunking solutions. The answer was: synchronous optical network (SONET) optical transport systems, designed and deployed as interconnected ring configurations, to provide "5 9's" availability and 50 millisecond protection.

Infrastructure Evolution

Network Layer	History	Today	Vision
Services	DS0	DS1, DS3, OC-N	OC-N, GE, SAN, λ
Service Network	DMS	DCS 3/1 & 3/3 DSX-1, DSX-3 FDF	TDM Switch Packet Switch
Transport Network	DCS 3/1 & 3/3 DSX-1, DSX-3	SONET/SDH DCS & ADM	Managed Photonic Layer
Optical Layer	1310nm Async FOTS	1310nm DWDM Fiber Relief	

Figure 4.1 Over time, the transport layer becomes the carrier's service layer (from Brent Allen, Nortel Networks).

> *Note:* References to SONET networks in this document apply equally to synchronous digital hierarchy (SDH) networks used in Europe and Asia.

A typical network configuration is depicted in Figure 4.2.

The SONET optical transport layer was designed to treat all traffic identically as "mission-critical." Typically, these networks took months to

Transport Network
Present Mode of Operation
ILEC/PTT CO
ILEC/PTT CO (Hub)
Metro Access
WB/BB DCS
3/1
3/3
Large Offices Only
DS1, DS3
DSX-3
OC-3/12/48
OC-N
Fiber PatchPanel
OC-N
Metro IOF Core
OC-48/192
DS3/OC-N
WB/BB DCS
OC-N
Fiber PatchPanel
OC-N
OC-N, GE, SAN
Service Switching & Testing
• DS-1s via 3/1 DCS
• DS-3s via 3/3 DCS if present, otherwise manual via DSX-3
• OC-N services via SONET DCS if present, otherwise manual via FDF
Adaptation, Grooming & Aggregation
• Services bundled into IOF trunks with appropriate destination

Figure 4.2 Today's service environment is primarily built on electrical DS1/DS3 services.

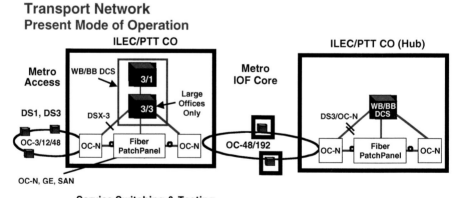

provision and provided static connectivity over periods of many months or years. Traffic growth beyond the installed ring capacity was managed by overlaying multiple rings or by deploying higher-speed rings. Responsibility for optical transport deployment generally rested with the carrier's network planning organization whose primary objective was to reduce the "cost-per-bit."

4.2.2 Traditional assumptions about optical transport no longer apply

The Internet explosion has diminished the predominance of voice traffic and private line traffic, and has invalidated many of the assumptions on which the original SONET networks were designed.[1] Indeed, the volume of data traffic has now surpassed that of voice (although not in revenue). Also, generally accepted notions of communities of interest and predictable traffic patterns no longer apply because the Internet has created a dynamic, global marketplace. As a result, the traditional approach of building SONET-based ring networks to handle traffic growth is inadequate in a number of ways:

1. Ring structures have long deployment times that result in lost carrier opportunities.
2. Equipment scaling requires large, step-function investments because complete rings must be added.
3. Operational costs spiral as carriers are forced to manage traffic across multiple, independent stacked rings.

A more dynamic and cost-effective network model is needed.

Indeed, not only has the volume of data traffic exploded, but also the bandwidth required by individual client devices at the edge of the network has grown. Core routers with SONET interfaces at OC48 and OC192 rates are now common. Storage devices now incorporate optical fiber channel interfaces to provide high-speed connectivity within local area networks (LANs) and metropolitan area networks (MANs). These devices support dozens of 1- and 2Gb/s interfaces and will move to hundreds of 10Gb/s ports over the next few years. Thus, the existing SONET-ring technology, which was designed to transport highly multiplexed voice and private line traffic, is being reexamined.

As optical transport gets closer to the end customer, the importance of providing differentiated optical services that better match the customer's application increases. With advances in optical technology and competition continuing to reduce the cost-per-bit, a carrier's optical investment will increasingly be measured against its success at generating additional revenue. Service providers require flexible service solutions to economically manage the service churn, service mix, and service growth characteristic of metro markets. Long-haul carriers also need service flexibility as they extend their optical networks to provide end-to-end integrated optical services.

4.2.3   *The optical transport network as an optical services network*

The preceding discussion strongly suggests the need to transform today's optical transport network into one that is more focused on facilitating evolving customer applications. We call this new network construct the *optical broadband services (OBS) network.* This optical network must not only respond to the scaling issues seen in the last few years but must also provide a platform for delivering a wide portfolio of emerging optical services. Unlike an optical *transport* network, the optical broadband *services* network will also have to support features such as SLA management, billing, service provisioning, customer network management, switching, and security.

The first steps toward this evolution are under way. Long-haul operators have opened their networks for wholesale, point-to-point OC-n, and wavelength services. Metro operators have deployed special-assembly dense wavelength division multiplexing (DWDM) networks for enterprise storage applications. The growing demand for these and other optical services require a service-oriented network evolution strategy. The following sections take a closer look at the emerging service set, followed by a proposed network strategy.

4.3   *Emerging applications and network requirements — metro*

4.3.1   *Service evolution*

The metro optical services market offers a myriad of services that cater to the high-speed, bandwidth-intensive demands of diverse market segments. These services vary in bandwidth granularity, security levels, protection schemes, provisioning speeds, protocol transparency, and pricing models. Figure 4.3 illustrates how metro service technologies have evolved over time. Optical Ethernet and DWDM are now being positioned as the technologies to deliver the next generation of broadband services.

4.3.2   *Enterprise applications driving metro optical services*

Table 4.1 provides a high-level industry segmentation and examples of business and technology trends that will influence the growth of broadband communications.

In addition, universal applications include e-mail, file transfer protocol (FTP), corporate intranet portals, electronic data interchange (EDI), customer relationship management (CRM), and enterprise resoue planning (ERP). Through the analysis of these vertical market segments, we gain insight into:

- The protocols that are driven by the applications
- The bandwidth that is dictated by the number of processing points and volume of information

Service Maturity and Evolution

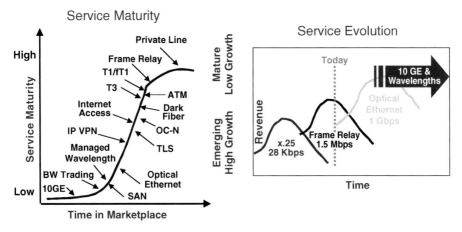

Figure 4.3 A combination of fiber availability, bandwidth demand, service pricing, and new technology are changing the nature of metro services.

Table 4.1 Vertical Markets and Applications That Will Influence the Growth of Broadband Communications

Industries and Segments	Trends and Applications Driving the Bandwidth Demand
Financial Services/Banking	
Banks, Insurance Firms, Brokerage Firms, Real Estate Companies, Credit and Financial Services Agencies	Cost of electronic transactions decreasing while number of transactions skyrocketing (e.g., e-commerce, online trading) Globalization trends (International Securities Exchange, Global Equity Market) Electronic Signatures Act (e.g., online contracts) Imaging and Video Applications: (e.g., online real estate services)
Manufacturing	
Engineering Companies, High-Tech Manufacturing, Publishing Firms	"Net impact in every link of the supply chain" (e.g., customers order online; components availability online; customers track orders online) Online design applications (CAD/CAM), allow for multisite collaboration Online publishing: personalized, customized

(continued)

Table 4.1 (continued) Trends That Will Influence the Growth of Broadband Communications

Industries and Segments	Trends and Applications Driving the Bandwidth Demand
Business Services/Utilities	
Service Providers, Advertising Agencies, Media Companies, Retail, Utilities	Online advertising (e.g., customized to audience) Online inventory tracking increasing efficiency in retail Deregulation and mergers driving changes for utility organizations (e.g., online energy trading; online billing; real-time online metering and bill processing)
Healthcare	
Hospitals, Pharmaceutical Firms, Laboratories, Insurance Providers, Home Healthcare Centers	Internet touches all sectors (e.g., online physician scheduling, insurance approval, and pharmacy) Standardized and electronic medical records Telemedicine and imaging records distribution Streamline paperwork processing and reduce spending
Education	
Education Boards, Universities, Libraries, Research Centers	Increasing student population and decreasing teaching time (e.g., distance learning, videoconferencing) Research and development based upon complex and data intensive computing
Government/Public Sector	
Federal, Provincial, Municipal Government, Public Agencies	Interconnectivity between government institutions, research, education institutions (e.g., file sharing, video depositions) Distance learning offers a cost-effective application for employee training
Transportation	
Airlines, Shipping, Trucking and Railroad, Travel Agencies	Internet allows for increased customer involvement and improved customer service (e.g., online reservations, ticketing, delivery tracking) Increased seasonal usage places a requirement for "bandwidth on demand"

- The distribution of bandwidth that is shaped by communities of interest
- Connectivity characteristics of applications (e.g., volume of traffic, messaging size; real- vs. nonreal-time nature)
- The maturity of the applications as well as potential emerging applications

Figure 4.4 provides an example of the new set of metro connectivity requirements that are not efficiently satisfied by traditional service offerings.

As an example, security regulators in the financial services segment have imposed strict guidelines on the integrity of databases that warehouse electronic transactions. In order to comply with these new regulations, financial institutions designed a new IT infrastructure with a disaster recovery site to support real-time backup of all electronic transactions. The majority of these transaction systems are based upon mainframe computing centers with storage devices utilizing enterprise connectivity (ESCON) and fiber channel protocols. Once financial institutions specified the protocol, channel counts, and site locations, they asked their carriers to develop broadband service offerings to address their distinct communication requirements. Managed fiber services (MFS) were developed and have been successfully deployed by numerous metro optical carriers. Carriers have also started to position this service into other vertical segments that have similar communication requirements.

Large Enterprise Connectivity Requirements

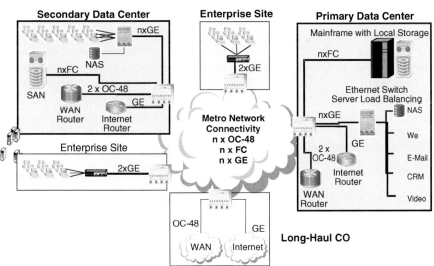

Figure 4.4 A diverse set of metro connectivity requirements that are not typically satisfied by existing transport networks.

4.3.3   Managed fiber services

Today's MFS are predominantly custom-built, with the underlying fiber and its associated optical systems dedicated to a single customer. An MFS offers customers a fully managed solution with the following values:

- Interface flexibility — this supports all commonly used proprietary interfaces
- Service transparency —— the carrier does not alter data and overhead bits
- Channel scalability — this is based on high-capacity DWDM systems
- Protection — unprotected options are available
- Low capital cost — the carrier assumes the initial investment
- Fully managed optical service — performance is guaranteed through SLAs and carrier management
- Rapid service extension — once the original MFS has been in place, additional wavelengths can be turned up in days
- Secure dedicated infrastructure

The MFS is based on DWDM systems that accept optical signals in native mode (from 16Mbit/s to 10Gbit/s) and map them onto a wavelength. As a result, these systems are highly forecast tolerant, easily accommodating "whatever the enterprise throws at them." For example, a change from fiber channel 100 Mbit/s to 200 Mbit/s, or from 100Mbit/s to 1Gbit/s Ethernet can be handled in the network without any hardware changes; only a simple software configuration change is required. Expensive extension and conversion devices that map ESCON onto Internet protocol (IP) or asynchronous transfer mode (ATM), or onto a slower speed facility are avoided. This flexibility shortens provisioning cycles, reduces engineering times, and reduces life cycle costs. The service is a custom build and, as such, the price for the service reflects the carrier's cost to build the dedicated network. The service usually requires the customer to sign a 3- to 5-year contract. In order to address a wider market segment that requires fewer channels, carriers are developing a wavelength-based service described in the next section.

4.3.4   Managed wavelength service

A managed wavelength service (MWS) is based on a shared network infrastructure wherein the fiber and optical systems support multiple customers in a metro area. Unlike an MFS where a customer first commits to the service, the deployment of a shared network requires that the service provider must first have a strong understanding of target customer demand and building locations. It is anticipated that carriers will move aggressively to MWS offerings when the volume of enterprise wavelength applications increases, allowing for more accurate forecasts.

Similar to MFS, the simplest definition for an MWS service does not tie the service description to any specific enterprise application. From a service provider's perspective, however, this has several service limitations including:

- Point-to-point connectivity only
- Limited performance and service monitoring
- Positioned as an intra-metro service
- Service tied directly to DWDM technology
- Difficult to develop realistic service forecast

The solution to these issues is to position specific MWS services that can leverage a carrier's DWDM infrastructure. The remainder of this section presents several near-term opportunities based on MWS, while the new OBS network, described in Section 4.5, suggests an evolution path to accomplish this.

4.3.4.1 *LAN connect — optical Ethernet over MWS*

LAN Connect allows enterprises to connect their LANs to their carrier's network using standard LAN interfaces based on optical Ethernet technology. This dramatically reduces the cost and complexity of inter-connection compared with SONET interfaces. Connecting through Ethernet allows the enterprise to use existing LAN-based methods of operation and expertise.

For enterprises, migration from the present mode of operation to *LAN Connect* service is graceful because off-site resources can be connected seamlessly to an existing LAN. It enables enterprises to re-think their computing models. For example, because off-site communication is viewed as complex and expensive, many enterprises have replicated and distributed servers, databases, etc. *LAN Connect* allows enterprises to move from highly distributed environments to more centralized models with dramatic cost reduction and productivity increases. It has been demonstrated that incremental expenditures on *LAN Connect* services are more than offset by savings in routers, computing systems, and network operations.

LAN Connect also reduces cost, because operational tasks such as provisioning are made much simpler. For example, adding bandwidth no longer involves truck-rolls but can be accomplished using software-based tools. Bandwidth can be managed from 1 to 1000Mbs in 1Mbs increments and fit into Ethernet LAN/WAN management structures. *LAN Connect* offers packet aggregation and switching with a new cost/performance factor.

4.3.4.2 *Storage connect — ESCON/FICON fiber channel over MWS*

Today, the availability and management of information have become integral to achieving corporate objectives. To keep pace with this data explosion, new information management approaches are rapidly being adopted that embody new storage networking models. For example, storage area networks (SAN), high-speed special-purpose networks that separate an enterprise organization's data from its clients and servers, are being embraced. In mainframe computing environments, SANs use ESCON and FICON protocols to interconnect servers and storage devices. The fiber channel protocol is used in more open systems environments, such as UNIX and Windows NT. Because these technologies have distance limitations (due in most part to latency and protocol timing issues), the task of inter-connecting SAN islands is one that is ideally suited to optical networking. By providing *SAN Connect*, service providers are discovering new opportunities to offer new services and distinguish themselves in today's competitive marketplace.

4.3.4.3   Video connect

Another connectivity service enabled by the MWS architecture is *Video Connect*. Producing, editing, and delivering digital broadcast material requires:

- File transfers between film and video postproduction houses
- Program previewing between postproduction houses and advertising agencies
- Program viewing between postproduction house and client
- Last-minute editing between postproduction house and broadcast facility
- Special previewing at designated venues
- Broadcasting from feeds and remote tape machines to cable head ends
- High-quality video signals that conform to SMPTE 259M standards

The ANSI (American National Standards Institute)/SMPTE (Society of Motion Picture and Television Engineers) standard 259M-1993 describes a serial digital interface, operating with 4:2:2 serial component signals. In order to conform to this standard, the *Video Connect* service must transport a one-way 270Mbps digital video signal point-to-point.

4.4   Emerging applications and network requirements — long-haul

4.4.1   Private line services

The transport of highly multiplexed, mission-critical traffic over long distances is the traditional wholesale service provided by long-haul operators. Typically, these services were DS-1 and DS-3 private lines (PL) sold to incumbent local exchange carriers (ILECs) or large enterprises. They required several months to provision and long-term contracts. As discussed previously, SONET-ring networks (actually many interconnected rings) were deployed to support these services. As the size and number of bandwidth pipes increased, direct SONET optical interfaces — initially OC12, and now OC48 and OC192 — were introduced at the handoff points.

4.4.2   Wholesale wavelength services

More recently, the emergence of new applications and a new generation of service providers have created a new set of service requirements. In contrast to the reasonably well-defined PL market described previously, a diversified wholesale market based on DWDM technology has emerged. DWDM technology, which was originally developed as a fiber-relief technology, has now found a place in transparent optical networking.

Many of these new entrants had neither the time nor inclination to build their own networks and, instead, looked to lease capacity from existing suppliers. These new business models generated new service requirements. For example, Internet service providers (ISPs) and application service providers (ASPs) require large bandwidth pipes to interconnect their core router networks, typically with optical OC48 interfaces and OC192 interfaces. These IP networks do not require the 5 9's protection of traditional PL, nor can they afford the cost. Instead, they preferred unprotected, point-to-point high-bandwidth services. The wholesale bandwidth market has expanded to include the following applications:

- Unprotected router interconnect or diversely routed connections
- OC48, OC192 for IP traffic
- Alternate routing for survivability
- Closing rings or providing extra capacity on saturated routes
- Interim bandwidth while carrier is building own facilities
- Increasing geographical coverage
- Enterprise connectivity services

Many of these applications also require that the wholesale carrier not modify the signal — a characteristic known as service transparency. This is either because the service protocol to be carried is not SONET (e.g., FICON for storage), or because the application requires that the wholesale carrier not modify its SONET overhead (e.g., in the case of ring completion). Again, these requirements are not easily satisfied by traditionally designed transport networks, so other technologies are being used.

Initially, the solution to meet these particular service demands focused on the long-term leasing of dark fiber under agreements known as indefeasible right to use (IRU); but with increasing demand and the advent of DWDM technology, leasing wavelengths within a fiber made more economical sense. These MWS services are deployed over point-to-point 2.5Gb/s or 10Gb/s DWDM links. As multiplexing is not involved, only minimal "SONET-like" overhead is used, and enhanced service transparency is provided.

Protection in a DWDM system is expensive, as a route-diverse wavelength must be dedicated as a backup. As a result, most MWS implementations are unprotected, and restoration is provided through other mechanisms (e.g., using Layer 3 or diverse routing through another carrier).

4.4.3   Next-generation wholesale service characteristics

Although wholesale PL and MWS services attend to many of the emerging service demands, others also need to be addressed. Several of these are discussed next.

4.4.3.1   Service granularity and flexibility

PL and MWS services are offered in distinct bandwidth blocks. Moving to a different speed generally requires interface changes and truck-rolls, creating significant cost barriers. New services such as optical Ethernet provide the opportunity for a wider range of speeds that can be customer-provisioned without equipment changes. As well, today services provide either 5 9's protection or no protection. Many new data applications would benefit from levels of protection somewhere between these two extremes (see Section 4.7.5).

4.4.3.2   Service transparency

As discussed previously, MWS services provide transparency, however, at the cost of dedicating a full wavelength. For example, fiber channel supports payloads of either 200, 400, or 800Mb/s, and hence significant bandwidth is wasted when using a 2.5Gb/s MWS service. New mapping techniques, coupled with virtual concatenation services, promise to mitigate this issue (see Section 4.6.3).

4.4.3.3   Service growth and provisioning time

Many applications require a more dynamic assignment of bandwidth. For example, an ISP may find traffic in the morning primarily originating on the East Coast and shifting westward during the day. The introduction of switched optical services controlled by the customer would be a powerful tool.

4.4.3.4   Point-to-point circuit

Circuits are provisioned on a point-to-point basis, while many retail service providers and large enterprises need to dynamically allocate bandwidth between multiple locations. An enterprise may focus on transaction mirroring within a regional network during the day while, during the night, needs to perform data-backup across the country. A managed network service that provides secure, on-demand connections between specified members is needed to support a wide range of multipoint/multi-site applications (see Section 4.7.4).

4.4.3.5   Pricing flexibility

Expanded billing options are needed to align billing with a customer's real usage (e.g., moving from flat billing to usage-based for scheduled and on-demand applications; see Section 4.7.3).

In addition to these service challenges, DWDM technology for high bandwidth, transparent services has been introduced as an overlay network to SONET for sub-wavelength services. This separation has caused undesirable inefficiencies and costs. A flexible OBS network is needed to address these issues. This network will provide a new level of service, control, and pricing flexibility to meet customer demands, all on a *single* network infrastructure.

4.5   Services architectural framework — evolution to the optical broadband services network

4.5.1   High-level architecture

This section proposes a new architectural framework to efficiently support the applications discussed in Sections 4.3 and 4.4, as well as providing a platform for future broadband services. The proposed architecture allows carriers to evolve their networks while protecting their current network investments and operational systems. The framework is based on a *"Service Adaptive" Edge* and a *"Service Agnostic" Core*. This separation of services from the core network is critical in ensuring that:

- New services can be quickly accommodated without churning the network infrastructure
- New core technology can be incorporated without affecting existing services

The focus of the network edge is service flexibility, service transparency, and service velocity to drive new revenue, while the focus of the core is minimization of capital and operational costs, as illustrated in Figure 4.5.

Optical Broadband Services Network

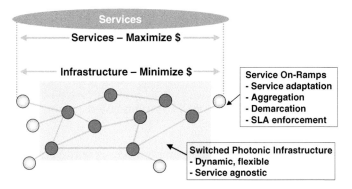

Figure 4.5 High-level optical broadband services architecture differentiates the edge from the core.

Another important feature of the architecture is the consolidation of wavelength and sub-wavelength services onto a single, manageable infrastructure. As part of the consolidation, the architecture adds a high degree of service monitoring and switching capabilities to MWS services, well beyond those of today's implementation.

A pragmatic three-phase evolution to implementing an OBS network is described next.

4.5.2 Phase 1: POP/CO consolidation through optical switching

The first step is the simplification of SONET ring management by using an optical switch (optical cross-connect [OXC]) to consolidate co-located, stacked add/drop multiplexers (ADMs). In this first step, the optical switch is used as a nodal element, automating the interconnection of existing stacked rings. Less physical equipment is required and consequently provides significant capital expense (CAPEX) savings. Operational expense (OPEX) savings come from:

- Lower space and power requirements
- Lower management overhead
- Elimination of truck-rolls and record keeping as patch-panel reconfigurations are replaced by remote service provisioning

4.5.3 Phase 2: Optical switching and intelligent control plane

The next step is to interconnect the optical switches to create a true optical network. Several standards bodies (ITU,[2] IETF,[3] and OIF[5]) are energetically working toward this objective. In particular, the current focus is on specifying a control plane over a switched optical layer that, on request from either the end user or network operator, will dynamically make connections across the optical layer. ITU's Recommendation G.807 (previously known as G.astn — automatic switched transport network) has emerged as the global framework for this "intelligent" switched optical network (see Figure 4.6).

The control plane operates over arbitrary network topologies (i.e., ring, mesh, and hybrid), and also connects to legacy networks. The control plane supports a range of restoration options and is designed to work over SONET (ring or mesh) and OTN networks. G.807 represents a logical framework so that, although the control plane is presented as a functionally distinct plane, it can be physically implemented as software within the OXCs themselves.

Note: The commonly used acronym, OXC, is used here to refer to an Optical Switch. In this context, OXC can either be an O-E-O or OO switch.

In Figure 4.6, the optical traffic layer consists of three important building blocks:

ITU G.807 (ASTN) Architectural Framework

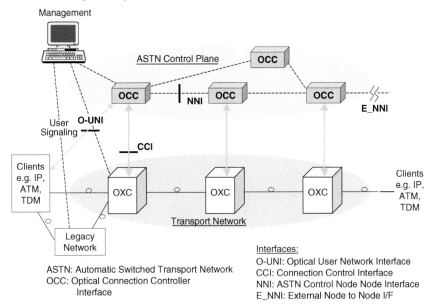

Figure 4.6 The reference architecture for the automatic switched transport network (ITU recommendation G.807).

1. High-capacity line systems approaching Terabits per second, with long all-optical reach of a few thousand kms between end city pairs
2. Optical switching platforms that will, over time, combine opaque and all-optical (photonic) switch fabrics, to provide unprecedented open-ended scalability and the ability to switch synchronous transport signal (STS) paths, wavelengths, bands of wavelengths, or entire line-system fibers[4]
3. Tunable devices (sources, filters, and receivers) offering flexible selectivity, as well as reduced inventory and operations savings

The control plane is overlaid on the traffic layer and performs a number of functions. First, unlike today's networks, the ASTN-controlled network auto-discovers the resources and topology of the optical traffic layer. This process begins with each OXC discovering its available resources, service capabilities, and local connectivity through the exchange of discovery signals between itself and its neighbors. Each OXC reports its local topology information to its associated OCC over the CCI interface. The OCCs, using the node-to-node interface (NNI), and a version of the open-shortest-path-first (OSPF) protocol extended for use on the optical network, collectively and automatically discover the entire transport network topology and available bandwidth.

This up-to-date topology database is maintained by each OCC controller and allows the OCC to compute complete paths through the network.

The OCCs use a modified multi-protocol label switching (MPLS) signaling protocol called Generalized MPLS, that is being developed in the Internet engineering task force (IETF) to dynamically signal for a lightpath to be established. In this way, lightpaths can be made, modified, or torn down in seconds. By distributing routing, the network becomes more robust and is able to support higher signaling performance.

The control plane also supports an optical user-network interface (UNI). This interface allows a client at the edge of the network to signal for a connection to be set up or torn down. Initially, the UNI capability will be used by network management systems to proxy-signal connections on behalf of clients. UNI functionality is being defined in the OIF.[5] The external NNI (E_NNI) provides the interface between ASTN networks that are under different network administrations. The E_NNI is a modified UNI with some NNI functions for exchanging address and topology summaries.

4.5.4   Phase 3: Service adaptation and management

The introduction of an optical control plane and direct client signaling through the UNI profoundly changes how optical networks will deliver services and how they will be managed. Today's management systems and network planning functions will have to adapt to a new paradigm where a substantial portion of the network control and network knowledge is no longer centralized in the management systems. Service control, which previously involved largely manual processes, will now be passed to automated processes in the network layer. As a result, network and service information will be distributed within the actual network elements.

Industry efforts have focused on intelligent networks that efficiently establish "A-to-Z" connections, not directly on the services themselves. Therefore, an additional mechanism is required at the ingress of the network to map a service into an appropriate set of network connection requests. Because of the importance of this mapping, the following section is devoted to its functional description.

4.6   Service and network adaptation

4.6.1   Objectives

Sections 4.3 and 4.4 highlighted the need for the optical network to support a wide range of new services, while Section 4.5 suggested a concurrent and equally dramatic evolution of its core technology and architecture. As a result, service providers require a mechanism that will allow them to respond to new service opportunities while they evolve their core network — service delivery must be independent of the network's core technology. *Service Adaptation (SA)* and *Network Adaptation (NA)* play this important role in the network, and will be the keys to future profitability. Together SA and NA isolate the delivery of services from the specifics of core's transport technology.

Service and Network Adaptation

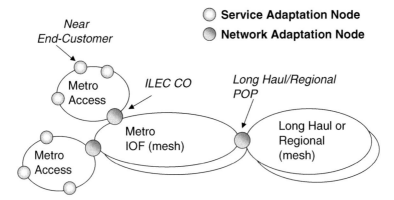

Figure 4.7 Service and network adaptation isolate services from the specifics of core transport technology.

Specifically, SA and NA allow the an optical service to be:

- Managed and monitored as a standard connection
- Efficiently packed and carried
- Carried through various optical switching fabrics
- Handed off to another carrier if required

Figure 4.7 highlights the roles of SA and NA in the network.

SA and NA consist of two functions: *mapping* and *wrapping*. *Mapping* transforms client signals into payloads that the network can transport, while at the same time providing the requisite level of service transparency. This is followed by *wrapping,* which encapsulates the user payload with overhead to support network management and forward error correction (FEC).

Two network wrappers are of interest. The first is SONET/SDH and is the dominant optical networking technology. The second is OTN, which is under development in the ITU (see ITU Recommendation G.709). Because SONET/SDH networks were originally designed for multiplexing and networking of lower-speed traffic and not for the transport of single wavelength payloads, current SONET implementations are considered too costly for wavelength services. As discussed in Section 4.3, DWDM technology has found a niche in serving this latter market, overcoming some of the perceived deficiencies with SONET networks. OTN adds the management capabilities missing in today's DWDM point-to-point, transparent wavelength solutions; however, OTN technology is in its embryonic stages and will not address sub-wavelength applications, while SONET technology is being enhanced to remove some of its current deficiencies. Many of these new SONET capabilities are presented below. Thus, given the proven manageability of and the staggering investment ($250B worldwide) in SONET networks, the discussion below suggests that both SONET and OTN will have a place in the future OBS network.

Four emerging technologies tied together provide the promise of effective SA/NA solutions: These technologies are described in Section 4.6.4.

4.6.2   Service adaptation requirements

Services at the network edge "initiate calls" that result in the network making a connection type that satisfies the SLA associated with the service. Thus, every service request results in a connection with characteristics appropriate to the service in terms of: destination, bandwidth, availability/CoS, delay, bit error rate (BER), jitter, etc.

Service Adaptation performs this function and has the following objectives and characteristics:

- SA provides a demarcation point into the network that allows the carrier to police and monitor customer traffic.
- SA is done at the optical edge and involves mapping services into formats that can be carried in *standard* network containers.
- SA delivers service transparency by using the appropriate encapsulation process.
- Ingress and egress SA ports must be compatible as they are service specific.
- Service performance measures (PMs) are managed at the egress and ingress ports, while only connection PMs need to be monitored within the network.
- SA can be a Layer 1 or Layer 0 function. In the medium term, SA will be a Layer 1 function.

4.6.3   Network adaptation requirements

Network adaptation allows sub-networks or different carrier networks to connect to one another and has the following objectives and characteristics:

- NA involves a handoff point of standard network containers (typically either OTN or SONET) from one sub-network to another.
- NA provides a demarcation point for business or administrative purposes.
- NA is used to provide a transition between:
  - Different technologies (e.g., wavelength plans, amplifier chains)
  - Domains of protection/restoration or grooming and muxing
- Metro access into interoffice facilities (IOF) typically involves NA due to technology differences, muxing, and management reasons.
- NA is typically required at the metro/long-haul (LH) boundary, driven by business separation.

- For the LH carrier, NA is also an SA, with the service being the transport of standard metro containers (i.e., SONET or OTN containers).
- NA can be a Layer 1 or Layer 0 function. In the medium term it will be a Layer 1 function.

4.6.4   SA/NA technologies

Four critical SA/NA technologies are discussed next.

4.6.4.1   Generic framing procedure (GFP)

GFP (ITU Recommendation G.7041) provides a multiservice adaptation into synchronous networks. GFP also encapsulates variable length payloads of specified data client signals into synchronous payloads for transport over SONET/SDH or OTN networks. GFP supports framed clients such as Ethernet, using *Frame-Mapped GFP*; and constant bit-rate, block-coded signals such as FICON using *Transparent GFP*. It provides simple robust and low-latency adaptation for data into optical transport payload containers.

The client services currently being defined include Ethernet, fiber channel, and FICON. GFP allows a wide variety of data services to be efficiently carried over the existing SONET network, as sub-wavelength service rates can be multiplexed into higher network line rates.

4.6.4.2   G.modem

G.modem is an ITU proposal for transparent mapping and wrapping synchronous payloads into STS payloads. G.modem (see ITU G.872 and D.282) allows any continuously clocked optical signal to be *transparently* carried across existing SONET/SDH network in such a way that the bit and frequency content of the received signal is preserved. It complements GFP for clocked traffic. The advantage of G.modem is that, once wrapped in this way, any service can be networked transparently through existing SONET networks. This is especially significant for client OTN-wrapped signals, because no OTN switching fabrics exist, and can only be provisioned as point-to-point connections. Indeed, G.modem's initial application is to carry OTN-wrapped signals on existing SONET networks.[6] The proposed G.modem adaptations for OTN signals are:

ODU1 → STS-51 or STS-3c-17v
ODU2 → STS-204 or STS-3c-68v

G.modem can also be used to transparently carry client SONET and wavelength signals across the SONET network:

Clear OC-48 or 2.5Gb/s wavelength → STS-51 or STS-3c-17v
Clear OC-192 or 10Gb/s wavelength → STS-204 or STS-3c-68v

G.modem adds service transparency capabilities to existing SONET networks.

4.6.4.3   SONET wrapping using virtual concatenation (VCAT) — for efficient service wrapping and networking

Once the client signal has been mapped using either GFP-mapped or G.modem, it can then be carried in a SONET or OTN network. In the former case, VCAT provides an efficient way of fitting the synchronous payload into the right number of SONET lower bandwidth payloads, typically nxSTS-1s or nxSTS-3cs.

VCAT is an SDH/SONET network technology that has been standardized by ITU-T, ANSI, and European Telecommunications Standards Institute (ETSI) and combines arbitrary numbers of individual virtual containers (VCs) to form a single, virtual higher-capacity link. It provides a more dynamic and granular provisioning mechanism than has been previously possible in traditional SDH/SONET-based networks. It produces "right-sized" transport links over an arbitrary network topology.

4.6.4.4   OTN wrapping for point-to-point DWDM link

Although OTN is not useful as a wrapper for networking (i.e., switching), it is useful as an internal network mechanism to transparently manage wavelength signals over point-to-point DWDM links. OTN wrapping adds signal overhead information to help in performance management and in extending link reach through FEC. The additional overhead in a 10Gb/s signal results in a 10.7Gb/s line rate signal and will be accommodated by most line systems. The OTN wrapper is added at the ingress DWDM transponder and removed at the egress.

4.7   New service opportunities on the optical broadband services network

The new OBS Network[7] will allow carriers to:

- Quickly introduce and deploy differentiated services to increase their subscriber base
- Provide new pricing paradigms (based on shorter durations) to attract new customers
- Lower OPEX through connection automation and customer self-management
- Continue to drive down CAPEX by using the network more efficiently

Some near-term opportunities include:

1. Rapid-service provisioning
2. CoS wavelength services
3. Bandwidth-on-demand (BoD) services
4. Optical virtual private network (OVPN)
5. Packet/optical inter-working

4.7.1   Rapid provisioning

The speed of provisioning (i.e., setting up a completely new connection on customer request) has become a critical competitive weapon. Provisioning involves three components:

1. Customer attachment to the network (e.g., network ingress and egress access)
2. Having sufficient resources within the network itself
3. Finding and connecting these resources

The OBS network facilitates the first and second operations.

A combination of auto discovery, new/enhanced planning tools, and mesh flexibility will increase the probability that the network has sufficient connection capacity:

- *Auto discovery* — This allows an accurate, real-time view of network capability. Often, misalignment occurs between the capability that the planning tools believe is in the network and the actual capability. This may result in delay because the planner cannot see available capacity, or because he or she sees capacity that is not actually available, causing a dropout later in the provisioning process.
- *Planning tools* — These tools allow accurate visibility of the network and can be used to facilitate just-in-time provisioning of additional capacity. This is not possible in the present mode of operation (PMO) because of the inaccuracy of the data. New planning tools can utilize real time network utilization information to more accurately predict where additional capacity will be required.
- *Mesh flexibility* — Often, capacity exists but it is not in the right place. For example, capacity is in the wrong timeslot, it is on the wrong link, or the ADM patch panel interconnect is not in place.

Introduction of OBS's auto-discovery mechanisms changes today's manually intensive and error-prone PMO to a request from the management plane to the control plane. Because the control plane has an up-to-date, accurate view of the network resources, connection across the network can be made reliably in seconds.

4.7.2   Connection granularity and class of service (COS) connection

OBS will provide a wide range of bandwidth options: 2.5Gb/s and 10Gb/s wavelength services evolving to 40Gb/s services in the near future, in addition to SONET-wrapped STS-1/3c/12c/48c/192c and unlimited virtually concatenation options. Routing can be accommodated on a wide range of criteria including least delay, least cost, nodal and link diversity, or other geographic considerations. A wide range of protection and restoration options is available, including protected (50ms), unprotected, shared mesh (100–200ms), reroute on failure, and pre-emptible.

4.7.3   Bandwidth on demand (BoD)

Unlike rapid provisioning, BoD allows customers to directly *set up/tear down/modify* connection requests to the control plane. The customer request can be initiated by either of the following methods:

> From a client device as an UNI request
> From a customer's network operations center (NOC) — in this case, the customer will have access to a secure Web portal that will validate the request and proxy-signal to the source OCC

The latter would be a more likely scenario in the near term because few client devices will have the intelligence to autonomously drive new bandwidth requests.

BoD is *not* the same as rapid provisioning. BoD is about delivering a new service type that bills on a per-use basis. In BoD service, the customer first negotiates a BoD contract through a management process. The customer's access to the network is then established and fixed; thereafter, he controls his connections on a demand basis. Rapid provisioning, as discussed earlier, is about easing operational issues in delivering services.

4.7.4   Optical VPN

OVPN provides the carrier and its customers with a mechanism for controlling, monitoring, and securing the flexibility provided by an OBS network. In an OVPN service,[8] all users, subject to a prearranged policy, freely share all available network resources. Bandwidth is provided and billed for on an as-needed basis. This is appealing to users with changing connectivity requirements, who do not have strict requirements on bandwidth availability and who have a reasonable level of trust in their provider's ability to set up connections only between valid members. Initial target users for this service are quite broad and include large enterprises and carriers/ISPs.

OVPN provides the following benefits to customers:

- *Reduced costs* — the end user gets the benefits of a private network without having to manage his own infrastructure. The service provider's infrastructure is shared among many virtual private networks. The end user pays for only what is essential to his business.
- *Flexibility* — the end user gets a view of his virtual private network that exactly matches his business needs. Within this view, he has a degree of control and visibility over his resources; for example, he can see faults, get bandwidth on demand, and monitor performance.

4.7.5   Packet/optical inter-working

As packet services are expected to be a major client of optical networks, a significant effort is under way toward developing complementary Layer 2/3 and Layer 1/0 products that deliver cost-effective, differentiated, open, and scalable networking solutions.

This work falls into three broad areas:

1. Delivering cross-layer, application-driven network management and services
2. Layer 2 and Layer 1 inter-working for network optimization and scaling
3. Low-cost router-to-optical interfaces

The fundamental building blocks to achieve this synergy are the UNI and the OVPN constructs described previously. OVPNs provide routers with a dynamic, secure environment to establish optical paths, while the UNI allows autonomous bandwidth assignment where and when it is needed.

Service opportunities being assessed include:

- On-demand route diversity for increased reliability
- Optical bypass of tandem router for increased performance, addressing scalability, cost, and latency
- Mitigating fan out exhaust issues by using channelized ports with UNI control
- Coordinated optical and packet layer quality of service
- Coordinated optical and packet layer traffic engineering
- Cross-layer restoration and protection optimization and coordination

4.8   Conclusions

This chapter has presented the drivers and requirements for evolving the current optical transport network into an *Optical Broadband Services (OBS) Network.* A number of key elements will need to be provided and integrated:

1. A dynamic optical transport layer
2. A standards-based signaling and routing control plane based on the emerging global G.astn framework
3. A flexible and standards-based service adaptation layer

Service adaptation is an area of growing importance because it provides the gateway for moving services onto an increasingly agile and photonic core. Several service adaptation technologies are being developed that will allow carriers to evolve their existing SONET networks into a new OBS Network.

Managed wavelength and new data services are shaping the next-generation networks, and will play significant roles as the new OBS network is introduced. OVPN is an important service building block for a variety of applications. OVPNs provide the appropriate closed and secure environments needed to manage the flexibility of the emerging intelligent networks. For end users, OVPNs better align services with their value; for service providers, OVPNs improve utilization of their resources.

Acknowledgments

The authors thank the following people who have contributed to the ideas presented in this paper: Louis-Rene Pare, James Goodchild, Brent Allen, Malcolm Betts, Steve Foster, Steve Harris, Raymond Aubin, Don Ellis, and Maja Jelaca.

References

1. Gruber, J., The Drive To IP on Optics, The Carrier IP Telephony 2000 Comprehensive Report, IEC, pp. 113–123.
2. Mayer, M., ASON Standardization within the ITU, IIR Conference, Barcelona, Spain, Nov. 2000.
3. Su, D. et al., Standards: standards activities for MPLS over WDM networks, *Optical Networks*, vol. 1, no. 3, July 2000, pp. 66–69.
4. Gruber, J. and Ramaswami, R., Moving toward all-optical networks, *Lightwave Mag.*, Dec. 22, 2000, pp. 60, 62–64, 66, 68.
5. Aboul-Magd, O. et al., User Network Interface (UNI) 1.0 Signaling Specification, Contribution Number: OIF2000.125.6, Sept. 2001.
6. McGuire, A. and Flavin, A.J., Interworking between SHD and OTN-based transport networks, *BT Technol. J.*, vol. 19, no. 3, July 2001.
7. Benjamin, D. et al., Optical services over the Intelligent Optical Network, *IEEE Commun.*, Sept. 2001, vol, 39, no.9.
8. Ye, E. et al., Service Requirements for Optical Virtual Private Networks, OIF Contribution, 2002.038.01, Jan. 2002.

chapter five

Multiprotocol label switching

Matthew N.O. Sadiku
Prairie View A&M University

Contents

5.1 Introduction

Due to the proliferation of personal computers that are connected to the Internet and the popularity of the World Wide Web (WWW), Internet use has gone beyond all expectations. The adoption rate of the Internet has surpassed all other preceding technologies such as telephone, radio, television, and computer. This phenomenal growth has made the Internet protocol (IP) suite the most predominant networking technology. With the rapid growth of Internet services and the recent advances in dense wavelength division multiplexing (DWDM) technology, an alternative to asynchronous transfer mode (ATM) for multiplexing multiple services over individual circuits is needed. The once fast and high-bandwidth ATM switches are not good enough because Internet backbone routers are outperforming them.

Multiprotocol label switching (MPLS) offers a simpler mechanism for packet-oriented traffic engineering. The term "multiprotocol" implies that its techniques are applications to any network layer protocol; however, in this chapter we regard IP as the network layer protocol.

MPLS is an extension to the existing IP architecture. It is the latest step in the evolution of routing and forwarding technology for the Internet core. It is a new technology being standardized by the Internet Engineering Task Force (IETF) designed to enhance the speed, scalability, and service provisioning capabilities in the Internet. As a technology for backbone networks, MPLS can be used for IP as well as other network-layer protocols. It has become the prime candidate for IP-over-ATM backbone networks.[1,2]

This chapter presents the basic features and architecture of MPLS. It also discusses its popular applications such as virtual private network, traffic engineering, and IP and ATM integration. The generalized MPLS (GMPLS) is also discussed.

5.2 MPLS basics

Cisco has been working within the IETF to develop a standard for invented tag switching since they first shipped it in 1998. That standard is MPLS. Thus, tag switching is a prestandard implementation of MPLS. MPLS provides a solid framework that supports the deployment of advanced routing services because it solves a number of complex problems. It addresses the scalability issues associated with the currently deployed IP-over-ATM overlay model and significantly reduces the complexity of network operation.

MPLS is a means of forwarding information at a very high rate. It combines the speed and performance of Layer 2 with the scalability and IP intelligence of Layer 3. It is based on the following key concepts:[3–5]

- It is a collection of distributed control protocols used in setting up paths in IP networks.
- It separates an IP router's function into two parts: forwarding and control.
- It provides a mean to map IP addresses to simple, fixed-length, protocol-specific identifiers known as "labels" (i.e., it separates forwarding information [label] from the content of the IP header).
- It uses a single-forwarding paradigm (label swapping) at the data plane to support multiple-routing paradigms at the control plane.
- It eliminates the need for interrogating the IP header of every packet at every intermediate router.
- It enables constraint-based routing, which allows the distribution of the traffic load on alternate routes when the shortest path is overloaded.
- It is an integration of Layer 2 and Layer 3 technologies. By making conventional Layer 2 features available to Layer 3, MPLS enables traffic engineering.

- It is independent of the underlying link-layer technology. It uses different technologies and link-layer mechanisms to realize the label swapping forwarding paradigm. It supports the IP, ATM, and frame-relay Layer 2 protocols.
- It provides flexibility in the delivery of new routing services.
- It supports the delivery of services with quality of service (QoS) guarantees and thereby facilitates voice, video, and data service integration.
- It facilitates scalability through traffic aggregation.
- It allows the construction of virtual private network (VPN) in IP networks.
- It offers a standards-based solution that promotes multivendor interoperability.

The aim of MPLS is to improve the scalability and performance of the prevalent hop-by-hop routine and forwarding across packet networks. Its primary goal is to standardize a technology that integrates the label switching forwarding paradigm with network layer routing.

5.2.1 MPLS protocol stack architecture

As depicted in Figure 5.1, a data network's architecture typically has four layers: IP for applications and services, ATM for traffic engineering, SONET/SDH for transport, and DWDM for capacity. Multilayer architectures like this suffer from cost-ineffectiveness and the need to manage complex network resulting from dissimilar technologies. The trend is to evolve core IP networks away from the overlay model and toward integrated solutions. This evolution is now feasible due to the developments in MPLS and optical internetworking systems. MPLS is a connection-oriented protocol that appears between the link layer and the network layer, as ATM or frame relay, but without a dependence on the link layer.

The concept of label switching is key to MPLS. The key function of MPLS is to perform traffic aggregation. In a conventional IP network, streams of traffic between two points in the network are divided into many IP packets with each packet having its own header: the source address and destination address. Along the way, each router inspects every packet header in order to route it. The less time routers spend inspecting packets, the more time

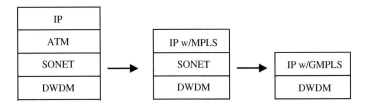

Figure 5.1 Evolution of technology toward photonic networking.

they have to forward them. It is a waste of resources to inspect every IP packet header when transferring a large number of packets destined for the same destination from the same source. MPLS recognizes such packets and adds a special label identifying them as a unified flow of information. In other words, a label is a short, fixed length, locally significant identifier used to identify a stream. Intermediate routers would then see the labels and forward them rapidly toward their destination.

At the ingress point of an MPLS network, the so-called *label edge router (LER)* adds a label to each incoming IP packet according to its destination and the state of the network. The label is embedded in the Layer 2 protocol header of the packet. The basic task of the routers is to forward packets as efficiently as possible from source to destination. To achieve this, each router needs information on the topology and status of the network, and where the egress of the respective destination is located. Label switching relies on the setup of switched paths through the network known as *label switching paths (LSPs)*. In other words, LSP is an established logical MPLS connection, which links an LER via a *label switched router (LSR)* to another LER, as illustrated in Figure 5.2. A key feature of MPLS is that one or more flows can be assigned to the same label or LSP. When a packet is sent on an LSP, a label is applied to the packet. On ATM links, for example, the label may be carried as the virtual circuit identifier or virtual path identifier applied to each ATM cell. A similar scheme has been proposed for SONET.

In an MPLS network, an LSP is set up for each path through the network. The router examines the header to determine which LSP to use, adds the appropriate LSP identifier (a Layer 2 label) to the packet and forwards it to the next hop. All the subsequent nodes forward the packet along the LSP identified by the label. LSPs typically follow the shortest path from source to destination. Once the traffic is within the MPLS network, only the label is used to make decisions on the next hop where the packet is to be sent, in contrast to traditional destination-based hop-by-hop forwarding used in IP networks. Each label has only hop-by-hop relevance. This feature allows for an efficient resource usage and provides high-speed forwarding.

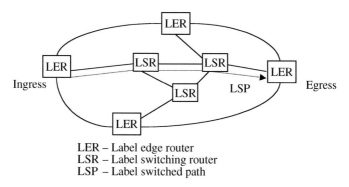

LER – Label edge router
LSR – Label switching router
LSP – Label switched path

Figure 5.2 MPLS architecture.

Bits 1	15	16	32	Variable	Variable
U	Message Type	Message Length	Message ID	Mandatory Parameters	Optional Parameters

Figure 5.3 The format of LDP messages.

Thus, an MPLS network consists of MPLS nodes called label switch routers (LSRs) connected by circuits called label switched paths (LSPs). The main component of MPLS architecture is the label distribution protocol (LDP), which is a set of protocols by which an LSR communicates with an adjacent peer by means of exchanging labels. To establish an LSP, a signaling protocol such as LDP is needed. Each LSP has a forwarding equivalent class (FEC) associated with it. The classification of packets into specific FECs identifies the set of packets that will be mapped to a path through the network. At the ingress of an MPLS domain, all incoming packets will be assigned to one of these FECs. Once a packet gets to the end of the last egress edge router, the labels are popped and the packet is routed according to the conventional IP rules.

As mentioned before, label distribution among the various LSRs is done by the label distribution protocol (LDP). The LDP protocol has four types of messages serving different purposes:[6,7]

- *Discovery messages* — to advertise the presence of LSRs
- *Session messages* — to establish and maintain LDP sessions
- *Advertisement messages* — to create, change, and delete label mappings for FECs
- *Notification messages* — to carry advisory and error information

The format of each LDP message is presented in Figure 5.3, where U (1 bit) denotes unknown messages and is used to notify the sender in case an LSR fails to recognize a message. The 15-bit message-type field is used to identify an LSP as one of the 10 defined types. The 16-bit message-length field specifies the total length of the message in bytes. The 32-bit message ID uniquely identifies the message. Mandatory and optional parameters employ a special coding called *type-length-value* (TLV). Within the IETF, two candidate protocols for the LDP are resource reservation protocol (RSVP) and constraint-based routing label distribution protocol (CR-LDP).

Labels exist within some transport media that MPLS nodes can use in making forwarding decisions. Examples include ATM's virtual path identifier/virtual circuit identifier (VPI/VCI) field and frame relay's data link connection identifier (DLCI). Other transport media, such as Ethernet and point-to-point links, must use a *shim label*. The MPLS generic shim label format is illustrated in Figure 5.4. It consists of the following fields:[4]

- Label (20 bits) — This is a locally significant ID used to represent an FEC during the forwarding operation.
- EXP (3 bits) — This "Experimental" field is currently being considered for QoS implementations.

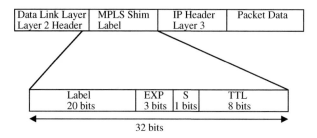

Figure 5.4 MPLS shim label format.

- S (1 bit) — This stack field indicates the presence of a label stack.
- TTL (8 bits) — The "time-to-live" field signifies the number of MPLS nodes a packet has traversed. It is decremented at each LSR hop and is used to throw away looping packets.

The 32-bit shim label is used to identify an FEC.

5.3   MPLS applications

Multiprotocol label switching can be used for the Internet protocol (IP) or any other network-layer protocols. It can be deployed in corporate networks as well as in backbone networks. Its greatest strength is perhaps its seamless coexistence with IP traffic and its reuse of proven IP routing protocols. It addresses today's network backbone requirements by providing a standards-based solution that supports network scalability, builds interoperable networks, and integrates IP and ATM.

MPLS may be regarded as a general-purpose tunneling protocol that enables several network services, which are distinguished by LDPs they use to build tunnels and what kind of traffic flows through the tunnels. Popular applications of MPLS are discussed next.[8]

5.3.1   Virtual private network (VPN)

MPLS supports virtual private network (VPN) services. VPNs share a single infrastructure of routers or switches between multiple independent networks. Using MPLS for VPNs is an attractive alternative to building VPNs using either ATM or frame relay permanent virtual circuits (PVCs), or various forms of tunnels, to interconnect routers at customer premises. MPLS-based VPNs have some advantages over a PVC-based model. They allow customers to choose their own addressing plans, and encryption is usually not required. They also enable the development of extranets using flexible policies. An MPLS-based VPN employs LSPs to provide tunnel-like isolation. The key benefits of MPLS-based VPNs include increased scalability, QoS support, and implicit data security mechanism.

5.3.2 Traffic engineering

MPLS is useful in the field of traffic engineering, which refers to the ability to control traffic flows in a network with the goal of reducing congestion and improving network utilization. Traffic engineering may also be regarded as the task of optimizing the use of network resources by provisioning LSPs as dedicated paths between two end points. Traffic engineering is essential for service provider and Internet service provider (ISP) backbones. Such backbones must support a high use of transmission capacity, and the networks must be able to withstand link or node failures. Traffic engineering is inherently taken care of in MPLS using explicitly routed point-to-point paths. Using explicitly routed paths is a more straightforward and flexible way of handling congestion of a complex network. MPLS traffic engineering provides an integrated approach to traffic engineering. LSPs can also be established by a variety of different control modules. In other words, MPLS uses explicitly routed LSPs for traffic engineering purposes. The MPLS traffic engineering control block performs all crucial functions such as resource discovery, network state, path computation, and route management. At the time a label is applied to a flow or session, predefined traffic engineering metrics can be programmed into the forwarding hardware guarantee levels of traffic bandwidth, jitter, and congestion control.[9–12]

5.3.3 IP and ATM integration

MPLS enables IP and ATM integration. In fact, MPLS is primarily a solution for IP-over-ATM backbone. MPLS enables an ATM switch to perform virtually all of the functions of an IP router. This is made possible by the fact that MPLS label swapping is the same as the forwarding paradigm provided by ATM switch hardware. MPLS is the Internet's best long-term solution to efficient, high-performance forwarding and traffic differentiation. Although MPLS can operate over any data link layer, ATM is by far the most suitable technology due to its ability to transport data, voice, and video with the offer of a defined QoS. MPLS has some advantages over a typical IP-over-ATM overlay model. First, it offers simpler integration of control information. It also avoids complex mechanisms to provide mapping from IP multicast addresses to multipoint virtual circuits. Second, the scalability of IP routing is improved relative to an overlay model. Third, it makes routing at the IP layer more likely to be optimal.

The selling point for MPLS is its ability to support LSPs from edge to edge. This allows sophisticated load balancing, QoS, and MPLS-based VPNs to be deployed by service providers.

Although MPLS is a technology designed for IP-based terrestrial networks, its capabilities can be made for satellite constellations where extensive routing is required.[13] It should be mentioned that, as great as MPLS is, not every network needs it. Only large networks such as wide area networks (WANs) with complex mesh topology must engineer traffic.

5.4   Generalized MPLS

The next logical evolutionary step toward all optical networks is to employ a variation of MPLS that supports different wavelengths for different data flows on the transport layer. This technique is already under investigation and it is called generalized *multiprotocol label switching* (GMPLS), which is also an enhancement of multiprotocol lambda switching (MPλS).[14,15] It is a suite of protocol extensions that support multiple switching types such as packet switching, time-division multiplexing (TDM) switching, lambda switching, and fiber switching.

The basic features of GMPLS include the following:

- It is a set of traffic engineering and optical extensions to existing MPLS routing and signaling protocols.
- It is an optical networking standard that represents an integral part of next-generation data and optical networks.
- It provides the necessary bridge between IP and optical layers, therefore enabling networks to be both interoperable and scalable.
- It brings more intelligence to optical networks; more intelligent optical devices help network operators do better restoration and better utilize bandwidth.
- It is designed to allow edge networking devices such as routers and switches to request bandwidth from the optical layer.
- It supports hierarchical LSPs and optical LSPs (or lightpaths).
- It is designed to handle multiple traffic types simultaneously. It creates a control plane to support multiple switching layers:
  - Packet switching for forwarding based upon packet or cell headers
  - Time-division switching for forwarding based upon the data's time slot in a repeating cycle (e.g., SONET/SDH, PDH)
  - Wavelength (lambda) switching for forwarding data based on the wavelength on which it was received
  - Spatial switching for forwarding data based on a position of the data in real-world physical spaces (e.g., incoming port or fiber to outgoing port or fiber)

The main focus of GMPLS is on the control plane of these various layers because each of them can use physically diverse data or forwarding planes. The goal is to cover both the routing and signaling portions of the control plane.

GMPLS extends MPLS to encompass time division, wavelength, and spatial switching. It extends the proven MPLS LSP mechanisms to create generalized labels and generalized LSPs. These extensions of MPLS include:[15]

- *Enhancements to signaling protocols* — The resource reservation protocol (RSVP) or constraint-based routing label distribution protocol (CR-LDP) is used as signaling protocol. GMPLS requires that an LSP start and end on similar types of devices. Although LPSs are unidirectional in MPLS architecture, they could be unidirectional or bidirectional lightpaths in GMPLS. GMPLS signaling enables an upstream node to suggest a label, although the suggestion may be overridden by a downstreams node.

- *Enhancements of routing* — GMPLS enables many improvements to routing. With GMPLS, explicit paths can be established across network layers. Open shortest path first (OSPF) or intermediate system to intermediate system (IS-IS) is used as interior gateway routing protocol (IGP).
- *Introduction of link management protocol (LMP)* — This is a new protocol designed to resolve issues related to link management of optical networks. LMP is what a router uses to decipher the availability of a link between itself and another router. It provides four basic functions for a node pair: control channel management, link connectivity verification, link property correlation, and fault isolation.
- *Additional functionality to address issues related to MPLS control plane* — GMPLS allows the control to be physically diverse from the associated data plane.

These extensions affect routing and signaling protocols for activities such as label distribution, traffic engineering, protection, and restoration. They also enable rapid provisioning and management of network services.

Although MPLS was designed exclusively for packet services, the primary goal of GMPLS is to provide a single suite of protocols that would be applicable to all kinds of traffic. By consolidating different traffic types, GMPLS permits simplification of networks and improves their scalability in ways unheard of until now. It offers the means by which networks can be scaled and simplified by deployment of a new class of network element designed to handle multiple traffic types simultaneously.

GMPLS network architecture offers a common mechanism for routing, data forwarding, and traffic engineering on transport networks with dense wavelength division multiplexing (DWDM). Such a future network will consist of elements such as routers, switches, DWDM systems, optical add-drop multiplexers (OADMs), optical cross-connects (OXCs), photonic cross-connects (PXCs), etc. that will employ GMPLS to dynamically provision resources and to provide network survivability using protection and restoration techniques. These optical network elements will have full control of the wavelengths and could create self-connecting and self-regulating networks. The technology behind GMPLS is micro-electric mechanical systems (MEMs), which can use micro-mirrors to redirect lambdas. This has opened the doors to a bandwidth explosion. This will eventually enable networks that require less management and overhead.

GMPLS offers several key benefits to service providers who migrate their services from their current model to the GMPLS unified control plane. First, GMPLS gives network operators the freedom to design their networks to best meet their specific objectives. Second, GMPLS provides accelerated service provisioning of any type of service, at any time, with any level of QoS, and to any destination. Third, GMPLS enables greater service intelligence and efficiency because it allows the network system to view the entire network topology, network node switching capability, and network resource status across all packet, TDM, and wavelength services. GMPLS will play a major role in the deployment of next generation networks; it will provide a common control mechanism across both the packet-switched and transport domain for establishing label switched paths.

5.5 Conclusion

The work on MPLS continues. The MPLS Forum, an international organization (www.mplsforum.org), advances the successful deployment of multivendor MPLS networks. It is hoped that optical MPLS over DWDM will also emerge from the MPLS standards.

References

1. T. Li, MPLS and the evolving Internet architecture, *IEEE Commun. Mag.,* Dec. 1999, pp. 38–41.
2. G. Armitage, MPLS: the magic behind the myths, *IEEE Commun. Mag.,* Jan. 2000, pp. 124–131.
3. J. Lawrence, Designing multiprotocol label switching networks, *IEEE Commun. Mag.,* July 2001, pp. 134–142.
4. A. Viswanathan et al., Evolution of multiprotocol label switching, *IEEE Commun. Mag.,* May 1998, pp. 165–173.
5. G. Hagard and M. Wolf, Multiprotocol label switching in ATM networks, *Ericsson Rev.,* vol. 75, no. 1, 1998, pp. 32–39.
6. T.M. Chen and T.H. Oh, Reliable services in MPLS, *IEEE Commun. Mag.,* Dec. 1999, pp. 58–62.
7. F. Holness and J. Griffiths, Multiprotocol label switching within the core network, *British Telecommun. Eng.,* vol. 18, pt. 2, Aug. 1999, pp. 97–100.
8. B. Davie and Y. Rekhter, *MPLS: Technology and Applications,* San Francisco: Morgan Kaufmann Publishers, 2000.
9. G. Swallow, MPLS advantages for traffic engineering, *IEEE Commun. Mag.,* Dec. 1999, pp. 54–57.
10. A. Ghanwani et al., Traffic engineering standards in IP networks using MPLS, *IEEE Commun. Mag.,* Dec. 1999, pp. 49–53.
11. X. Xiao et al., Traffic engineering with MPLS in the Internet, *IEEE Networks,* March/April 2000, pp. 28–33.
12. D.O. Awduche, MPLS and traffic engineering in IP networks, *IEEE Commun. Mag.,* Dec. 1999, pp. 42–47.
13. T. Ors and C. Rosenberg, Providing IP QoS over GEO satellites systems using MPLS, *Int. J. Satellite Commun.,* vol. 19, no. 5, 2001, pp. 443–461.
14. A. Banerjee et al., Generalized multiprotocol label switching: an overview of routing and management enhancements, *IEEE Commun. Mag.,* Jan. 2001, pp. 144–150.
15. A. Banerjee et al., Generalized multiprotocol label switching: an overview of signaling enhancements and recovery techniques, *IEEE Commun. Mag.,* July 2001, pp. 144–151.

chapter six

Dynamic synchronous transfer mode

Matthew N.O. Sadiku
Prairie View A&M University

Contents

6.1   Introduction

Computer networks such as the Internet have traditionally provided asynchronous communication, which uses packet-switching and store-and-forward techniques. Telephone networks, on the other hand, have provided real-time communication using circuit-switching and time-division techniques. Asynchronous transfer mode (ATM) has been introduced to provide all kinds of services; however, it has been demonstrated again and again that ATM cannot cope with those services that require high quality of service (QoS) guarantees such as voice, video, and hi-fi audio in an economically satisfactory manner. Internet protocol (IP) over ATM has been found to be a less elegant combination of protocols. A need exists for a transfer mode that can accommodate real-time as well as nonreal-time traffic. Also, with the large amount of data transfer capacity offered by the current fiber networks, the processing and buffering at switch and access points on the network is causing a bottleneck problem. Against this background, dynamic synchronous transfer mode (DTM) was developed.[1-3]

DTM is a new transport network technology designed specifically for the foreseen explosion of real time media in the next-generation networks. It is a broadband network architecture conceived in 1985 and developed at the Royal Institute of Technology, Sweden. It is an attempt to combine the advantages of both asynchronous and synchronous media access schemes. It is a networking scheme designed to fully utilize the capacity of optical fiber as a physical medium by emphasizing simplicity and avoiding computation-intensive policing, queuing, buffering, and control mechanism. This is achieved through the technology's inherent characteristics, which include dynamic bandwidth allocation, low propagation delay, almost zero delay variation, full traffic isolation between channels, and high-speed transmission.

DTM was developed to meet Layer 2 requirements for advanced IP services. Although Layer 2 requirements can get rather extensive, they include:[4]

Simple scalability — As the demand on networks has increased, the need for easily scalable resources has grown exponentially.

High availability — To ensure this, the network must be designed with high link, node, and switching redundancy at low cost.

Support for integrated services — To ensure high-quality IP network, a Layer 2 platform should provide real-time support with near jitter-free transport.

Resource management — This important feature provides the network operator with billing information.

6.2   DTM basic features

DTM is based on circuit switching augmented with dynamic reallocation of time slots. It is basically a time division multiplexing (TDM) scheme. It is a switching as well as a transmission scheme that can serve as a substitute for ATM. DTM is similar to SONET/SDH. It may run on top of SONET/SDH or work alone. It is compared with other transport network technologies in Figure 6.1. DTM guarantees high transport quality even over large-scale networks and it works well over DWDM.

DTM uses a unidirectional or simplex medium with capacity shared by all connected nodes (i.e., multiple access). The medium (called link) can assume different topologies such as dual bus, folded bus, or ring. A typical bus structure is presented in Figure 6.2. Although DTM is based on TDM, the entire capacity of a fiber link is divided into small fixed-size 125 μs frames. Each frame is further divided into 64-bit time slots. The frame length of 125 μs with 64 bits per slot is to enable easy adaptation to digital voice.

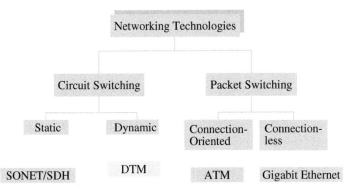

Figure 6.1 A comparison of network technologies.

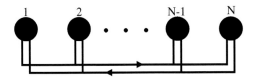

Figure 6.2 A DTM network with N nodes.

The number of slots per frame depends on the bit rate. For example, a bit rate of 2.5 Gbps produces approximately 4800 slots per frame. As illustrated in Figure 6.3, a slot is either a control slot or data slot and, if necessary, control slots may be converted to data slots or vice versa. The few control slots are used as broadcast control channels between the nodes. They are used to set up and tear down connections and to reassign data slots when a node requires a change in bandwidth. The majority of the slots are data slots, which constitute bandwidth and are dynamically assigned to nodes. At initialization, data slots are allocated to nodes according to a predefined distribution. A node needs to own a slot to transmit. DTM employs a distributed algorithm for slot reallocation, where free slots are distributed among the nodes.

The service provided by a DTM network is based on channels. A channel consists of a set of time slots with a sender and an arbitrary number of receivers. It gives guaranteed capacity and constant delay, making it suitable for delay-sensitive applications with real-time requirement such as video

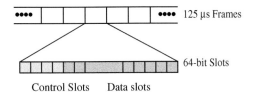

Figure 6.3 DTM multiplexing format.

and audio. Channels on the physically shared medium are realized by the TDM scheme, as depicted in Figure 6.3. A node forms a channel by allocating a set of data slots for the channel and by sending a control message for channel establishment. A channel may take one of the following forms:

Simplex — a channel that is set up from sender to receivers
Duplex — a channel consists of two channels, one in each direction
Multirate — a channel consisting of an arbitrary number of data slots
Multicast — a channel having several receivers

Traditionally, a circuit is a point-to-point connection between a transmitter and a receiver. DTM employs a shared medium, which inherently supports multicast because a slot can be read by several nodes on a bus. DTM channels can support the asymmetric patterns of Internet traffic. They can combine dynamically to optimize capacity.

Thus, DTM is based on some basic ideas that can be summarized as follows:[5]

- Pre-allocate bandwidth (slots) to all nodes and provide a mechanism for fast reallocation if necessary.
- Employ circuits to concentrate all processing in a connection establishment phase and make data transfer phase simple.
- Connect each node to more than one fiber and provide switching between different dual fiber buses.
- Employ wavelength multiplexing to increase the usage of each node.

DTM is optimized as either a direct voice, video, or data connection as well as a broadband carrier network. It does not use hop-by-hop decision making. Instead, it is based on source-routing and on-the-fly connection setup (i.e., channels are rapidly set up, broken down, or adjusted in size as needed). Using source-routing avoids complicated routing processes at the transit node and yields a simple high-speed backbone switch. With the dynamic resource management of DTM, network operators have excellent network control, much greater flexibility, higher utilization, and simplifier maintenance and administration. The features and benefits of DTM are summarized in Table 6.1.

Table 6.1 DTM Features and Benefits

Feature	Benefit
Dynamic allocation of resources	High bandwidth utilization
Distributed switching	Scalable in space and capacity
Efficient multicasting	Secure transfer of confidential information
Optional redundancy	Support of new services
Monitoring of bandwidth usage	High availability
High logical connectivity	Service with real-time guarantee
Smooth path for real-time applications	

6.3 DTM technology

DTM is a fast circuit switching that guarantees latency, almost zero jitter, traffic isolation, and flexible resource reservation. The technology includes switching, control signal for setup, routing, and network management. One may think of DTM as a next-generation SONET with switching features and flexible, configurable, and on-demand bandwidth allocation.

The DTM is a multiservice, integrating technology that may be divided into different parts: network architecture, signaling, transport, mapping, and channel management.[6]

6.3.1 Architecture

DTM operates at the lower three layers of the OSI model (i.e., physical, data link, and network layers). The physical layer deals with the transport of slots between nodes, bit encoding/decoding, slot synchronization, and framing. The data link layer contains functions and protocols required to construct, maintain, and transport data over a data link. The network layer handles the setup, modification, and removal of channels operating across multiple data links. This layer also handles switching and performs the distribution of synchronization information in the network. The DTM network is based on the DTM protocol suite in Figure 6.4. The protocol suite provides a complete communication architecture including the medium access control (MAC) layer, the synchronization scheme, routing and addressing of the ports. The MAC protocol provides a synchronous service. The DTM MAC protocol does not require optical termination in each node, in contrast to distributed queue dual bus (DQDB). Because a DTM channel provides a

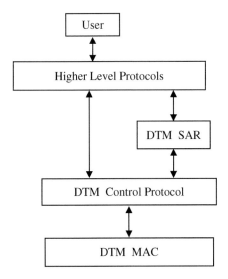

Figure 6.4 The DTM protocol suite.

synchronous transfer of data, providing asynchronous service requires using a DTM segmentation and reassembly (SAR) protocol, which is similar to the SAR protocols in DQDB and ATM.

6.3.2   Signaling

This describes the core signaling that is responsible for channel establishment, resource management, link state, routing, synchronization, and channel termination.

6.3.3   Transport

This describes the transport mechanism in the network. The implicit specification covers the switching elements, while the explicit specification deals with the physical link protocols required to transport DTM frames over physical media.

6.3.4   Mapping

This describes the mapping of various non-DTM technologies onto the DTM-core service so that the DTM system can transport these streams. This includes mapping of PDH over DTM, mapping of SONET/SDH over DTM, mapping of Ethernet over DTM, mapping of ATM over DTM, and mapping of multiprotocol label switching (MPLS) over DTM. DTM is a network architecture for integrated services. It offers different types of traffic: IP over DTM, DTM local area network (LAN) emulation (DLE), pleisiochronous digital hierarchy (PDH) transport, SDH/SONET tunneling, and transport of studio video. DLE is used to connect LANs by transferring Ethernet packets over DTM channels and making the attached nodes feel they are part of a LAN. The DTM only provides the conventional CBR (constant bit rate) service.

6.3.5   Management

Channel management handles the setup, modification of capacity, and removal of channels throughout the network. DTM channels can be managed in two ways: traffic-driven channel management and controlled channel management. The traffic-driven channel management type provides plug-and-play management and high resource utilization. Controlled channel management is executed by explicit reservations created by network management; it provides a basis for supporting applications with QoS demands. Both traffic-driven and controlled channels may coexist in a DTM network.

6.4   Comparison of DTM and ATM

ATM was developed in the mid 1980s, before the widespread use of IP. ATM has been found to be an inefficient transport for IP traffic and cannot scale to the high speeds some networks need. DTM, on the other hand, is designed to complement IP and is optimized to efficiently carry IP traffic. Similar to ATM, DTM network can adapt to variations in the traffic and can divide its bandwidth between hosts according to their demand. It can serve as a substitute for ATM because it offers better bandwidth management than ATM and scales easier up to higher bandwidth.

DTM avoids the major drawbacks of ATM:

1. The space overhead is low for DTM because it uses time division switching so that no packet header is needed. In contrast, the space overhead for ATM is about 10% (100 x 5/53).
2. DTM provides good QoS because it is based on TDM. To get similar performance from ATM, it has to be implemented over SONET/SDH, which requires additional, expensive equipment.
3. Because DTM is based on circuit switching, traffic management is simplified. In contrast, ATM is based on statistical multiplexing and so cell loss and cell transfer delay jitter (i.e., the variation in the delay) may occur.
4. A DTM network uses source-routing and on-the-fly connection setup (i.e., connections are created and released dynamically on a burst-by-burst basis). This makes DTM suitable for transferring fairly large data like the World Wide Web (WWW) traffic because such transfer does not require a preestablished connection.

ATM does not suit WWW traffic because it is connection-oriented and it incurs large connection setup delay when the network becomes large.

ATM has its merit with provision of a complete QoS support for a variety of traffic but it has to do per cell processing and queue controls at the switches, thereby creating performance problems. The DTM is a fast circuit switching that guarantees a constant delay transmission, low jitter, and bandwidth directly proportional to the number of slots that are on the channel and low loss rates.

6.5   Standardization and deployment

DTM technology used to be proprietary and most of its patents were owned by Ericsson and Net Insight (both headquartered in Sweden). The standardization of DTM should enable its many benefits to be shared across vendor solutions. Net Insight and Dynarc are mainly involved in the standardization process for DTM. In March 2001, DTM was recognized as a standard for real-time media networking technology by the European standards organization ETSI (European Telecommunications Standards Institute). The DTM multipart standard (ES 201 803) details the system architecture, physical interfaces, system signaling, the mapping of various services (SONET/SDH, ATM, frame relay, Ethernet, MPLS, etc.) over DTM, and management information. Although standardization is an ongoing process and it is expected that additional parts of the standard will be completed in the near future, this major milestone makes DTM technology an open standard and should speed up its development and deployment.

DTM technology has been proposed for building metropolitan area networks (MANs). A typical MAN consists of an optical fiber ring surrounding a large city with businesses (nodes) connected to the ring at various points. Users can use a MAN to get access to other networks.

The DTM has met some success in Europe and Scandinavia. In North America, DTM faces a lot of competing homegrown technologies that are perceived to be better. Also, because of the heavy investment already made on ATM and SONET equipment, it will take a while for DTM to be deployed in the United States. DTM is better suited for carriers who can afford the luxury of building a new network than for carriers with legacy equipment.

6.6   Conclusion

DTM arose due to the recent explosive popularity of IP for critical applications. It combines the advantages of guaranteed throughput, channel isolation, and inherent QoS found in SONET/SDH with the flexibility of packet-based networks such as ATM and Gigabit Ethernet.

DTM is a switching as well as transmission technique. It is designed with the purpose of fully utilizing the capacity of optical fiber as a transmission medium. It has been developed and is undergoing standardization. It is well suited for future high-speed backbone networks. Although the technology is new, it is gaining wide acceptance.

References

1. C. Bohm et al., The DTM gigabit network, *J. High-Speed Networks,* vol. 3, no. 2, 1994, pp. 109–126.
2. N. Yamanaka and K. Shiomoto, DTM: dynamic transfer mode based on dynamically assigned short-hold time-slot relay, *IEICE Trans. Communication,* vol. E82-B, no. 2, Feb. 1999, pp. 439–446.
3. C. Bohm et al., Fast circuit switching for the next generation of high-performance networks, *IEEE J. Selected Areas in Commun.,* vol. 14, no. 2, Feb. 1996, pp. 298–305.
4. O. Schagerlund, Understanding dynamic synchronous transfer mode technology, *Computer Technol. Rev.,* vol. 18, no. 3, March 1999, pp. 32, 36, 54.
5. L. Gauffin et al., Multi-gigabit networking based on DTM: a TDM medium access technique with dynamic bandwidth allocation, *Computer Networks and ISDN Sys.,* vol. 24, 1992, pp. 119–130.
6. Dynamic synchronous transfer mode (DTM); Part 1: System description, ETSI Standard ETSI ES 201 803–1, 2001.

chapter seven

A survey on fair bandwidth allocation for multicast over the Internet

Hooman Yousefizadeh
Florida Atlantic University
Ali Zilouchian
Florida Atlantic University
Mohammad Ilyas
Florida Atlantic University

Contents

7.1   Introduction

With fast growth in the speed and processing power of computers and network bandwidth, applications such as video conferencing and distance learning will become more feasible and popular. These applications face a new challenge of how to use network resources and how to allocate network bandwidth on links that are being shared by different flows. With the increase of these kinds of applications, multicast sessions will be seen more in the network. Considering the heterogeneity of the Internet (including optical Internet), an important issue pertaining to multicasting video is the congestion of some links, while other links may be underutilized. On the other hand, fair scheme addresses the case where each receiver is provided with video quality commensurate to its bandwidth capacity, regardless of the limitations of the other receivers of the same session. The definition of the fairness is a challenge, as is design of protocols to achieve this fairness. This chapter discusses a survey of different works to address the problem of multicast over the Internet.

7.2 Fairness aspects

In video unicast over the Internet, the sender sends video commensurate to the bottleneck link on the path from the sender to the receiver. In video multicast, however, due to heterogeneity of the Internet, it is not that simple. If the sender sends video with a rate commensurate to the receivers with the low bandwidth available in their paths from the sender, the receivers that have higher bandwidth available are not using the network resources efficiently. On the other hand, if the sender transmits video commensurate to high-capacity receivers the links in the path of low-capacity receivers will get congested, and as a result the quality of video will degrade for such receivers. So a very important challenge is to find a way that each receiver receives the video commensurate to available capacity in its path from sender. We adopt the definition of "fairness" from Rubenstein et al.,[11] as follows (for simplicity, we assume that utility is equal to the rate.):

> *Fully-Utilized-Receiver-Fairness:* A receiver's, "r's," rate is fully-utilized-receiver-fair if either its rate is equal to the maximum desired rate or at least one link exists in its path where all the other receivers from any session that are sharing that link has a rate less than or equal to the rate of r.
>
> *Same-Path-Receiver-Fairness:* A pair of receivers r and r' are Same-Path-Receiver-Fair if the route from the sender to them is exactly the same and either they have equal rates or the one with the smaller rate has that rate due to its session constraint on desired rate.

Researchers took different approaches to achieve fair bandwidth allocation, mainly in two categories.

1. *Receiver-Driven Approaches* — The sender sends video in a number of layers and lets receivers decide how many number of layers they are capable of receiving. These decisions are usually made through some join experiments to different layers of video. These algorithms have good performance regarding *intra-session fairness* (i.e., fairness

between different receivers of the same session); however, they do not guarantee good *intersession fairness* (i.e., the fair allocation of the capacity between different sessions in the common links). In fact, the sessions that start at a later time usually receive a lower share of the network bandwidth.

2. *Sender-Driven Approaches* —The receivers' feedback information about the capacity of the network and the path from the sender to them and based on that, sender decides at which rate the video should be transmitted to the receivers. The potential problem of these approaches is the *feedback implosion* problem. This problem occurs when all the receivers try to send information about their desired rates — they add to the traffic on the network and, as a result, the network becomes overwhelmed. In other words, this feedback brings network to the congestion.

Much research is in progress to find a protocol that can achieve both intersession and intra-session fairness. Different methods are being considered to send rate information back to the senders, and at the same time. In this chapter, we first introduce the important receiver-driven protocols and then some of the sender-driven approaches.

7.3   Receiver-driven approaches

Two major categories of receiver-driven approaches to achieve intra-session fairness are well known by the research community:

1. Replicated video multicast
2. Layered video multicast

7.3.1   Replicated video multicast

In this method, proposed by Jiang et al., the sender sends a number of video streams carrying the same video but with different qualities, and therefore requiring different rates.[2] The authors have introduced destination set grouping (DSG) protocol. In DSG, a set of receivers with capabilities close to each other will group together. Therefore, a number of multicast trees exist, each targeted to receivers with different levels of capacity. When a new receiver wants to join the multicast group, it will choose the group that is close to its capacity. For each group of receivers, the sender uses a feedback-controlled mechanism to adjust the rate of the stream, within some limits. If, due to a change of traffic in the network, a receiver wants to receive a quality higher than the higher limit or lower than the lower limit of the current stream, it can switch to another group. When this happens, the multicast trees will be updated.

Figure 7.1 presents an example of replicated video multicast using DSG protocol. Each line style depicts a different stream for the same video

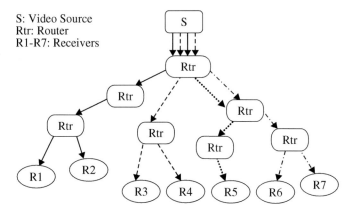

Figure 7.1 Replicated video multicast using DSG protocol.

with a different quality. This method of using replicated streams has the disadvantage of redundancy. The same video stream is being sent with different qualities.

7.3.2   *Layered video multicast*

This method takes advantage of video layering (see Figure 7.2). In this method, although video is being sent in a number of streams, similar to the previous method, different streams are not independent from each other. All receivers will receive the first layer, which is the base layer. Receivers can join to more layers commensurate to their capacities.

The greater the number of layers, the better quality of video. These streams, in spite of replicated video, are augmented to each other to give a better quality to the video overall. If L1 < L2, receivers cannot join the layer L2 multicast tree without joining layer L1. In fact, layer L1 without layer L2 does not give any meaningful information.

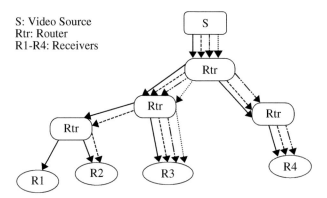

Figure 7.2 Example of layered video multicast.

Two well-known protocols that have been suggested for layered video multicast are receiver-driven layered multicast (RLM) and layered video multicast with retransmission (LVMR), which will be explained next.

7.3.2.1   *Receiver-driven layered multicast*

McCanne et al. first introduced the RLM protocol.[3] In this protocol, considering the fact that different receivers with heterogeneous capacities are available, the sender sends the video in the different layers, and receivers decide how many layers of video they shall receive. In fact, receivers keep joining with more layers of video to increase the quality of the video they are receiving. This process continues until either they subscribe all layers or congestion occurs. The protocol works based on join experiments performed by receivers. Receivers join a new layer, then wait for a time period (detection time), and monitor the network and its packet loss. If packet loss exceeds a threshold, they deduce that congestion has occurred and drop the layer. The protocol also considers the fact that when a join experiment fails, the other receivers can learn from it. The protocol does not repeat the experiment if it is likely to fail.

In this protocol, the concept of shared learning is introduced. Any receiver that desires to conduct a join experiment notifies all receivers in the group by multicasting a message including the experimental layer. Therefore, if an experiment fails, all the networks can increase their join timer and try the experiment after a longer period of time. Also, to correlate congestion to the corresponding join experiment, if one receiver is conducting a join experiment for a layer, other receivers wait until after a predefined detection time. This period can be readjusted when congestion is detected. If the other receivers want to join to higher layers, they wait for detection time before they conduct their join experiment. Therefore, by limiting the interference of join experiments, the probability that a receiver interprets congestion as a result of its join experiment (although it is not the case) decreases. In addition, if two receivers are conducting join experiments for different layers (L1 and L2, when L1 < L2) and congestion occurs, the receiver conducting the join experiment for L1 does not drop the layer L1 unless the second receiver drops all layers L2 to L1 + 1 and the network is still in congestion.

7.3.2.2   *Layered video multicast with retransmission*

Li et al. pointed out two issues in RLM that make the protocol inefficient.[4,5] First, it is not necessary for every receiver to know about each experiment and its result. Second, this extra information, like the beginning of experiment and its result, are consuming additional bandwidth. To address these two issues, they introduced the layered video multicast with retransmission (LVMR) protocol. This protocol takes advantage of several observations. One is that the join experiments on the layers lower than or equal to the current maximum layer in the subnet do not affect other receivers. Join experiments of the layers above this level and lower than maximum layers in a domain do not affect the receivers that are outside the domain. Another observation

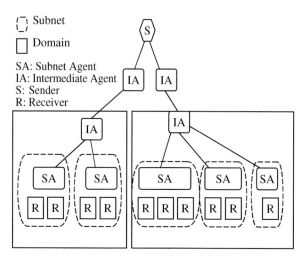

Figure 7.3 Example of hierarchical order of agents in the network.

is that a join experiment on layer L cannot affect receivers that are already receiving layer L. If receiver A's join experiment on level L causes congestion in receiver B, then receiver A itself should also be congested. Second, receiver B's join experiment on level L will also cause congestion in receiver A.

LVMR works as follows: at every domain, an intermediate agent (IA) exists, and in each subnet within a domain a subnet agent (SA) exists. These agents will collect the status within their domains. Every higher-level agent will receive the information from lower-level agents and pass it to lower-level and higher-level agents. Therefore, information goes from receivers to SAs and from SAs to IAs and so on until it reaches the sender. Also, the flow of compiled information goes from the sender to the receivers through IAs and SAs. The details of this protocol can be found in Li et al.[4,5] Figure 7.3 illustrates the hierarchical architecture in LVMR.

In this protocol, three conditions are defined as follows:

Congestion Condition — when packet loss rate exceeds a threshold or percentage of frames delayed exceeds a threshold
Unload Condition — when packet loss rate is less than a threshold and percentage of frames delayed is below a threshold
Load Condition — when we are not in congestion or unload conditions

In the following discussion, T_u^i is the time period that the receiver needs to stay in an unloaded state before it tries adding the next layer. T_c^i is the time period that the receiver needs to stay in a congestion condition before it drops the highest layer. Three "states" are related to the previous three conditions and a *temporary state* for a time period T_t, whenever it drops or adds a layer. Receivers move from one state to another when the condition changes. If the receiver has been in an unloaded state for a time period and it is still in the unload condition, it will add a layer, and if the receiver stays in a congested state for a time period and it is still in a congestion condition, it will drop the highest layer. In both cases, the receiver goes to a temporary state and remains there for T_t; then, based on the condition, it moves to a congested, loaded, or unloaded state. When a join experiments fails, the receiver increases T_u^i exponentially by factor α to avoid oscillation due to an add-layer experiment that occurs too soon. When it succeeds, T_u^i is decreased by factor α, and $T_u^{i+1} = T_u^i$ T_c^i is set to an initial value; but after each add-layer experiment of adding layer i, it is set to a small number in order to detect congestion quickly and drop the layer in this case. Also, if the receiver that is in a congested state finds another congested receiver on the same subnet, it increases T_c^i by one. This is due to the fact that it is more likely that congestion is caused by the other receiver. The details of work can be found in Li et al.[4]

Other important issues are mutual confusion and third party confusion. If receivers A and B conduct an experiment simultaneously and congestion occurs, they cannot know who caused congestion. Also, a third receiver with a current level of subscription lower than these two receivers cannot know who caused congestion. To solve these issues receivers, before conducting an experiment, inform SA and wait for its response before they start add experiment. SA in turn informs IA. If IA receives more than one request, it will make the decision about which one or multiple of the experiments can be conducted. Based on the results, IA builds some kind of knowledge base that helps it figure out the effect of these experiments on each other. If they are independent, it will allow both of the experiments (or maybe more). If they are not independent, IA will only let the lower-level experiment run. Examples can be found in Li et al.[4]

One important point that is clear about the mentioned protocols is that they assume that a number of layers are sent and each layer has a rate. With these given groups and group rates, the receivers should decide to how many layers they want to join; however, some questions are left unanswered. In how many should the sender provide the data? And what rate should be chosen for each group to have the best performance over the network? The sender-driven approaches try to answer these questions. Notice that in sender-driven approaches, the sender should have the information about network traffic along the paths in the multicast tree. Therefore, in these algorithms the task of receivers and routers should also be stated.

7.4   Sender-driven approaches

In these approaches, receivers send feedback to the sender about their desired rates. The sender then decides on how many different rates to use for sending the video and what the values of those rates should be. Some research has been conducted recently in this area. We will introduce one of the early works in this area next.

Yang et al. introduce an algorithm that with the use of dynamic programming finds the optimal partitioning of the rates.[12] This is the optimal number of multicast groups and rate of transmission in each group. Before we explain this algorithm, we introduce the isolated rate. Isolated rate in a multicast session is defined as the reception rate of the receiver if no constraints are imposed from other receivers of the session. This is the reception rate in case this receiver belonged to a unicast session. In other words, this rate is only limited by the receiver itself, or by the capacity of the path leading from the sender to this receiver. The parameters that are needed for each receiver are *loss tolerance* and *isolated rate*. Authors use the inter-receiver-fairness utility function for the receiver (introduced by Jiang et al.) or the rate utility function defined as follows:[1]

$$U_{IRF}(r,g) = \frac{\min(r,g)}{\max(r,g)}$$

where r is the receiver's isolated rate.

$$U_{rate}(r,g) = \min(r,g)$$

and g is group rate.

The group utility function is defined as the sum of utilities of all the receivers in the group, and session utility is the sum of all of the utilities of all of the groups in the session. Clearly, group utility is a function of the receiver isolated rates and group rate. So in this algorithm receivers, upon receiving the *calculate-rate* message from the sender, calculate their isolated rates and send them back to the sender. They also proved that the maximum session utility occurs for group rate equal to isolated rate of one of the receivers in the group. Therefore, upon receiving all of the isolated rates, the sender calculates the optimal group rate by simply calculating group utility function for all of the isolated rates and choosing the one that gives the maximum utility. Notice that if the group rates are ordered, the calculation of this optimum rate is even easier. Then, the sender transmits group rates to the receivers, and they choose the rate that is close to their isolated rate. For more information about the role of routers as well as more detail about this algorithm, see Yang et al.[12]

Clearly, the algorithm first assumes the predetermined number of groups; however, still the question of the number of groups is not answered. To answer this question for a given set of receivers, they have studied the ratio (Q_K) of optimal session utility when we have K groups $V^*(K)$ to the session utility when we have N (number of receivers) groups $V^*(N)$, for different values of K. It is clear that $V^*(N)$ is maximum, because it can choose the one rate to each receiver. Therefore, it has the highest possible level of granularity. But on the other hand, it will add to the overhead, so we want to have a compromise between overhead and utilization. They studied 300 receivers when receiver isolated rates follow three different probability distribution functions; namely, uniform, normal, and bimodal. Results show that with four or five groups, Q_K for both U_{IRF} and U_{rate} will reach about 80%. Therefore, it appears that the best number of groups would be about four or five. For more groups, increase of utility is not enough to justify the complexity and overhead that is being added to the network. Earlier works in this area can be found in Jiang et al. and Cheng and Ammar.[2,13]

7.5   Concluding remarks

In this chapter, issues regarding video multicast were introduced. Several methods of receiver-driven layered video multicast are presented. DSG protocol for replicated video multicast is introduced. Recent works in this topic can be found in Cheng et al.[6] RLM and LVMR are also presented; these are the bases of the most advanced receiver-driven protocols. Please see Lai and Pan, Gopalakrishnan et al., and Li et al. for recent developments in this topic.[8–10] Then, a sender-driven approach is introduced. A protocol to find the best rates for each group, suggested in Li et al., is explained.[10] We did not, however, concentrate on the routers and their roles in the video multicast. This was done to avoid lengthening the chapter, and by no means were we trying to imply that the router has a less important role in the protocols. Several other protocols are proposed to find the best rate for each outgoing link in the routers. See Graves et al., and Sarkar and Tassiulas (2000, 2001) for a few examples.[14–16] Yousefi'zadeh offers an analytical algorithm for finding group rates, and then finding the partitioning that results in maximum utilization.[17] Authors solve the problem in two steps. In the first step, they find the best rates for each group. In the second step, they find the new partitioning for the receivers.

References

1. T. Jiang, M. Ammar, and E.W. Zegura, Inter receiver fairness: a novel performance measure for multicast ABR sessions, *Proc. ACM SIGMETRICS 1998*, Madison, Wisconsin, June 1998.
2. T. Jiang, M. Ammar, and E.W. Zegura, On the use of destination set grouping to improve inter-receiver fairness for multicast ABR sessions, *Proc. IEEE IN-FOCOM*, Tel Aviv, Israel, pp. 42–51, March 2000.
3. S. McCanne, V. Jacobson, and M. Vetterli, Receiver-driven layered multicast, *Proc. ACM SIGCOMM 1996*, Stanford, California, September 1996.
4. X. Li, S. Paul, and M. Ammar, Layered video multicast with retransmissions (LVMR): evaluation of hierarchical rate control, *Proc. IEEE INFOCOM 1998*, pp. 1062–1072, San Francisco, California, March 1998.
5. X. Li, S. Paul, P. Pancha, and M. Ammar, Layered video multicast with re-transmission (LVMR): evaluation of error recovery schemes, *Proc. NOSSDAV 1997*, St. Louis, Missouri, May 1997.
6. S.Y. Cheng, M. Ammar, and X. Li, On the use of destination set grouping to improve fairness in multicast video distribution, *Proc. IEEE INFOCOM 1996*, San Francisco, California, pp. 553–560, March 1996.

7. X. Li, S. Paul, and M.H. Ammar, Multi-session rate control for layered video, *Proc. Symp. on Multimedia Computing and Networking,* San Jose, California, January 1999.

8. W. Lai and C.Y. Pan, Achieving inter-session fairness for layered video multicast, *Proc. IEEE Conf. on Commun.,* vol. 3, pp. 935–939, June 2001.

9. R. Gopalakrishnan, J. Griffiioen, G. Hjalmtysson, and C.J. Sreenan, Stability and fairness issues in layered muliticast, *Proc. NOSSDAV 1999,* June 1999.

10. X. Li, M.H. Ammar, and S. Paul, Video multicast over the Internet, *IEEE Network Magazine,* vol. 14, no. 2, pp. 46–60, 1999.

11. D. Rubenstein, J. Kurose, and D. Towsley, The impact of multicast layering on network fairness, *IEEE Trans. on Networking,* vol. 10, no. 2, pp. 169–182, 2002.

12. Y.R. Yang, M.S. Kim, and S.S. Lam, Optimal partitioning of multicast receivers, *Proc. 8th Int. Conf. on Network Protocols,* pp. 1–12, November 2000.

13. S.Y. Cheng and M. Ammar, Using destination set grouping to improve performance of window-controlled multipoint connections, *Computer Commun.,* vol. 19, pp. 723 –736, 1996.

14. E.E. Graves, R. Srikant, and Don Towsley, Decentralized computation of weighted max-min fair bandwidth allocation in networks with multicast flows, *International Workshop on Digital Communication,* Italy, 2001, LNCS 2170, pp. 326–342, 2001.

15. S. Sarkar and L. Tassiulas, Back-pressure-based multicast scheduling for fair bandwidth allocation, *Proc. IEEE INFOCOM,* pp. 1123–1132, 2001.

16. S. Sarkar and L. Tassiulas, Distributed algorithms for computation of fair rates in multirate multicast trees, *Proc. IEEE INFOCOM 2000,* vol. 1, pp. 52–61, 2000.

17. H. Yousefi'zadeh, H. Jafarkhani, and A. Habibi, Layered media multicast control (LMMC): rate allocation and partitioning, Submitted to *IEEE/ACM Trans. on Networking.*

chapter eight

Emerging optical network management

Imrich Chlamtac
University of Texas at Dallas
W.R. Franta
GATX Capital
Jason P. Jue
University of Texas at Dallas

Contents

0-8493-1333-3/03/$0.00+$1.50
© 2003 by CRC Press LLC

121

8.1   Introduction

In this chapter, we argue that carrier networks are changing through an evolutionary, not a revolutionary, migration process. As a result, end-to-end networks with SONET, optical, and packet sub-networks will exist for some time to come. We examine the consequences of this evolutionary migration process on the end-to-end provisioning and network control processes, discuss the protocols and net management structures being developed, and discuss their use in a technology mixed networking world. We begin with a description of the reasons for migration and end with an examination of algorithms relevant to provisioning and control in a mixed packet, SONET, and optical networking infrastructure.

During 1999 and 2001, the trade press had us believing that all-optical networking would soon revolutionize the data communications world and the networks of all the PTTs, CLECs, RBOCs, IXCs, etc. During 1999 and 2000, the venture capital community invested $29 billion and $52.8 billion, respectively, in networking technology and service companies. This data comes from a study conducted for *Network World* by PriceWaterhouseCoopers' Money Tree, in Partnership with VentureOne Survey.

By our count, by early 2001 over 300 companies focused on optical (photonic) equipment development. Networking product stalwarts were also heavily engaged in product development or company acquisition. Collectively, these companies were producing optical components, optical modules, optical switches, optical cross-connects, and optical add/drop multiplexer equipment. The optical switch, optical cross-connect, and optical multiplexer companies all produced accompanying "management" software for the provisioning and control of their equipment. Many of the optically focused companies were on rather long development cycles, 2 years or more, and were targeted to have product at the end of 2001 or the first half of 2002.

A number of other service companies were focused on the delivery of next-generation optically driven services, as exemplified by the emerging IXCs and CLECs in North America. These companies were intent on revolutionizing networking, making sweeping changes in the widely supported need to shift from voice-based to data-based networking infrastructures.

The rationale for all the "optical" enthusiasm was the promise of significantly more data-carrying bandwidth, new optically based service provider revenue-generating service opportunities, and simplified network infrastructures that greatly reduced service provider capital expense (CAPEX) and operational expense (OPEX) costs. Equipment provider PowerPoint presentations commonly made pledges that ranged from 30% (minimum) to over 70% or more savings in CAPEX, OPEX, or both.

New optical equipment promised to reduce or eliminate the costly electronic regeneration of optical signals through use of Erbium Doped Fiber or Raman optical amplifiers that allow optical signals to be transmitted up to 400 km to 6000 km without "electronic" (meaning solely in the optical domain, with no optical signal conversion to electronic signals necessary) 3R retrofit. New optical switching and mux equipment promised to ease the burden of "path" provisioning, "path" monitoring, and "path" restoration of optical data carrying "paths." We will more formally define paths a bit later.

Several materials were being investigated for optical components (chips), including SiGe and InP for optical muxes and demuxes, GaAs for amplifiers, and lithium niobate and InP for modulators. A number of optical switching technologies were being developed with tiny moveable mirrors based on micro-electro-mechanical systems (MEMS) becoming the leading candidate to the production of small to very large (1024×1024) port optical switches. Dense wavelength division multiplexing (DWDM) was promising to provide substantially more bandwidth over a single fiber, with advances carrying the technology from 4 or 5 channels (wavelengths) to over 160 channels. While all this was occurring, optical transmissions were migrating from 2.5Gb/s to 10Gb/s per wavelength, with the promise of 40Gb/s wavelength transmissions just around the corner. The basic premise of the *optical revolution* was that *by doing everything optically* instead of electronically, networks became much cheaper to install and operate, while simultaneously providing infinitely more power and flexibility of operation. With promises like these, the optical revolution could not be stopped, or so many thought.

Certain optical technologies have, indeed, gained customer traction, by having proven themselves both economically and technologically. Among the proven technologies are 2.5Gb/s and 10Gb/s DWDM transport systems, together with certain associated EDFAs. Optical switching, including fiber switching, wavelength switching, wavelength bundle switching, etc. have not gained customer traction.

Then came the "photonic nuclear winter" in early 2001. (Although this phrase is not original to the authors, they do not remember where they first heard or saw it used.) Beginning in the first quarter of 2001 the economy slowed, infrastructure capital became unavailable, carriers began to fail, nonessential optical infrastructure spending ceased, and the optical revolution slowed. Optical equipment provider attitudes shifted from thoughts of rapid growth to thoughts of survival.

In many cases, technological barriers were more difficult to overcome than developers had originally thought. With 40Gb/s transmissions, for example, the effects of chromatic dispersion, polarization-mode dispersion, fiber nonlinear effects, and component costs proved to be more severe than anticipated. More time was needed anyway to refine these technologies and to further the development of tunable lasers and all-optical wavelength converters. More time was needed to refine the distinguishable different needs in equipment cost, switching, speed, and amplifier needs, etc., of the access, metro core, regional core, and core (including long-haul) networking infrastructures. More time was needed to develop a uniform (standards-based) approach to the management of optical network infrastructures, for the wealth of product developments had also allowed a number of product-specific and incompatible network management systems to be created.

During the photonic nuclear winter, those companies that can survive will become stronger, with their products refined and certain competition eliminated. During this time the needs of the optical market segments will continue to evolve, with product solutions for each market segment becoming clearer and products being refined to use more highly integrated optical components and modules — moves that will tend to reduce product cost and increase product functionality.

During 2002 we have come to believe that with some exceptions the optical revolution, as it was envisioned during 2000 and 2001, will not occur. *The revolution called for the rapid replacement of electronically oriented networks by all-optical networks, with the retention of electronics only at points of user ingress and egress.* Instead, some form of *optical evolution will occur.* This evolution to optical networks will occur at a slower, more rational pace than was expected for the revolution. Evolution also sadly implies that not all of the over 300 extant optical equipment providers can thrive, and massive consolidation in equipment providers is inevitable. Evolution implies the refinement of optical network management standards and trial deployment of these standards prior to widespread deployment. Evolution implies that "out with the old and in with the new," will not occur, but that side-by-side existence of old and new will be much the norm.

Most projections, including those for optical components, the demand for bandwidth, etc., suggest a significant "up-tick" in carrier optical equipment spending beginning in 2003; spending that we believe will follow the revised, not the original, deployment model. Further, because of the photonic nuclear winter and the continued development of optical technologies during the photonic nuclear winter, the forms of products deployed in 2003 and beyond may be more integrated, and different, from the forms that would have been installed during 2001 and 2002 had the market prospered. This observation also suggests that optical technology is now well ahead of the carrier's ability to utilize it.

8.2 Optical networking today

Without returning to prestandard service provider networking era, we may say that we are entering the third generation of carrier (service provider)-based networking. The first generation is based on SONET networks, with T1/E1, etc. "tributary" or "feeder" lines. First-generation networks were voice traffic oriented.

Second-generation networking is based on SONET networks with DWDM point-to-point transmission systems and, in many cases, high-speed routers or asynchronous transfer method (ATM) switches in the network core. Second-generation networks carry data, even if that data is PCM digitized voice. In some cases, second-generation networks carry digitized voice.

Third-generation networks are exclusively data transmission oriented, with voice being carried as one data type. Third-generation networks, according to our principal of evolution given previously, are networks based on "old" SONET equipment, "new" (e.g., this implies new features such as virtual concatenation, which we will discuss later in the chapter) SONET equipment, local area network (LAN) and ATM switches and routers as well as on optical add/drop multiplexers, optical switches (fiber and wavelength), and (where needed) all-optical long-haul transmission systems. Third-generation subscriber access may well be optically based, with SONET or Ethernet as the core access technology. A third-generation network may well contain equipment from at least two, if not three, of these network generations. Correspondingly, third-generation network operation must be based on industry standards-based network management systems that bind together three generations of equipment into some kind of seamless whole.

8.2.1   The current network management situation

Current carrier network infrastructures are populated with SONET, router, LAN switch, ATM switch, time division multiplex (TDM), recently with DWDM equipment, and, finally and still, with some older TDM equipment. These technologies are incompatible in network management terms and, additionally, are mismatched in data payload transfer capabilities and other technology-related matters.

TDM equipment is managed by a vendor-independent protocol called TL1. TL1 dates from a 1984 Bellcore design. TL1 is based on a command line interface representative of a man-to-machine protocol that is designed to transmit commands to machines and receive messages from machines. It is designed to control network elements (e.g., blades) within a platform and to convey alarm and fault information. Although TL1 was applied to SONET, it never gained widespread use beyond TDM equipment.

SONET equipment is principally managed by an ISO-developed and ITU-adapted protocol known as the Common Management Information Protocol (CMIP). CMIP was designed to be a machine-to-machine protocol, and as with TL1, was intended to be vendor neutral. CMIP dates from 1988. CMIP has found use for SONET equipment management and has also appeared, together with SNMP (see below), in some ATM (switch) platforms. CMIP never gained traction beyond SONET and has not, for example, been used in the management of LAN networks.

LAN equipment, including routers, switches, and most ATM switches, use the Simple Network Management Protocol (SNMP) as the management vehicle. SNMP was developed by the IETF.

Discounting TL1, CMIP and SNMP are basically incompatible in operation. CMIP uses a connection-oriented (telco and ATM-oriented) model of operation, wherein either the networking platform or the management client initiate execution of a handshake sequence to establish contact, each with the other, prior to the exchange information and commands. SMNP uses a (LAN-oriented) connectionless model with no connection sequence prior to the exchange of commands and messages. SNMP is indeed simple, and employs only four communication primitives (*get, get-next, set, and trap*), whereas CMIP uses six (*get, set, action, create, delete, and even-report*). CMIP is more complex, even though the *get, set, trap, and event-report* primitives are similar in operation. Because of the connection model differences that exist between the SNMP and CMIP protocols, it is not surprising to find that each uses a different suite of support protocols. SNMP uses the connectionless UDP transmission protocol and Internet protocol (IP) packets (as defined as part of the TCP/IP suite of protocols), while CMIP uses the connection-oriented TP transport protocol and CLNP, the ISO counterpart of the Internet's IP protocol. Although both use ASN.1 (Abstract Syntax Notation One, a data formatting standard)-based "database" or data tree standards, differences exist between the two.

8.2.2   Other considerations complicating unified network management

SONET and LAN equipment represent the majority of all installed networking equipment. Under the evolutionary model advocated earlier, networks based on equipment from these two major classes of equipment types will persist, and all-optical networking infrastructures will be added to, and piece-wise replace, these extant architectures. Therefore, other SONET and LAN equipment differences and deficiencies, which affect (make more complex or restrict the effectiveness of) next-generation networks, *and next generation equipment "integrating" network management systems*, must also be mentioned. The most significant of these differences and deficiencies include the following.

SONET remains a single wavelength networking architecture, meaning that the management of DWDM transmission systems is treated as a second, separate entity. Correspondingly, and not surprisingly, SONET does not incorporate all-optical switches and add/drop muxes.

SONET has, for good reasons, a hierarchical structure. The hierarchical structure and *synchronous timing* used in SONET networks, allows for the rapid and correct muxing of several smaller (slower) data streams into a single larger (faster) data stream, and provides equally for the reverse demuxing process. (Each data stream level has a corresponding frame format for carriage of the stream data.) The SONET hierarchy is built up from the base Level-1 stream (with a speed of 51.8Mb/s), with the hierarchy containing stream levels of (multiples) of 3, 12, 24, 48, 192, and soon 768. It also defines four virtual tributaries (feeder streams) with rates from 1.5Mb/s (T1s) to 6.3Mb/s. Under this hierarchy, three Level-1, 51Mb/s streams are concatenated to make up one Level-3, 155Mb/s stream. Then, 4 Level-3, 155Mb/s, streams can be concatenated to realize one Level 12, 622Mb/s, stream, and so on up the hierarchy. LAN structures are not inherently hier-archically organized (packets are packets), although 10Mb/s Ethernet, to 100Mb/s Ethernet to 1000Mb/e Ethernet may give that appearance. With LAN structures *asynchronous timing* rules. In SONET, the notion of "time slot interchange" (TSI) is used. TSI allows a SONET node to remove a stream of level X from a higher level (speed) stream (e.g., level Y) and insert the X-level stream into another stream of level Y, or lower than Y, depending on the numeric relationship between X and Y. Some SONET equipment (generally installed in North America) does not support TSI, and the absence of TSI capability complicates provisioning requests. For future reference, we note that concatenation and TSI may be considered elements of data stream *"grooming,"* a subject to which we will return later.

SONET operates under the reliability principle of *protection,* which is used to *restore* failed service. LANs operate under the principal of service *availability.* SONET reserves secondary path resources (e.g., fiber links) to take over in the case of primary resource failures. The "switch-over" to secondary resources (restoration of service) is guaranteed to occur in less than 50 milliseconds. Restoration is guided by the "overhead" information carried in each SONET level-X (for all supported X values) frame that reports on the health and well-being of SONET resources along the "path." LAN availability allows SNMP traps to signal failures, with routing algorithms (such as the Open Shortest-Path First routing algorithm and protocol [OSPF], which is the most widely used routing protocol and the successor to the earlier Routing Information Protocol [RIP]) used to determine new paths (routes) for packets to reach destinations via new routes. No "availability" restoration time is guaranteed.

SONET was designed to carry digitized voice in frames, and not to carry data packets. Methods of carrying data packets in SONET streams were invented more recently and are now widely used. These protocols bear names like Packet Over SONET (POS), which utilizes the Point-to-Point Protocol (PPP), and the now ancient HDLC framing format. LANs were designed to carry data packets, and not digitized voice. New LAN protocols have been invented to carry digitized voice over LANs as data packets. These protocols are just coming into widespread use, and are collectively known as supporting Voice over IP (VoIP) protocols.

SONET networks are organized using hierarchically arranged notions of sections, lines, and paths. Sections are physical links between signal ampli-fiers and regenerators. Lines are the physical paths between two muxes (and may be made up of multiple sections), and a path connects two communi-cating end points (and is perhaps composed of several lines). LAN networks are not inherently hierarchically organized. In LANs, each device is a peer with all other devices in the network, and peer devices are interconnected by communication links (that may be WAN lines, Ethernet links or segments, fiber optic links, etc.).

The SONET network topology is inherently one of interconnected *rings* (and occasionally collapsed rings that appear as star [hub] topologies or point-to-point topologies), with a ring made up of SONET mux and cross-connect platforms connected by point-to-point physical links. LAN networks are generically *mesh* topologies (some of which may also be collapsed to point-to-point or hub structures).

The SONET asynchronous timing, hierarchical framing structures, restoration technology, and other data correction technologies, allow SONET networks to guarantee *Quality of Service* (QoS) for transported data stream frames. All of these guarantees are provided by physical layer equipment. LAN networking technologies were initially best effort packet delivery oriented. The lower layers could lose or discard packets, and it remained the responsibility of the two communicating end points to retransmit lost or damaged packets. Recently QoS technology has been incorporated into LAN equipment (technology with names like Diffserv, Class of Service, Multiprotocol Label Switching [MPLS], etc.) for the most part intended to ensure the availability of bandwidth to serve the declared needs of a packet stream. QoS levels delivered by LANs only approximate the QoS guarantees given by SONET.

SONET uses multiplexing, but *multiplexing* based on *reserved bandwidth*. LANs use *statistical multiplexing*, generally, without bandwidth reservation. As a consequence, a SONET frame will always find bandwidth available for its carriage, while a LAN packet may have to wait to receive bandwidth for its carriage.

Due to all of the preceding differences, SONET equipment is generally more expensive than LAN equipment (DWDM aside), although in stating so we realize we are comparing an apple to an orange.

8.2.3   *Where does all this leave us?*

None of the material presented in the previous section was purported to provide direct support for the management of optical switches, optical add/drop multiplexers, or any other form of all-optical platforms (network node); both SONET and LAN equipment perform electronic switching and adding/dropping of frames or packets, respectively. Further, as pointed out, integrated management support for DWDM transmission systems was also not provided.

The previous section does, however, lead us to certain conclusions. First, SONET is inherently, in some sense, more reliable than is the LAN. SONET uses *physical layer* protection paths to provide network *reliability*, and employs a physical layer network management architecture. LANs use *network layer* alternate routed paths to provide network *availability*, and use a network layer management architecture. SONET restoration is likely faster than LAN alternate path convergence.

Second, the network QoS guarantees for the two technologies are quite different, although LANs now provide for bandwidth reservation in some fashion. ATM equipment straddles the LAN–SONET world with respect to these measures. ATM is hard to provision, like SONET, and that is why permanent virtual circuits, not dynamic virtual circuits, are used. ATM's QoS attributes are closer to those of SONET than they are to those of LAN equipment, although LAN equipment is closing the gap.

Third, LANs were developed for data delivery under best effort or class of service guarantees. SONET was developed for guaranteed delivery of frames containing digitized voice. As a result, SONET has *limited flexibility,* exhibits *inefficiencies of bandwidth utilization,* exhibits *slow and difficult provisioning* of "circuits," and is realized from expensive platforms when applied to the transport of data.

Slow and difficult SONET provisioning results from the extensive use of manual- and people-intensive provisioning tools, techniques, and processes. The other two limitations result from the hierarchical "strength" of SONET itself when we employ that hierarchy to the carriage of LAN generated and consumed packets. The following table summarizes the situation, and the entries in the table result from the fact that a LAN packet stream can <u>only</u> be inserted into a frame of some level in the SONET hierarchy. For all cases listed, we must assume the unused bandwidth to be wasted.

Packet Service Rate	SONET Frame Level Required	% Bandwidth Utilization
10 Mb/s Ethernet	51Mb/s level-1 frame	20
100Mb/s Ethernet	155Mb/s level-3 frame	42
1000Mb/s Ethernet	2448Mb/s level-48 frame[a]	42

[a] We shall not explain why level-24 frames are inadequate.

Source: Adapted from *Delivering Ethernet over SONET Using Virtual Concatenation,* by Nilam Ruparelia, appearing in CommsDesign.com.

LAN equipment is rapidly improving to provide SONET- and ATM-like QoS and reliability features, but at least from a perception viewpoint it trails both in these measures. LANS are fairly easy to provision as so much is based on self-discovery and self-regulation.

8.2.4   Adding pure optical equipment

To the mix of considerations given in the last few sections we must now add those associated with pure optical networking. All SONET and LAN equipment performs add/drop and switching functions electronically. Many consider either of these classes of equipment to be optical simply because they employ DWDM point-to-point links. We restrict optical networking to networks that use equipment that employs "pure" *optical add/drop* or *optical switching technologies.*

With LAN/SONET equipment, adding, dropping, or switching a single wavelength or a bundle of wavelengths from a composite DWDM optical signal requires three steps. The first requires conversion of the incoming DWDM optical (O) signal to electronic (E) form, after which the individual wavelengths can be switched electronically, followed by recombination, that is, by the conversion of the electronic signals back to combined optical (O) signals for continued carriage. This sequence is known as the O-E-O conversion process. With pure optical equipment these functions are all performed in the optical domain, giving rise to the concept of O-O-O conversion. Optical switches have been developed that switch the complete DWDM contents of an incoming fiber to another outgoing fiber. Other switching devices switch individual wavelengths from an incoming fiber to another outgoing fiber.

As we stated in the introduction, fiber switches that contain 4×4, 8×8, 64×64, or 1024×1024 port switching matrices, as determined by the switching metro, to backbone core network switching application are being (or have been) developed. (Larger matrix switches have been discussed, but not built.) The most developed form of switching matrix technology is MEMS (tiny movable mirrors).

Devices that switch individual wavelengths are often optical add/drop muxes, although other forms of wavelength switching, including those that combine switching with wavelength conversion, also have application. Under the add/drop mux scenario, wavelengths are added to, or extracted from, a composite DWDM signal without conversion of the nonabstracted or added wavelength signals to electronic form. Unlike fiber switches, the number of add/drop ports is generally small in number.

Optical switches of either class are generally surrounded with additional optical equipment that amplifies the wavelength or composite signals optically, or that compensates for chromatic or polarization mode dispersion, or a number of so-called fiber-based nonlinear (but signal distorting) effects. These attendant optical components may be integrated into the switch or mux equipment or attached to it. Although optical switching is often said to be *transparent* to both the speed of the optical signal (in Mb/s) and the format of the signal (packet header, packet form, etc.), in some cases the attendant optical equipment compensates for signal distortions that increase (nonlinearly) with optical signal speed.

Optical devices place new requirements on network management systems. Switching functions must be performed and optical performance must be monitored. An optical switch cannot interrogate the contents of a packet or frame header in the optical domain as an electronic switch can in the electronic domain. It is therefore optical (bit stream) signals that an optical device switches, and not packet or frame streams. Network management, therefore, must determine the switching patterns of optical switches. Occasionally net management must also select settings for the secondary optical devices associated with an optical switch, although these activities are becoming governed by feedback loops in the devices themselves. With respect to a collection of switches in an all-optical sub-network, net management software is required to:

- Find a fiber or wavelength *path**
- From some add/drop starting point where a bit stream signal is to be "added"

- Through a series of switches and/or add/drop muxes
- To some destination add/drop mux where the bit stream signal is to be "extracted"

This task requires knowledge of optical network topology, fibers, switches and switch ports, wavelength, and fiber usage, etc., and algorithms that use all of this information to compute an acceptable (possibly bidirectional) optical bit stream path between the two "end" points. Often the computation of an acceptable path must take into account the nature and quality of individual fibers, the number of wavelengths on a fiber, the signal losses incurred during optical switching, wavelength conversion possibilities, etc. In some instances, the net management must adjust amplifier or other optical device parameters to provide for good quality of transmission along the computed optical path.

The first class of net management system-required information is semi-*static* (e.g., topology), while the second class of information is time-varying or *dynamic* (e.g., usage). The network management system must be able to set up and terminate appropriate paths based on information of both classes that it periodically monitors and collects. It can be seen, even though simplified, that optical networks place new and complex needs on network management systems.

In all the optical networking presentations we have seen, the impetus for all-optical networking is always:

- Transparency, as discussed previously
- Reduced Capex costs due to the elimination of massive amounts of expensive electronics for O-E-O conversions
- Reduced Opex costs due to the use of automated provisioning techniques, as broached previously
- The increased bandwidth that optical DWDM channels can provide

Clearly, the differences and deficiencies described earlier must be either tolerated or improved upon or overcome and the new requirements of optical networks must be accommodated, for nothing prevents and everything demands the efficient and reliable end-to-end carriage of data (IP) packets. Quite clearly, the evolutionary model imposes a carriage that likely begins and ends on Ethernets, with intervening passage through pure optical and SONET network-based network infrastructure segments. The trick is to quickly and easily provision the carriage capability and efficiently use the network resources along the provisioned route. A number of network management-related protocols are being developed to address these needs.

8.3 Third-generation network management goals

The change of networking focus from voice transport to data transport has sharpened the need for new network management features that integrate the SONET, LAN, and all-optical network equipment classes and provides for the rapid provisioning of, and efficient utilization of resources in, these integrated networks. With our migration to optical and integrated networks, time, trial, and trouble have taught us that we want networks in which the management software must *automatically*:

- *Discover* what resources (links, link capacities, switches and switch ports, wavelengths, etc.) are available and useful to any provisioning request
- *Construct* (i.e., identify or compute) the proper data stream path, observing any desirable constraints
- *Manage* path setup and path maintenance, including restoration in the face of failure, as well as path termination

It is the automatic execution of the discover, construct, and manage tasks by the "control plane" of the network management hardware and software that (among other things):

- Allows for the introduction of new services, including those wherein the subscriber ultimately only pays for resources used
- Reduces provisioning time from days to weeks, to minutes or milliseconds
- Allows carriers to more fully use available network resources
- Eliminates the detrimental effects of lost inventory (network resources)
- Allows fine-tuning of network growth plans, as the control plane (almost as a byproduct of its essential tasks) monitors and reports the utilization of resources

which, collectively, facilitate more rapid development and alteration of new services while providing substantial CAPEX and OPEX savings. This is accomplished through the elimination of long, hard, craftsperson-provisioning tasks and the "truck-rolls."

In short, what is needed and desired is a way to more:

- Automatically increase bandwidth utilization
- Reduce provisioning times
- Support multivendor (configuration) interoperability
- Support multi-equipment class including at least selective use of OADM and OXC equipment
- Support multidomain networks
- Reduce new service definition and activation times
- All in an end-to-end manner
- While maintaining support for security, billing, and settlement requirements, as well as service reliability

8.4 An introduction to recent control plane protocol developments

Third-generation management protocols are being developed by a number of organizations, including the Internet Engineering Task Force (IETF), the International Telecommunications Union (ITU), the American National Standards Institute (ANSI), the Institute of Electrical and Electronic Engineers (IEEE), and the Optical Inter-Networking Forum (OIF). The protocols being developed by these standards development bodies include:

- The Generic Framing Procedure (GFP) or the virtual concatenation protocol, a grooming standard for the more efficient carriage of Ethernet (or any other) packet stream over SONET
- The Optical Transport Network (OTN) architecture standard, the successor to SONET which integrates DWDM, and its associated management architectures into the architecture
- The Generalized Multiprotocol Label Switching (GMPLS) architecture and its supporting protocols, standards that integrate the provisioning of TDM, SONET, optical, and LAN-integrated end-to-end network infrastructures

8.4.1 Generic framing procedure (GRP)

GRP uses a concept called virtual concatenation to improve efficiency in the carriage of packets over SONET by improving the utilization figures presented in the earlier table. GRP addresses one aspect of *grooming,* namely the intelligent optimization of bandwidth throughout a network. Recall that SONET carries data in frames, which come in a variety of *fixed* sizes and which collectively define the SONET frame (and speed) hierarchy. The purpose of GRP is to create variable-sized frames, sized to better fit the packet or data they are intended to carry. From another perspective GRP allows the *fragmentation* of SONET data streams for insertion into several frames, while *concatenation* provides for the combination of lower-level frames to form higher-level frames. Notice that with fragmentation, a 100Mb/s Fast Ethernet packet stream fits into 2 level-1 SONET frames (representing 102.2Mb/s of bandwidth), while a 1000Mb/s Gigabit Ethernet packet stream fits into 7 level-3 frames. A 10Mb/s Ethernet packet stream fits into 7 virtual tributary frames (each at 1.544Mb/s). Given this approach, the utilization figures are given in the following table.

Packet Service Rate	VC SONET Framing	% Bandwidth Utilization
10 Mb/s Ethernet	6 VT1.5 frames	95
100Mb/s Ethernet	2 level-1 frames	97
1000Mb/s Ethernet	7 level-3 frames	91

Source: Adapted from N. Ruparelia, *Delivering Ethernet Over SONET Using Virtual Concatenation,* appearing in CommsDesign.com.

Not only does virtual concatenation (fragmentation) much improve bandwidth utilization, it leaves the operation of SONET rings intact. For, although all of the SONET frames containing the fragments of a packet data stream must end up at the same destination end point, the virtual concatenation "reassembly" information is carried in the path overhead bytes of a frame and need only be examined and used by the communicating end points and not by any of the intermediate SONET equipment nodes.

Implementation of virtual concatenation can be linked to the GMPLS protocol as described next. Virtual concatenation would be associated with new (or upgraded) SONET equipment deployed where packet streams can enter or exit a ring.

8.4.2   Optical transport network (OTN)

OTN, a development of the ITU and ANSI, is intended to address the shortcomings of SONET. The OTN architecture places three optical sub-layers beneath the SONET/ATM layer. The three sub-layers provide for:

- End-to-end networking over a single wavelength
- Networking of a multiwavelength (DWDM) signal
- The transmission of wavelengths on a fiber span

These three layers eliminate the constraint of a one-wavelength SONET architecture. Although useful, these three sub-layers alone do not reflect the structure of OTN. The architecture of OTN follows that of SONET with the hierarchy of optical channels, optical multiplex sections, and optical transmission sections (i.e., these three elements are the three sub-layers previously named) paralleling the SONET hierarchy of section, line, and path. Connections between two end points at any level of the hierarchy are called trails. As with SONET, each layer contains "overhead" information for the management of that layer. The OTN optical channel, much like the SONET path, transports an optical bit stream between the two end points. Unlike SONET, OTN is asynchronously timed like LANs.

OTN also has parallels to the SONET level (frame/speed) hierarchy outlined earlier. Just as the SONET hierarchy is based on multiples of the Level-1 (51.1Mb/s) stream, the OTN hierarchy is built up using multiples of the optical transport module (OTM). To accommodate DWDM, the OTM is specified in terms of a number of optical wavelengths and the speeds associated with the wavelengths. These associations appear as OTM — nu.sp, where nu provides information on the number of wavelengths used to support the data stream and sp provides information on the speeds of the individual wavelength channels, which need not be all the same. Three speeds are supported, namely 2.5Gb/s, 10Gb/s, and 40Gb/s. Speeds are designated by the digits 1, 2, and 3, respectively. Thus, conventional interpretation would say that an OTM-3.23 reflects a data stream that employs up to 3 wavelengths, with the individual wavelengths operating a speed of either 10 or 40Gb/s. An OTM-3.2 would reflect a stream utilizing three wavelengths, each operating at a speed of 10Gb/s. There is also a frame structure associated with OTMs that we will not explore.

The management aspects of OTN are captured in the specifications for the Automatic Switched Transport Network (ASTN) and the Automatic Switched Optical Network. ASTN applies to SONET and OTN; ASON applies only to OTN. It is intended that the management of OTN-based networks be automated according to the precepts given in the previous section on management goals.

8.4.3   Generalized multiprotocol label switching (GMPLS)

Clearly, OTN represents an extension or revision of the networking model that SONET is built around. It represents thinking that comes from the telco community. Different, yet unifying, management models are also being produced by the LAN community.

Our first reaction may be to look to the management associated with ATM as the best base for a unified network management system, as ATM, and its standardized Network-to-Network Interface (NNI), called ATM PNNI, is widely used in carrier networks, and ATM is also used in LANs. This does not appear, however, to be the path the industry is taking. Instead, it has been noted that extensions to a LAN protocol called multiprotocol label switching (MPLS) can be applied to the pursuit of a unified management schema. The MPLS extensions are now known as the GMPLS.

MPLS uses the open shortest-path first protocol with traffic engineering extensions (OSPF-TE) to provide topology and resource information. It also uses similar extensions to the border gateway protocol (BGP-TE). Two protocols, resource reservation protocol (RSVP) or constant routing (CR)-LDP are also used for the signaling of provisioning requests. These two protocols are known as label distribution protocols (LDPs).

MPLS also simplifies packet forwarding, by allowing packet switching devices to look only at a layer two "label," and not an IP and/or packet header to determine forwarding decisions. The forwarding mechanism becomes a simple single lookup that determines packet forwarding. Specifically, if a packet arrives on port X and contains the layer-2 label Y, then the lookup determines that the packet is to be forwarded on port Z with label W. This process is called *label switching* (ATM also uses the process on cells) and, because of its simplicity, is much faster than IP packet header-based forwarding determinations.

It is also true that label switching separates the switching criteria from the packet contents (except for the label), and sense it is, any kind of mapping of packets to labels can be used in the forwarding process. The separation is thus between the control plane and the data plane. With this separation it has become apparent that the concepts can be extended to the control of SONET, optical, and TDM devices, which do not make forwarding, or switching, decisions based on packet content but instead on time slots, wavelengths, fibers (physical ports), etc. In its current form, GMPLS is packet-switch capable, TDM capable, wavelength-switch capable, and fiber-switch capable. With GMPLS, an end-to-end path of appropriate resources can be established through a number of sub-networks of different and varying technologies.

8.4.4   Other issues

Consistent with the evolutionary model, our first observation is that we need not soon expect a single, complete, tightly integrated, efficient network management system. Network management systems will themselves evolve over time. From the preceding discussion emerges the perception that the route to a single all-encompassing net management architecture is one based on linking current technologies and protocols. This leads to the notions of "*a over b*" networking scenarios and *multilayered networking*. For example, we introduced the notion of packet *over* SONET. More generally, we have IP over ATM networks over SONET networks over DWDM optical networks. All "over scenarios" increase networking complexity and create the need for protocols that support the layering of protocols. The industry direction, therefore, is toward a flattening of the number of layers ultimately concluding with IP *over* optical, using MPLS and GMPLS in ASTN. The architecture may require the management of a single layer, or perhaps two layers — one for packets and one for transport.

Second, we expect some evolution in the mechanisms used to realize provisioning or the reservation of networking resources for a packet stream flow. Two models are contending for the provisioning honors; namely, the *overlay* and the *peering* request models, which *make* and *ask for* network resources, respectively.

The *overlay model* allows edge data devices (e.g., data stream ingress routers) to request resources from the carrier transport network through a User Network Interface (UNI) to the transport network. With the UNI, the requester need not have any visibility of the resources within the transport network, nor to the topology of that network. The UNI approach is historically consistent with telco network operation. A UNI model has been developed by the OIF. So far, this provides primitives for the endpoint connection, connection termination, and connection status over SONET networks, although it appears that it can be extended to include DWDM and optical switching devices. Among the benefits of the overlay model are its ability to keep secret certain topology and resource information, accommodate older technologies, and allow for dissimilar sub-networks that can evolve independently.

Under the *peering model,* the requester is a peer with all the resources in the transport network and, as such, can directly communicate with them. The peering model is consistent with data network operations, and in particular with Internet operation. Under the peering model, the edge (data) equipment is responsible for the allocation of network resources, whereas under the UNI model, agents hidden behind the UNI are responsible for resource allocation. Any of the relevant LAN-oriented provisioning LAN protocols, including MPLS, can be used under the peering model.

Third, we note efforts to replace the previously described MIB, SNMP, UDP, IP, and the GDMO CMIP, TP, CLNP protocol stacks with the IDL, COBRA, TCP, IP protocol stack. These efforts are based on the purported better scalability, and "independence" of COBRA. COBRA, along with Java, is enjoying use in a number of business applications. Here, COBRA stands for the common object request broker architecture, and our mention signals a watch-for-developments situation.

8.4.5   Protocols for bandwidth management in third-generation networks

The rapid growth of the Internet as a medium for global multimedia communications has resulted in a need for scalable network infrastructures which are capable of providing large amounts of bandwidth at varying granularities, on demand, and with quality-of-service guarantees, to support a diverse range of emerging applications. As pointed out previously, these requirements mean that the DWDM network bandwidth in emerging wavelength-routed systems needs to be managed, or provisioned, under the previous constraints. Optical wavelength-division multiplexing (WDM) communication systems, which are currently capable of providing over 1 Tbps of bandwidth over a single fiber link, are now being deployed in telecommunications backbone networks. Existing systems implement WDM on a point-to-point basis, with the optical signal being converted back to electronics at each intermediate switching node. Emerging wavelength-routed systems will utilize all-optical cross-connects (OXCs) to enable the establishment of all-optical circuit-switched connections, or lightpaths,[5,6] between source nodes and destination nodes. Lightpaths eliminate electronic processing requirements at intermediate nodes, thereby reducing the overall delay for packets traversing the lightpath and also reducing network costs.

In these all-optical networks, new protocols are needed to provision resources for lightpaths. When a connection request arrives to the network, a connection management protocol must find a route and a wavelength for the lightpath, and must provision the appropriate network resources for the lightpath.

8.4.6   The current situation

In most existing core networks today, provisioning is a manual process that can be very time consuming. Provisioning times on the order of days or weeks are not uncommon. As traffic becomes more dynamic and as the rate of connection requests increases, automated provisioning methods will be required.

Given the limitations of existing provisioning solutions, there has been much recent effort to improve provisioning times and resource utilization by developing protocols and standards to facilitate the automated, on-demand provisioning of optical connections. The Optical Domain Service Interconnect (ODSI) consortium is working to standardize interfaces that would allow client networks and devices to interact with the optical network. The ODSI framework does not specify how the wavelengths are actually provisioned within the optical network, but simply defines how a client would request or release an optical connection from the optical network.

Automated provisioning schemes may either be centralized, in which a single management entity in the network handles all provisioning requests, or distributed, in which all nodes are capable of participating in the provisioning process. Centralized approaches are simple to implement; however, they are not scalable under very dynamic traffic conditions, and may potentially be a single point of failure in the network. On the other hand, distributed approaches are scalable in nature but require a greater degree of coordination among network nodes.

The Internet Engineering Task Force (IETF) is currently working on GMPLS, a generalized control framework for establishing various types of connections, including optical wavelength connections, in a distributed manner by utilizing IP-based protocols. Currently, there is much work being done by the IETF to modify existing IP routing and signaling protocols in order to support GMPLS.

Most emerging GMPLS solutions are expected to be based on IP routing which utilizes the OSPF protocol or some variation of this protocol. In the OSPF protocol, global state information must be maintained at each node. This state information is updated through periodic broadcast messages. For an optical network with DWDM links, each node maintains the availability status of each wavelength on each link. In very dynamic systems, maintaining global information in a distributed manner can be bandwidth intensive and can lead to state inconsistencies and conflicts. Such conflicts may lead to thrashing, in which a lightpath request will attempt to reserve a resource which was previously available but has actually been reserved by another request. At the same time, the original lightpath request will have reserved resources that may block other requests. The overall effect of this phenomenon is that an increased number of blocked requests will lead to a situation in which resources are reserved for an increased amount of time, but never utilized. This increased reservation of resources leads to further blocking as well as decreased utilization. The effect becomes more pronounced as the rate of provisioning requests increases, as the hop distance of connections increases, and as the network size increases. An alternative approach to maintaining distributed link-state information is to maintain global information at a centralized server, and to have the centralized server handle all connection requests. This approach eliminates the problem of outdated state information; however, the centralized server may become a processing bottleneck and may not be able to handle a large number of dynamic requests in a timely manner.

In addition to the efforts to improve provisioning times, much interest also lies in providing differentiated QoS in order to support various emerging multimedia applications. QoS services, which use MPLS to recognize, classify, and prioritize IP traffic, are starting to emerge. With these types of services, carriers and service providers will be able to offer varying QoS service levels for different classes of traffic. The use of MPLS will reduce the need for information technology (IT) managers to set policies on their IP networks for every user, application, or piece of equipment. The hope is that QoS for IP networks will jump-start voice-over-IP and video applications, including teleconferencing, and that QoS will be available as a service that IT managers can buy from the carriers or Internet service providers (ISPs). The new carrier services will typically provide at least three levels of service to reduce latency, delays, or jitter on IP networks, and will help businesses fine-tune traffic to give specific-application traffic different levels of priority. The problem is that it is still too early for most providers to tie QoS and service level agreements to these emerging service offerings, largely because this technology is too new. Most carriers are only now implementing MPLS on their networks and must work with customers to establish guaranteed service levels over time.

The static provisioning of bandwidth at the wavelength granularity provides some level of service guarantee for fairly static, constant-bit-rate applications; however, as the Internet continues to evolve, an increasing number of applications are expected to be very "bursty" in nature with on-demand, variable bit-rate requirements. These applications may lead to inefficient network utilization for the case of statically provisioned bandwidth, and will require optical layer services which are better able to handle dynamic changes in traffic. By providing burst-level provisioning of lightpaths, the optical network will be able to offer a higher degree of multiplexing, which will result in higher network utilization for bursty traffic; however, such burst-level reconfiguration may also lead to an increase in connection blocking probability due to the higher degree of uncertainty with regard to the network. Thus, it may be desirable to develop a network which is capable of providing both long-term as well as burst-level lightpath provisioning services in order to support a wide range of applications.

We claim that in the new third-generation optical networks, the design of the control protocols for dynamic lightpath establishment becomes a critical issue. These protocols must be able to determine routes and reserve resources for lightpaths in a manner that is scalable and provides a low degree of blocking for connection requests. The protocols must also be able to support highly dynamic and bursty traffic through the rapid and automated on-demand provisioning of lightpaths. The automated on-demand provisioning for providing DWDM network resources is a key element for the success, if not the survival, of operators and providers. It is believed that dynamic on-demand wavelength provisioning will enable service providers to respond quickly and economically to customer demands. The ability to activate users and services in seconds will not only provide an immense advantage over manual-allocation-based competitors, but will also enable a more efficient use of bandwidth and other network resources by providing the ability to dynamically allocate additional bandwidth by simply lighting up another wavelength when needed and releasing the wavelength when it is no longer needed. This dynamic provisioning ability will lead to significantly higher operational margins and will also provide the opportunity to support new services that are based on short duration connections and bursty traffic environments. With fast, automated, millisecond provisioning, it further becomes possible to dynamically control bandwidth costs, to provide protection and restoration dynamically on the fly without the need to preprovision bandwidth resources, and to provide prioritization for emerging QoS services. An automated provisioning system for all-optical networks is therefore an opportunity to control the future world of DWDM networking. It is therefore no surprise that the question for most carriers is not whether to migrate to end-to-end all-optical networks, but when.

8.4.7   Protocols for lightpath establishment and management

In order to establish lightpaths in a wavelength-routed network, algorithms and protocols must be developed to select routes and assign wavelengths to the lightpath, as well as to reserve network resources. The lightpath establishment problem may either be static or dynamic. In the static lightpath establishment problem, the connection requests are given and the objective is to either satisfy all requests while minimizing the network cost or to maximize the number of requests that are established over a given network. In the dynamic lightpath establishment problem, connection requests arrive to the network over time and only remain in the network for a limited duration before departing. In this problem, the objective is to minimize the probability that a connection request will be blocked.

More formally, the problem of finding a route for a lightpath and assigning a wavelength to the lightpath is often referred to as the routing and wavelength assignment (RWA) problem. When applied to static lightpath establishment, the objective of the RWA problem is to route lightpaths and assign wavelengths in a manner which minimizes the amount of network resources that are consumed, while at the same time ensuring that no two lightpaths share the same wavelength on the same fiber link. Furthermore, in the absence of wavelength conversion devices, the RWA problem operates under the constraint that a lightpath must occupy the same wavelength on each link in its route. This restriction is known as the wavelength-continuity constraint. The optimal formulation of the RWA problem is known to be NP-complete; therefore, heuristic solutions are often employed.

In the case of dynamic lightpath establishment, the objective is to dynamically find a route and select a wavelength in a manner that minimizes the overall connection blocking probability. The routing protocol for dynamic lightpath establishment may either be fixed or adaptive. In fixed routing schemes, routes are predetermined and do not change with changes in the network state. Examples include fixed shortest path routing and fixed alternate path routing. Adaptive routing protocols make routing decisions based on current network state information.

Dynamic routing and wavelength assignment protocols can furthermore rely either on global state information, as in link-state routing, or on local information. It has been demonstrated that protocols that utilize global state information generally outperform protocols in which no global information is available; however, global-information-based schemes require a high degree of control overhead in order to maintain the state information. If the number of nodes and links in the network is large or if connections are arriving and departing at a high rate, then it becomes increasingly difficult to maintain global information and it becomes highly likely that some nodes will have outdated information. In such cases, it may be preferable to utilize protocols that do not rely on global information. Since the first work on lightpath definition and establishment[7] and the first work on lightpath establishment with wavelength conversion,[8] a large number of lightpath establishment algorithms have been proposed. This chapter is too short for an exhaustive treatment of the topic, but a good survey of RWA techniques may be found in Zang et al.,[9] and an overview of lightpath establishment in WDM metropolitan area networks may be found in Xiao et al.[10]

8.4.8 Provisioning schemes

In addition to routing and wavelength assignment, a provisioning protocol is required to exchange control information among nodes and to reserve resources along the path. In many cases, the provisioning protocol is closely integrated with the routing and wavelength assignment protocols.

The provisioning protocol may either be centralized, in which requests are sent to a centralized entity which calculates the routes and attempts to reserve the resources, or it may be distributed, in which case each individual node calculates the routes and initiates the signaling for reserving resources.

In distributed provisioning methods for lightpath establishment, a control signal traverses each node along the route sequentially. The provisioning method may rely on either global information or only local information to make wavelength assignment and reservation decisions at each node. Existing provisioning approaches have typically been classified as being either source-initiation reservation (SIR) policies, or destination-initiated reservation (DIR) policies. In source-initiated schemes, wavelength resources are reserved as a control message traverses along the forward path to the destination, while in destination-initiated schemes, wavelength resources are reserved by a control message heading back toward the source node.

A general framework for classifying various provisioning methods can be constructed built from the following considerations. In general, a provisioning scheme may require a number of different types of signaling messages. We will refer to these messages as PROBE messages, RESERVE messages, RES_ACK messages, and RELEASE messages. A PROBE message gathers resource information along a selected route without actually reserving resources. The message contains a set of possible wavelengths that can potentially be utilized for a connection. Examples of PROBE messages are the PATH message in RSVP and the LABEL REQUEST message in CR-LDP. The RESERVE message attempts to reserve one or more wavelength resources on each link that it traverses. The RESERVE message also contains a set of resources to be reserved on each link. The RES_ACK message is used to confirm that the resources have been reserved along a path. When the source node receives a RES_ACK message, it may begin transmitting data. For the case in which all reservations are made in the backward direction, the RES_ACK message may be combined with the RESERVE message. The RELEASE message releases one or more wavelength resources on each link that it traverses. The set of wavelengths that are available on each and every link in the path traversed by a PROBE or RESERVE message will be referred to as the *available set*.

Provisioning methods may be characterized by the following parameters:

- The *reservation-initiating set* is the set of nodes at which reservations are initiated. The reservation may be initiated at the source node, the destination node, or an intermediate node. Reservations may also be initiated at more than one node in the path.
- The *direction of reservation* from an initiating node indicates that reservations may be initiated in the forward direction, the backward direction, or both the forward and backward directions.
- The *set of wavelengths* to be reserved by an initiating node allows the node to choose to reserve a single wavelength, a fixed set of wavelengths, or a variable set of wavelengths.
- The availability of *state information* at the source node means that the source node may have global state information regarding wavelength availability on each link in the network, or local information regarding wavelength availability on its own outgoing links.

SIR would be classified as a provisioning method in which reservations are initiated at the source node, and reservations are made in the forward direction. If global information is available at the source node, then a single wavelength may be reserved in the forward direction. If no global information is available, then either the available set of wavelengths or a subset of the available set of wavelengths will be reserved.

DIR would be classified as a provisioning method in which reservations are initiated by the destination node, and reservations are made in the backward direction. In these schemes, a PROBE message is sent from the source to the destination along the path. The set of wavelengths in the PROBE message may either be a single wavelength or multiple wavelengths depending on how much information is available at the source node. The provisioning scheme in the emerging GMPLS standard is an example of DIR.

8.4.9 Measures of success

A key measure of performance in dynamic wavelength-routed networks is the blocking probability, or the probability that an incoming connection request will be denied. One source of connection blocking is insufficient network resources. If a route with sufficient capacity cannot be found between the source node and destination node, then the connection request must be blocked. Furthermore, if no wavelength converters are present in the network, then the lightpath for the connection must utilize the same wavelength on each link in the path between the source node and the destination node. If no such wavelength is available, then the connection will be blocked even if capacity is available.

Connection blocking may also occur when routing and wavelength assignment decisions are made based on outdated network state information. State information may be outdated if state updates have not yet propagated throughout the network or if multiple simultaneous connection requests interfere with each other. When a lightpath is being established, control messages must propagate along the route of the lightpath in order to reserve network resources. It is possible that resources along the route, which were available when the connection request was first initiated at the source node, will no longer be available by the time the control message reaches the desired resource. In such a case, the request will be blocked. For the case of DIR, the PROBE message may find available resources in the forward direction, but the RESERVE message traveling in the backward direction may find that the resources have been taken by other connections.

For current and emerging systems in which provisioning is fairly static and lightpaths are established for long periods of time, most of the blocking will be due to insufficient resources. As network traffic continues to scale up and become more bursty in nature, a higher degree of multiplexing and flexibility will be required at the optical layer. Thus, lightpath establishment will become more dynamic in nature, with connection requests arriving at higher rates and with lightpaths being established for shorter time durations. In such situations, blocking due to conflicting connection requests may become an increasingly significant component of the overall connection blocking probability.

The blocking probability is a function of a number of factors, such as network load, the connection arrival rate, the method by which resources are reserved, and the amount of resources reserved by each connection request.

8.4.10 A short survey of signaling protocols

We now survey some of the more relevant approaches that have investigated signaling protocols for lightpath establishment in wavelength-routed optical networks. In Ramaswami and Segall, a distributed signaling protocol is proposed in which global information is available to all nodes.[11] The source node initiates a provisioning request by first finding a route and then signaling each node in the route to reserve resources. After receiving confirmation of the reservations from each node, the source node sends out another control message to each node in order to configure the cross-connects. When acknowledgments are received by the source node from each of the nodes on the route, the connection is established.

The work in Yuan et al. evaluates various hop-by-hop signaling protocols in which resources are reserved along the path in either the forward direction as a control message travels toward the destination, or in the backward direction as a control message travels back toward the source node.[12]

In SIR schemes, the detailed method of reserving wavelengths depends on how much information is available to the source node and to individual nodes. In the extreme case when the source node is maintaining complete state information, it will be aware of which wavelengths are available on each link. The source node may then send a connection setup message along the forward path, reserving the same available wavelength on each link in the path. For the case in which each node is only aware of the status of its own outgoing links, the source node may utilize a conservative reservation approach, choosing a single wavelength and sending out a control message to the next node attempting to reserve this wavelength along the entire path; however, there is no guarantee that the selected wavelength will be available along every link in the path. If the wavelength is blocked, the source node may then select a different wavelength and reattempt the connection. The limitation of this approach is that it may result in high setup times because it may take several attempts before a node can establish a lightpath.

An alternate approach that maximizes the likelihood of establishing a lightpath in a forward reservation scheme is to use an aggressive reservation scheme which over-reserves resources.[12] Multiple wavelengths may be reserved on each link in the path with the expectation that at least one wavelength will be available on all links in the path. In a greedy approach, all feasible wavelengths will be reserved at every link in the path. The redundant wavelength channels on every link along the path could be released when the request reaches the destination and a wavelength channel is selected. Alternately, the wavelengths can be released on a hop-by-hop basis as the control message traverses each node in the forward direction. That is, wavelengths that were reserved on previous links but found to be unavailable on the forward path are released immediately, so that these wavelengths can be reserved by some other transmission requests. The aggressive reservation scheme could maximize the possibility of establishing the current request; however, over-reserving some network capacity during the lightpath setup procedure could unnecessarily block some other transmission requests that arrive to the network at a later time.

Another approach to limiting the number of wavelengths that are reserved is to divide the wavelengths into groups. When reserving wavelengths on a link, a node will reserve only those wavelengths that belong to a specific group.[13] The choice of group is made at the source node and is based on the number of available wavelengths in each group. In such case, the size of the group is a critical parameter. If the group is too large, then too many resources will be reserved, but if the group is too small, then the likelihood of establishing a lightpath will be small.

To prevent the over-reservation of resources, reservations may be made after the control message has reached the destination and is headed back to the source. Such reservation schemes are referred to as *backward reservation* (destination-initiated) schemes. By reserving wavelengths in the reverse direction, the reserved wavelengths are idle for less time than if the wavelengths are reserved in the forward direction. Another advantage is that the connection request message can gather wavelength usage information along the path in the forward direction. The destination node can then use the gathered information in order to select an appropriate wavelength for reservation. It is demonstrated in Yuan et al. that backward reservation schemes generally outperform forward reservation schemes for the case in which no wavelength conversion occurs.[12] One possible drawback of a backward reservation scheme is that if multiple connections are being set up simultaneously, it is possible for a wavelength that was available on a link in the forward direction to be taken by another connection request, in which case the wavelength will no longer be available when the reservation message traverses the link in the reverse direction.

The work in Zang et al. evaluates a signaling protocol in which a link-state approach is utilized to maintain state information, and another protocol in which a distance-vector approach is utilized to maintain the state information.[14] Much work has also been done in the Internet community to extend existing IP-based protocols to support lightpath establishment. Currently, the most popular approach involves using a GMPLS control plane to establish connections, with OSPF protocols to maintain state information and CR-LDP and RSVP protocols to provide the basic messaging functionality for the signaling. In the GMPLS approach to signaling, the OSPF-TE is used to distribute global state information to all nodes. The source node determines the route for a lightpath by running a constraint-based shortest-path first protocol (CSPF). Upon finding a route and selecting a wavelength on the route, the source node sends out a label request message to the destination along the selected route. The label request message specifies the wavelength (as a suggested label) and the path. At the destination node, a label assignment message is sent back toward the source node and reserves the desired wavelength resource on each link in the path. The label request and label assignment messages are exchanged using either the CR-LDP or RSVP-TE protocols.

In Jue and Xiao, an analytical model was developed to evaluate the blocking probability characteristics of various SIR and DIR schemes.[15] The model considered blocking of connections due to insufficient resources as well as blocking due to outdated state information. The model assumed fixed routing with a random wavelength assignment policy. In Lu et al., a more accurate model was developed to measure the blocking probability of a DIR scheme.[16]

8.4.11 Implementation considerations

Although proprietary implementations can be found and are typical of current provisioning solutions, it is widely expected that within 3 years the provisioning protocols will be implemented within a GMPLS framework. In order to provide a high level of interoperability and to facilitate the integration of provisioning protocols on various types of equipment, a COBRA-based interface can be utilized to obtain hardware state information and to configure the switches and cross-connects. The provisioning protocol module must also be able to interface with other operational support system (OSS) modules. To facilitate integration of various modules, the ITU has developed the telecommunications management network (TMN) specification, an architectural model for networks that defines basic functional blocks and the associated interfaces between blocks in the OSS. Provisioning protocols may be developed within these specifications in order to provide standards-compatible interfaces with other OSS modules; however, whether or not the provisioning modules will be implemented in a TMN framework remains an open issue.

Lastly, link state information at each node can be monitored using protocols such as the link manager protocol (LMP) or the network transport interface protocol (NTIP).

References

1. D. Greenfield, *The Essential Guide to Optical Networking*, Prentice Hall, Upper Saddle River, N.J., 2002.
2. S. Clavenna, *Optical Signaling Systems*, www.lightreading.com/document.asp?doc = id7098, January 8, 2002.
3. Various, Section titled Optical networking: Signs of maturity, *IEEE Commun.*, vol. 40, no. 2, February 2002.
4. Various, Section titled Intelligence in optical networks, *IEEE Commun.*, vol. 39, no. 9, September 2001.
5. I. Chlamtac, A. Ganz, and G. Karmi, Lightpath communications: A novel approach to high-bandwidth optical WANs, *IEEE Trans. on Commun.*, vol. 40, no. 7, pp. 1171–1182, July 1992.
6. I. Chlamtac, A. Ganz, and G. Karmi, Lightnets: Topologies for high-speed optical networks, *IEEE J. on Lightwave Technol.*, vol. 11, no. 5, May 1993.
7. I. Chlamtac, A. Ganz, and G. Karmi, Purely Optical Networks for Terabit Communication, *IEEE INFOCOM*, 1989.
8. I. Chlamtac, A. Farago, and T. Zhang, Lightpath (wavelength) routing in large WDM networks, *IEEE J. on Selected Areas in Commun.*, jointly with *IEEE/OSA J. Lightwave Technol.*, vol. 14, no. 5, June 1996.
9. H. Zang, J.P. Jue, and B. Mukherjee, A review of routing and wavelength assignment approaches for wavelength-routed optical WDM networks, *SPIE/ Kluwer Optical Networks Mag.*, vol. 1, no. 1, pp. 47–60, Jan. 2000.
10. G. Xiao, J.P. Jue, and I. Chlamtac, Lightpath establishment in WDM metropolitan area networks, *SPIE/Kluwer Optical Networks Mag.*, special issue on Metropolitan Area Networks, 2003.

11. R. Ramaswami and A. Segall, Distributed network control for optical networks, *IEEE/ACM Trans. on Networking*, vol. 5, no. 6, pp. 936–943, Dec. 1997.

12. X. Yuan, R. Melhem, R. Gupta, Y. Mei, and C. Qiao, Distributed control protocols for wavelength reservation and their performance evaluation, *Photonic Network Commun.*, vol. 1, no. 3, pp. 207–218, 1999.

13. A.G. Stoica and A. Sengupta, On a dynamic wavelength assignment algorithm for wavelength-routed all-optical networks, *Proc. SPIE/IEEE/ACM OptiComm 2000*, Dallas, Texas, pp. 211–222, Oct. 2000.

14. H. Zang, J.P. Jue, L. Sahasrabuddhe, R. Ramamurthy, and B. Mukherjee, Dynamic lightpath establishment in wavelength-routed WDM networks, *IEEE Commun. Mag.*, vol. 39, no. 9, pp. 100–108, Sept. 2001.

15. J.P. Jue and G. Xiao, Analysis of blocking probability for connection management schemes in optical networks, *Proc. IEEE Globecom 2001*, San Antonio, Texas, Nov. 2001.

16. K. Lu, G. Xiao, and I. Chlamtac, Blocking analysis of dynamic lightpath establishment in wavelength-routed networks, *Proc. IEEE Int. Conf. on Commun. (ICC) 2002*, New York, Apr. 2002.

chapter nine

Optical network resource management and allocation

Ding Zhemin
Hong Kong University of Science and Technology
Mounir Hamdi
Hong Kong University of Science and Technology

Contents

9.1   Introduction

9.1.1   The big picture

The significant growth of data traffic in the past two decades, mainly driven by Internet-based services, has been shifting the telecommunication industry from the voice-centric to the data-centric world. The new next-generation Internet network is taking shape. Many issues and challenges still need to be addressed before the next-generation network is to be truly set up. Many people believe the core technology to meet those challenges is the wavelength division multiplexing (WDM) technology, considering its huge bandwidth capacity: a single mode fiber's potential bandwidth can reach up to 50 Tb/s;[1] low signal attenuation, low signal distortion, low power requirement, low material usage, small space requirement, and low cost.

WDM is a very promising technique in which multiple channels (wavelengths) are operated along a single fiber simultaneously, each on a different wavelength. These channels can be independently modulated to accommodate different traffic streams. Thus, WDM divides the huge bandwidth of an optical fiber into channels whose bandwidths are compatible with peak electronic processing speed. WDM-based optical networks have already been deployed and tested in the United States (MONET, NTONC projects), Europe (RACE, ACT projects), and other countries.

For our own purposes, the term optical network is not merely concerned with the physical layer but an optical network supporting high-speed wavelength routing, providing protection and restoration techniques and effective bandwidth management at the optical layer.

9.1.2   Enabling technology

One form of wavelength add/drop multiplexer (WADM) is illustrated in Figure 9.1. The WADM adds and drops each wavelength on a pair of inbound and outbound unidirectional links. The receiver does not need to demultiplex because only one wavelength reaches each receiver.

The greatest advantage of WDM is the increased network flexibility achievable with wavelength routing, which allows operators to provide network node-pairs with end-to-end optical channels, known as lightpath. The intermediate optical cross-connects (OXCs) route the lightpath from sources to destination, simplifying network management and processing.

It is a natural thought to combine the huge bandwidth capacity provided by WDM optical technology with the flexibility of other network forms. Different logical layers with various user services are set up in the WDM optical networks. Without entering into the details of the logical network structure, we show the general network architecture as in Figure 9.2. The

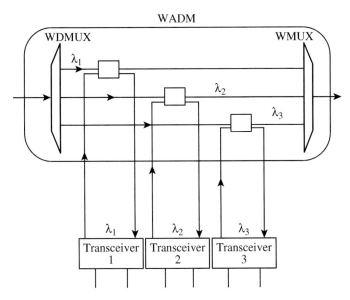

Figure 9.1 WADM structure. Three wavelengths, $_1$, $_2$, $_2$, are in the fiber.

virtual topology may take many specific forms, depending on the structure of logical layer. Three important types of networks are considered. They are SONET networks, ATM networks, and IP networks.

The rest of the chapter is organized as follows. Some basic elements of WDM networking are described in Section 9.2. They are RWA problem, traffic grooming, and optical networks survivability and optical opaque networks. In Section 9.3, several topics that have a very close relation with the optical network reconfiguration and quality of service (QoS) are

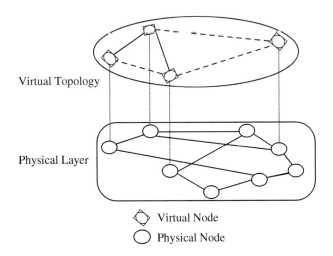

Figure 9.2 Multilayered network hierarchical architecture.

presented. First, we consider load balancing in the optical network. Then, a new switching technology called optical burst switching (OBS) is introduced. We also discuss the MPLS and the possible combination of OBS and MPLS to form an IP over WDM network that has the capability to provide QoS. A new resource allocation methodology called blocking island is given because we believe it could be potentially applied into many aspects of WDM optical networks. Concluding remarks are presented in Section 9.4.

9.2   Basic elements of WDM networking

This section focuses on basic issues in the reconfigurable optical networks.

9.2.1   Routing and wavelength assignment (RWA)

In WDM transmission, each data channel is carried on a unique wavelength (or optical frequency) and a single optical fiber has many different wavelengths. In a wavelength-routed WDM network, a lightpath (e.g., wavelength continuous path without processing in the intermediate nodes) is established between two communication nodes. A lightpath must occupy the same wavelength on all the fiber links it traverses if no wavelength converter exists. This property is known as *the wavelength continuity constraint*. Networks that use optical cross-connects to route lightpaths through the network are referred to as wavelength-routed networks. In order to satisfy a lightpath request in wavelength-routed WDM networks, we not only need to consider routing but the wavelength selection as well. Given a set of connection requests, the problem of setting up a lightpath by routing and assigning a wavelength to each connection is called RWA problem. It can be formulated as a combinatorial problem known to be nondeterministic polynomial time (NP)-complete.[2] Many heuristic algorithms have been proposed and evaluated under different assumptions. The traffic assumptions generally fall into one of two categories: static traffic, where all the connection requests are already known; and dynamic traffic, where connection requests arrive in a dynamic order and lightpaths are set up on demand. Most previous work focuses on single-fiber network where each node pair is connected by a single fiber link; while recently, more research has been carried out on multiple-fiber networks. The benefit of adding wavelength converters has also been studied. For a recent survey on RWA problems, please see Yoo and Banerjee[3] and Zang et al.[4]

The RWA problem is usually partitioned into two sub-problems: route selection and wavelength assignment. The routing schemes can be cataloged into static routing (including fixed-alternate routing) and dynamic routing.[5,7–9] The wavelength assignment policies include random assignment, most-used assignment, least-used assignment, and first-fit assignment.[10–12] These two sub-problems can also be solved jointly.[5,13]

9.2.1.1   Route computation

Shortest path (SP).   Shortest path algorithms find the shortest route from a given source to a destination in a graph.

K-shortest path (K-SP).   K-shortest path algorithm finds more than one route for each source and destination pair.

Adaptive routing (AR).   In adaptive routing, the route from a source node to a destination node is chosen dynamically, depending on the network state. The network state is determined by the set of all connections currently in progress. One form of adaptive routing is least congested-path (LCP) routing.[14] The congestion on a link is measured by the number of wavelengths available on the link. Links are considered to be more congested if they have fewer available wavelengths. The congestion on a path is indicated by the congestion on the most congested link in the path. A variant of LCP is proposed in Li and Somani, which only examines the first K links on each path, where K is a parameter.[9] Chan and Yum, and Li and Somani demonstrate that LCP performs much better than fixed k-shortest routing.[9,14]

9.2.1.2   Wavelength selection

The wavelength assignment problem is the selection among available wavelengths to maximize the wavelength utilization.

Random wavelength assignment (RWA).   For one route, a wavelength is chosen from the all-possible wavelengths set of this route randomly.

First-fit (FF).   In this scheme, all wavelengths are numbered. Pick up the first available wavelength in numerical order. It does not need the global information.

Least-used (LU)/SPREAD.   This scheme attempts to balance the load among all the wavelengths. Its performance is even worse than RWA in the evaluation so is not preferred in practice.

Most-used (MU)/PACK.   This scheme attempts to allocate the most utilized wavelength first. It outperforms LU and FF significantly while computation cost is similar to LU.[5,10]

For static traffic an integer linear program, which is modeled with the multi-commodity flow problem, can get us the exact answer. But the formulation of ILP is NP-complete. Some heuristic solutions are proposed. *Genetic algorithms (GA)*[15] are standard techniques for hard combinatorial optimization problem. The basic idea is to simulate evolution of genotypes and natural selection. *Simulated annealing algorithms (SA)*[15] is another standard technique for hard combinatorial optimization problem. The idea is to simulate annealing of some object in order to overcome a local minimum point in a sense of iterative improvement.

The restriction imposed by the wavelength continuity constraint can be avoided by the use of *wavelength conversion* (also referred to as wavelength translation). A wavelength converter is a device that takes a data channel modulated in wavelength i and outputs the data channel modulated in another wavelength j. If all optical nodes in the network have the full wavelength conversion capability, this optical network is equivalent to traditional circuit-switched networks. Wavelength converters thus result in improvements in network performance; however, the introduction of wavelength converters — whether they are all-optical or opto-electronic — is expected to significantly complicate the design of an optical cross-connect and therefore increases the cost greatly.[6]

In most network topologies, wavelength converters generally provide only modest improvements (approximately 10 to 40%)[46] in the traffic that can be supported for a given blocking probability.[6] Considering the limited improvements and expensive cost of converters, limited number of wavelength converters in Lee and Li, and Parys et al., and wavelength-converters-in-selected-nodes in Brackett are proposed and evaluated.[47–49]

Multicasting provides an efficient way of disseminating data from a source to a group of destinations. Supporting *multicast at the WDM layer* has several potential advantages:[50]

1. More efficient multicasting is possible with the knowledge of the physical (e.g., optical layer) topology.
2. Some optical switches are inherently capable of light splitting, which is more efficient than copying packets in electronics.
3. WDM multicast can alleviate the electronic processing bottlenecks.

To support multicasting, the nodes in a WDM network need to have light (optical) splitting capability. A node with splitting capability can forward an incoming message to more than one output link. A split capable node is very expensive due to its complex architecture.[50] People usually consider a network with a subset of nodes having the light-splitting capability. A network with a few split capable nodes is called a network with *sparse splitting capability.*

It has been demonstrated that finding a minimum Steiner tree for a multicast session, where the members are only a subset of the nodes in a network with an arbitrary topology, is an NP-complete problem.[51] Many heuristic algorithms have been proposed. They can be roughly classified into two categories.[50] The first contains algorithms based on the shortest-path heuristic (SPH), which minimizes the cost of the path from a multicast source to each of the members, while the second contains algorithms based on the minimum Steiner tree, which attempt to minimize the total cost of a multicast tree.

The problem of multicast routing in an optical network where all the nodes have splitting capability is dealt with in Sahasrabuddhe and Mukherjee.[52] The multicast tree generation problem in a network with sparse splitting has been dealt with in Zhang et al. and Sreenath et al.[50,53] Ali and Deogun propose a new switching architecture supporting multicasting and also design a corresponding multicasting algorithm based on the new switching architecture.[54]

In a network with sparse splitting capability, only a subset of nodes is multicast capable (MC). The other nodes are treated as multicast incapable (MI). Because of the limited MC nodes and wavelength continuity constraint, it may not be feasible to cover all nodes in a multicasting session in one light-tree (using only one wavelength). Zhang et al. propose a new multicast medium called *light-forest*, consisting of one or more light-trees (rooted at a multicast source).[50]

Sreenath et al. argue that the delay in optical networks is normally very low so its algorithm aims to minimize the cost of the forest than minimizing the delay on individual paths.[53] The algorithm proposed in Sreenath et al. is based on a concept called *virtual source (VS)*.[53] A virtual source is a node having both splitting and wavelength conversion capabilities.

9.2.2   *Traffic grooming*

Traffic grooming is a term used to describe how different (low-speed) traffic streams are packed into higher speed streams. Usually we study an optical WDM ring network that consists N nodes labeled 0,1, ... , N-1 in the clockwise direction, interconnected by fiber links. A virtual connection needs to be added and dropped only at the two end nodes of the connection. Because the connection will bypass the intermediate nodes, it may be possible to have some nodes on some wavelength where no add/drop is needed on any time slot; therefore the total number of transceivers will be reduced. That also means the total cost of the network will be reduced. Modiano and Chiu demonstrated that the general traffic-grooming problem is NP-complete.[23]

The traffic-grooming problem in WDM ring networks can be described as follows: given a certain number of wavelengths in a fiber link and an incoming traffic matrix, how we can pack those low-speed streams into different wavelengths and assign established connections to circles of the ring so that the number of ADMs (add/drop multiplexors) is minimized? Usually for one lightpath, we need to put an ADM at the source node and one at the destination node, respectively.

Let us look at a simple numerical example of grooming for a single traffic matrix in Berry and Modiano.[17] Suppose we have a ring with N = 5 nodes and a granularity of g = 4 (e.g., OC-12's on an OC-48). Consider the traffic requirement R = {1–2, 1–2, 1–3, 1–3, 1–4, 1–4, 1–5, 1–5}. The minimum number of wavelengths required to support R is 2. By using two wavelengths and dropping each wavelength at each node, R can be supported using 10 ADMs. Consider grooming the traffic so that {1–2,1–2, 1–3, 1–3} are placed on one wavelength and {1–4, 1–4, 1–5, 1–5} are placed on the second wavelength. In this case the traffic can be supported using only 6 ADMs (see Figure 9.3) In this example, the topology that minimized the

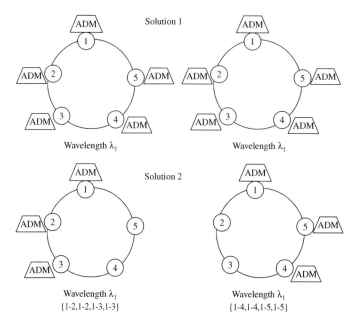

Figure 9.3 Traffic grooming example.

number of ADMs also minimized the number of wavelengths; however, in general this is not true.[18,19]

Much research has been conducted in order to solve this problem. Gerstel et al. provide a network design to overall network cost in WDM rings.[20] The overall network cost includes the number of wavelengths per fiber and the number of transceivers in the network. The authors argue, because the cost of terminating equipment is much more expensive than the cost of the number of wavelengths in the networks. The authors assume the network is circuit-switched and transceivers are nonblocking. Six-WDM ring network architectures are considered and analyzed. They are full optical ring, single-hub, point-to-point WDM ring, hierarchical ring, and incremental ring.

Zhang and Qiao discuss the cost-effective design of SONET over WDM ring networks where the traffic is uniform or nonuniform.[21] The authors claim the algorithms proposed, whose goal is to minimize the number of wavelengths W and the cost of transceivers D, are "generic in that they can be applied to both unidirectional and bi-directional rings having an arbitrary number of nodes under both uniform and nonuniform traffic with an arbitrary grooming factor." The basic idea of the approach is divided into two steps. First, construct as few circles as possible to include all request connections so as to minimize W where each circle consists of multiple nonoverlapping connections; second, after the circles are constructed, another heuristic algorithm is used to groom up to m circles on to a wavelength ring while trying to overlap as many end nodes belonging to different circles as possible so as to result in a small D.

Wang et al. provide comprehensive mathematical definitions for the various traffic-grooming problems in unidirectional and bidirectional rings by employing integer linear programs (ILP).[22] Because the ILP problem is very complex and we already know the general traffic-grooming problem is NP-complete,[23] a heuristic simulated-annealing-based approach is proposed.

In Wang and Mukherjee, traffic grooming in interconnected multi-WDM rings is examined.[16] In typical telecommunication networks, multiple rings are interconnected to provide geographical coverage. Wang and Mukherjee propose several WDM-ring interconnection strategies.[16] Two heuristic traffic-grooming algorithms are proposed for the cases of optically interconnected and hierarchical network architectures. It is found that the optically interconnected strategy achieves the best transceiver saving; however, the hierarchical architecture is suitable for general physical network topologies and it provides moderate transceiver savings. A fully optical scheme uses fewest wavelengths among all proposed interconnection strategies but requires more transceivers.

Most early work on traffic grooming was focused on SONET rings. The objective is to minimize the number of SONET ADMs used in the ring network. With optical network topologies evolving from rings to meshes, traffic grooming in other topologies is becoming increasingly important. The traffic-grooming problem in star networks is studied in Dutta.[55] A few articles consider traffic grooming in mesh networks. Among them, the study by Sasaki and Lin assumes the incoming traffic to be incremental.[56] That is, the individual traffic streams are of equal bandwidth and the wavelength capacity is an integral multiple of that bandwidth. In Konda and Chow, the traffic-grooming problem is formulated as a special case of the multi-commodity flow problem.[57] The integer linear program (ILP) formulation in Zhu and Mukherjee tries to minimize the network throughput; two heuristics, namely MST and MRN, were proposed.[58]

Given a certain set of traffic, the traffic should be groomed in such a way that the number of wavelengths and transceivers are minimized. Notice that minimizing the number of transceivers is equivalent to minimizing the number of lightpaths, because each lightpath requires the appropriate equipment for receiving and transmitting. In general, we call this equipment transceivers; the problem is known as traffic grooming in mesh networks. Notice that we have two parameters to optimize: minimizing the number of wavelengths and minimizing the number of transceivers. We may not always be able to find an optimal solution for both parameters.

Consider a 5-node mesh network topology as presented in Figure 9.4. Assume it is a single-fiber WDM optical network, with each fiber having one wavelength channel. The capacity of each wavelength channel is two units of traffic streams. Three connection requests occur: (A, B), (B, C), and (A, C), with each requiring one unit of traffic stream. Two different methods are used for aggregating traffic and assigning lightpaths. The first assignment is illustrated in Figure 9.4a. Three lightpaths AB, BC, and $AEDC$ are set up; six transceivers are needed. The second solution is to groom the traffic of (A, C) into lightpaths AB and BC, as shown in Figure 9.4b. Let (A, B) share

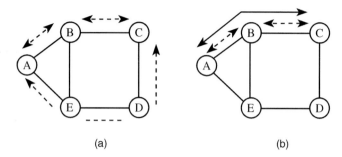

Figure 9.4 The illustrative example of traffic grooming in WDM optical mesh networks.

the lightpath and transceivers with (A, C) from A to B. Similarly, let (B, C) share the lightpath and transceivers with (A, C) from B to C. The second solution requires only two lightpaths to be set up, and four transceivers are needed. This example demonstrates that if we can groom the traffic and assign the lightpath appropriately, the cost of the network will be decreased.

9.2.3   Optical layer survivability

WDM networks are prone to failures of components such as links, nodes, and wavelength multiplexers/demultiplexers. Because these networks carry high volumes of traffic, failures may have severe consequences. Therefore, it is imperative that the networks be *fault-tolerant.* Fault tolerance refers to the ability of the network to reconfigure and reestablish communication upon failure, and is also known as the *survivability* of a network. The technique that uses preassigned capacity to ensure survivability is referred to as *protection,* and the technique that reroutes the affected traffic after failure occurrence by using available capacity is referred to as *restoration* in ITU-T Recommendation G.872.[24]

The basic types of network failure generally considered are link and node failures. Link failure usually occurs because of cable disruption; node failure is due to equipment failure at network nodes. Channel (wavelength) failure is also possible in WDM optical networks. A channel (wavelength) failure is usually caused by the failure of transmitting or receiving equipment operating on that channel (wavelength).

Most of the previous research on the survivability in WDM optical networks has focused on the recovery from a single link or node failure. This is primarily due to the fact that it is easier to reroute the failure of one piece of equipment at a time, at most. Also, the possibility of multiple equipment failures occurring at the same time is quite low. The key ideas and approaches used for single link failure can be extended to handle node failures and multiple component failures. Channel failure scenarios have received little attention so far.

In the following sections, we first introduce some techniques used in today's optical networks, as well as those possible for the WDM networks in the future. Then some lightpath restoration methods in WDM networks are discussed (see Figure 9.5).

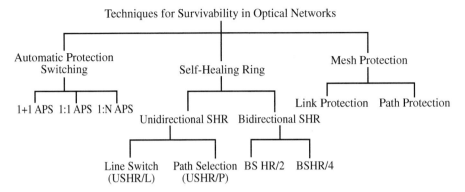

Figure 9.5 Techniques for survivability in optical networks.

9.2.3.1   Automatic protection switching

APS is typically used to handle link failures. It has three main architectures: 1 + 1, 1:1, and 1:N. Dedicated protection is commonly referred to as 1 + 1 and 1:1 protection. In *1+1 protection* the source node transmits on both the working and protection links simultaneously. If one link fails, the other link will be used to receive signals. In *1:1 protection*, transmission occurs on the working link only, while the protection link may be either idle or used to transmit low-priority traffic. Upon failure of the working lightpath, the protection lightpath will then be used. *Shared protection*, also referred to as *1:N*, allows spare wavelengths to be shared by a number of working lightpaths. In this scheme, N working links share a single protection link that provides protection against the failure of any one of the N working links.

9.2.3.2   Self-healing ring (SHR)

SONET SHR is a technique for survivable optical networks. Unidirectional SHR (USHR) and bidirectional SHR (BSHR) are two types of SHRs in SONET systems. The difference between these two categories is the direction of the traffic flow under normal operation. In USHR, the normal traffic flow goes around the ring in one direction. Any traffic routed to the protection ring because of a failure is carried in the opposite direction. In BSHR, working traffic flows in both directions.

By using the SHR technique, the network can get fast and 100% protection guarantee at the cost of a large amount of spare capacity. Generally, in a point-to-point link, APS is the best solution, while SHR offers the most benefits in a ring topology.[25] The disadvantage of the two schemes is the scalability. When a large mesh network is to be survivable, the spare capacity requirement becomes inordinate if the two schemes are employed. In this situation, dynamic restoration appears to be a feasible solution to do the job.

9.2.3.3   *Lightpath restoration*

The lightpath that carries traffic during normal operation is known as the primary lightpath. When a primary path lightpath fails, the traffic is rerouted over a new lightpath known as the backup lightpath.[26]

When an existing lightpath fails, a search is initiated to find a new lightpath that does not use the failed components. The successful recovery is not guaranteed because the attempt to establish a new lightpath may fail due to resource shortage at the time of failure recovery. To overcome the shortcomings, backup lightpaths are proposed, which yields a 100% restoration guarantee. A method is either *link-based* or *path-based*.[27,28] The link-based method selects a new path when a link fails. This method is unattractive for several reasons:[27] the choice of backup paths is limited, and because of the wavelength constraint the backup path must use the same wavelength as the working path. In the path-based restoration method, a backup lightpath is selected between the end nodes of the failed primary lightpath. This method employs better resource utilization than the link-based restoration method.

In a dynamic traffic scenario, the *primary backup multiplexing* can further improve the resource utilization. This technique allows a primary lightpath and one or more backup lightpaths to share a channel. Some factors can explain why it is feasible for the dynamic traffic. First, failures do not occur frequently enough in practice to warrant full reservation. Second, every lightpath does not necessarily need fault tolerance to ensure network survivability. Third, at any instant of time, only a few primary lightpaths critically require fault tolerance.

9.2.3.4   *Multiple layer protection*

The use of optical switches and all-optical components constitutes a new network layer, called optical layer or WDM layer, into the layered structure. The WDM layer supports different higher-layer services, such as SONET connections, asynchronous transfer mode (ATM) virtual circuits, and IP-switched datagram traffic. Researchers are becoming increasingly more interested in employing IP control plane within optical networks to support dynamic provisioning and restoration of lightpaths. Specially, it is believed that IP routing protocols and multiprotocol label switching (MPLS) signaling protocols could be adapted for optical networking needs.

According to the layered structure of a network, survivability can be offered in the WDM layer or the higher layer. WDM layer survivability has many advantages: it is much faster for recovery at the WDM layer, it makes more efficient use of restoration capacity because of resource sharing among different service layers, and it provides transparency — the protection is independent of the protocols used in higher layers.

Due to the fact that the full survivability merely in the WDM layer proves very expensive, multiplayer protection schemes are proposed to provide the cost-effective performance as well as the differentiated QoS for survivability to different users. Usually, the combination of IP and WDM restoration problem is solved by assuming that the traffic routed on any line of the network can be made tolerable to faults by using either an IP restoration scheme or a WDM protection scheme. For each line, the multilayer scheme chooses the right scheme that has the minimum network cost.

9.2.4 Opaque vs. transparency

Full optical transparency means that every kind of signal (analog or digital and with whatever format) can be transmitted through the network. No device in the network must depend on the signal characteristics, but must allow for different transmission properties. The transparency property is considered one of the most important optical network characteristics; however, full transparency implies that no digital signal processing is present and that the network performs only analog functions like filtering, space or wavelength switching, wavelength conversion, and so on, although several problems are raised by the transparency property. It is difficult for the optical nodes to control the transmission impairment with the hug transmission capacity provided by WDM networks;[29] the full transparency in the optical networks will require a wide and detailed set of standards to provide full compatibility in different vendors.[30]

9.3 Optical network reconfiguration and quality of service (QoS)

This section discusses some important techniques in handling optical network reconfiguration and QoS; namely load balancing, blocking island, and optical burst switching.

9.3.1 Load balancing

In wavelength-routed networks, the passive or configurable optical nodes and their fiber connections constitute the physical topology of the network. The logical topology of the network is decided by the lightpath. Using optical cross-connects (OXCs), the lightpath can be dynamically changed, so the logical topology is also changed. When incoming traffic patterns are changing, we may need to reconfigure the network logical topology to achieve the load balancing.

The goal of load balancing is different. In Narula-Tam and Modiano, the goal is to minimize the network delay by reconfiguring the logical topology,[31] while in Banerjee and Mukherjee, it is to maximize the network throughput (by minimizing the packet average hop distance).[32] Narula-Tam and Modiano only consider the ring topology and assume wavelengths are always available.[31] They try to gradually adjust the topology, avoiding disrupting too many lines. Banerjee and Mukherjee present an exact integer linear programming (ILP) formulate for the complete logical topology design.[32] Although ILP is very time-consuming, the authors claim that the first few step iterations can lead to very good results. Some studies assume that the new logical topology is known *a priori* and were concerned with the cost and sequence of branch-exchange operations to transform from the original virtual topology to the new virtual topology.[31,33–35] Banerjee and Mukherjee try to obtain a new optimal topology from the ILP formulation, and the topology is "closest" to the previous topology.[32] "Closest" means the constituent lightpaths and the routes for these lightpaths of the new topology are largely similar to the previous one.

In general, we can reduce the load on each link through network reconfiguration. For example, we have a traffic matrix: (0,1) = 1/4, (1,2) = 1/4, (2, 3) = 1/4, (3, 0) = 1/4. We consider routing this traffic on a unidirectional ring physical topology with four nodes connected in the clockwise direction. Figure 9.6a demonstrates that the virtual topology of the network is also configured with clockwise direction, and the load on each logical link is 3/4. If the virtual topology is configured into counter-clockwise ring, as illustrated in Figure 9.6b, the load on each logical link is 1/4, reducing the load by a factor of 3.

The problem of determining optimal logical configuration so as to minimize the maximal link load consists of two sub-problems. One is the lightpath connectivity problem; the other is the traffic routing problem. In order

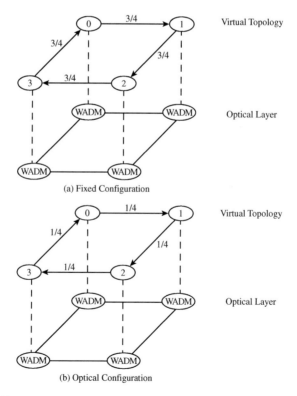

Figure 9.6 Different configurations of the logical layer.

to achieve the optimal topology, the two sub-problems must be solved jointly. Bienstorck and Gunluk, however, have proved that this problem is NP-complete,[36] so it is usually solved separately.[35] In the process of reconfiguration, rapid changing traffic may bring unpredictable reconfigurations that sometimes disrupt the normal network operation. Although the traffic changes dynamically, so does the optimal network configuration. Notice this reconfiguration takes time. At the rate of one gigabit per second, even a few milliseconds' delay may cause tens of megabits of traffic to be rerouted or buffered at each node that is reconfigured.

Labourdette et al. suggest multi-step (branch exchange sequence) strategies to move from the current logical topology to the optimal logical topology.[34] The disadvantages of this approach are that it takes quite a long time to finish reconfiguration, and an intermediate disconnected network is formed in the process. Other authors propose different cost schemes, arguing that the reconfiguration only takes place when its benefits outweigh the costs of reconfiguration.[56,58,59]

Narula-Tam and Modiano assume that the unlimited wavelength in each fiber and each node has a small number of transceiver ports. It evaluates the proposed algorithms based on three randomized traffic models:

1. The traffic between each pair of nodes is independent and identically distributed (i.i.d) with a uniform distribution between 0 and 1.
2. Significant proportions of the traffic flow from a single source to multiple destinations or from multiple sources to a single destination.
3. Traffic is formed by selecting a random ordering of the network nodes i1, i2, i3..., and generating uniform traffic between nodes I_j and $I_{(j+1)mod\ N}$. All other entries in the traffic matrix are zero.

Based on their previous research in 2000,[31] Narula-Tam and Modiano further explore the impact of adding wavelength restriction in ring networks.[39] If the physical topology is a bidirectional ring and the logical topology has, at most, one lightpath between each source and destination node pair, it can be demonstrated that a minimum of P(N-P) wavelengths are required to implement all possible logical topologies. P is the number of transceivers per node, and N is the number of nodes in the network. Simulations, which demonstrate that the lower bound on wavelength requirements is very tight, verify this claim.

9.3.2 Optical burst switching

Usually, optical networks only provide circuit-switching capacity. With the rapid growth of data traffic, many current optical networks need to be upgraded to support packet-switched data traffic. Thus, IP over WDM, which studies how to support predominantly IP traffic over WDM optical paths, has grown into a very important area. IP routers perform four main tasks: routing, forwarding, switching, and buffering.[40]

How to define the function of optical packet switching is still an issue. One strategy is to execute transmissions and switching in the optical domain, while routing and forwarding are carried out electronically. This strategy efficiently permits optical packet layer to support a range of network protocols while harnessing the power of WDM transmissions.

Generally, an optical packet switch has three principal functions:[44] switching, buffering, and optionally routing and forwarding. Among different schemes, photonic packet switching appears to be a strong candidate[42] because of the high-speed, data rate/data format transparency and configurability it offers. Unfortunately, optical packet switching still faces many cost and technological difficulties. One major challenge is the current lack of optical random access memory. Another major challenge is the stringent requirement for synchronization, both between multiple packets arriving at different input ports of an optical switching fabric, and between a packet's header and its payload.

Optical burst switching (OBS) is a promising solution for the high-speed optical routers. The concept "burst switching" means at the ingress several packets with same destination and similar characteristics are assembled into one entity. After a control message (burst header packet [BHP] or burst control packet [BCP]) reserves the resource on the path for the transmission, the entity is forwarded to the egress port. OBS can provide improvements over wavelength routing in terms of bandwidth efficiency and core scalability via statistical multiplexing of bursts. Furthermore, the synchronization requirement will be less stringent than in optical packet switching due to looser coupling between control signals and data in OBS.

The concept of burst switching was proposed for voice communications in the early 1980s. Three major differences exist between burst switching and circuit and packet switching:

1. A burst has an intermediate granularity when compared with circuit and packet switching, which are a call and a packet, respectively.
2. In burst switching, bandwidth (for a burst) is reserved in a one-way process. That is, a burst can be sent without an acknowledgment for a successful reservation. In circuit switching, bandwidth for a call is reserved in a two-way process.
3. In burst switching, a burst will cut through intermediate nodes (or switches) without being buffered, whereas in packet switching, a packet is stored and forwarded at each intermediate node.

9.3.2.1 Optical burst-switching techniques

A burst-switching technique called reserve-a-fixed-duration (RFD), which is based on close-ended resource reservation and distributed control, was first described by Yoo and Qiao.[43] Two other burst-switching techniques, tell-and-go (TAG) and in-band-terminator (IBT), are also for distributed control but based on open-ended resource reservation.

A burst is usually defined as a talk spurt or data message. A burst formed at an edge switch/router may consist of megabyte of data. The three techniques all use a one-way process, and a burst can cut through the intermediate switches while the way they release bandwidth is different.[45]

1. *IBT* — A control message is sent as a header, followed by a burst that has the label of the end of the burst. Bandwidth is reserved and switches are configured as soon as the control message is processed, and released as soon as the label IBT is detected. One problem of IBT scheme is that the IBT must be optically processed.
2. *TAG* — This is similar to fast-circuit switching. A source node first sends a control (or setup) packet to reserve bandwidth, and then a corresponding data burst is transmitted without waiting for the acknowledgment of reservation until the source node sends a release packet to explicitly release the reserved bandwidth. Another method is to keep sending refreshing packets in the processing; if no refreshing packet exists, the bandwidth will be released after a timeout period. The problem is obviously the signaling overhead.
3. *RFD* — In this scheme, bandwidth reservation is close-ended in that at each switch, bandwidth is reserved for a duration specified by each control packet. This not only eliminates signaling overhead (associated with bandwidth release) but also facilitates the more intelligent and efficient resource utilization.

We can still see that OBS is a very good solution of switching in the optical networks for its low setup time, high (moderate) bandwidth utilization, and very good adaptability, while the implementation difficulty is much lower than the photonic packet switching. OBS allows switching of data entirely in the optical domain by performing resource allocation in the electronic domain.

9.3.2.2   MPLS

MPLS is the convergence of connection-oriented forwarding technique and Internet routing protocols. With MPLS, it is possible to switch data through a routing hierarchy without compromising hierarchical requirements. It can also associate a wide range of forwarding granularities with a label, ranging from all data destined through an "exit" router to a host-to-host application flow. MPLS has largely evolved from the need to use the high speed of existing label-swapping technologies, such as ATM. In the evolution of MPLS, Ipsilon's IP switching, IBM's ARIS (aggregate route-based IP switching), Cisco's TAG switching and Toshiba's cell switch router architectures are among the most important techniques in bringing the ATM high-speed switches and IP routing protocols together.

MPLS uses labels to make forwarding decisions at the network nodes called label-switching router (LSR). In MPLS, the space of all possible forwarding options is partitioned into forwarding equivalence class (FECs). For example, all the packets destined for a given egress and having the same quality of service (QoS) may belong to the same FEC. The packets are labeled at the ingress depending on the FEC to which they belong. Each of the intermediate nodes uses the label of the incoming packet to determine its next hop, and also performs label swapping by replacing the incoming label with the new outgoing label that identifies the respective FEC for the downstream. We all know that in IP routing (hop-by-hop, destination-based packet routing), each packet's next hop and output port are determined by a longest-prefix-match forwarding table lookup, with the packet's IP destination address as the key. The table is established and maintained by IP routing protocol(s). In the label-based forwarding, each packet's forwarding treatment is entirely determined by a single index lookup into a switching table, using the packet's MPLS label as the key. The switching table is loaded *a priori* with unique next-hop label, output port, queuing, and scheduling rules for all current MPLS label values. This mapping information is established and managed by the management engine in response to external requests for a labeled path through the LSR, and is only modified when a new label needs to be activated or an old label removed. The actual signaling for setting up, tearing down, and maintaining LSPs can be done either using label distribution protocols (LDPs) or via Resource Reservation Protocol (RSVP) extensions.

MPLS provides constraint-based routing. The ingress node can establish an explicit route through the network. Rather than inefficiently carrying the explicit route in each packet, MPLS allows the explicit route to be carried only at the time the label switched path (LSP) is set up. The subsequent packets traversing this path are forwarded using packet labels. Constraint-based routing is potentially useful for traffic engineering.

Recently, multi-protocol lambda switching (MPLS) was proposed,[71] which treats optical cross-connects (wavelength routers) as LSRs and wavelengths as labels to provision wavelength-routed paths.

9.3.2.3   A framework for IP over WDM using OBS and MPLS

We first introduce an IP network operating over the optical backbone using OBS and MPLS technologies[44] (see Figure 9.7).

Incoming IP packets are assembled into data bursts at the ingress IP router and transported across the optical core to the destination egress IP router. Either no buffering or very limited optical buffering (using fiber delay line) occurs. Each cross-connect in the optical backbone will have label-swapping information about the precomputed routes, which are stored in the label switching table (LST). The LST can be set up by using standard routing protocols with traffic engineering extensions to distribute optical resource information (available bandwidth per wavelength, number of wavelength per fiber) and LDP or RSVP to distribute labels. Whenever an ingress router has a data burst to transmit, it refers to its LST to determine the proper label. This label is set up with the control message passing routers before the data burst. The control message provides information about the length and offset of the data burst. If necessary, the QoS parameters derived from the control message can support the differentiated service. The major advantage of integration under such a framework is that using MPLS control will reduce many of the

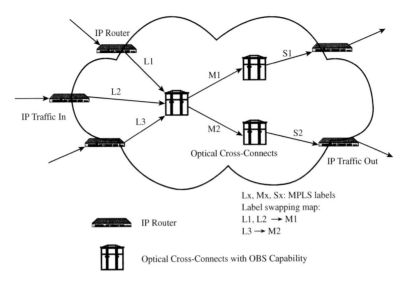

Figure 9.7 An IP-over-OBS WDM transmission backbone using MLPS.

complexities associated with defining/maintaining a separate optical (burst switching) layer, such as concerns over interface definitions, address assignment/resolution, interworking with higher-layer management, and so on.

9.3.3 Blocking island (BI) paradigm[59]

Developed from artificial intelligence, namely constraint satisfaction and abstraction and the theory of phase transition, BI provides an efficient way of abstracting resource (especially bandwidth) available in a communication network. Although it has no direct relation with the optical network, we believe its idea is very important to the optical resource allocation and management.

We assume all demands are unicast and the only QoS parameter taken into account is bandwidth. The network physical topology consists of V nodes arbitrarily connected by L bidirectional links. We model it as a network graph $G = (V, L)$.

BI clusters parts of network according to the bandwidth availability. A β-BI for a node x is the set of all nodes of the network that can be reached from x using links with at least β available bandwidth. For example, Figure 9.8 depicts a 40-BI for node V_1.

β-BI has some very useful properties. Next, we list a few without proof (for a proof, see Frei[59]).

> *Unicity* — One and only one β-BI exists for a node. Thus, if S is the β-BI for a node, S is the β-BI for every node in S.
>
> *Partition* — β_j-BI induces a partition of nodes in a network.

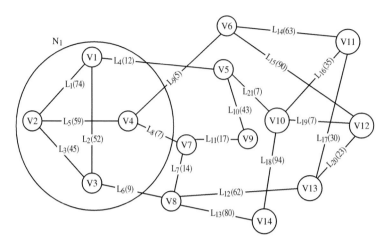

Figure 9.8 The topology of NSFNet. N_1 = {V1, V2, V3, V4} is the 40-blocking island (40-BI) for node V1.

> *Route existence* — Given a request $d_u = (x_u, y_u, \beta_u)$, it can be satisfied if and only if the node x_u and y_u are in the same β_u-BI.
> *Inclusion* — If $\beta_i < \beta_j$, the β_j-BI for a node is a subset of the β_2-BI for the same node.

Using the concept β-BI, we can construct a recursive decomposition of BI graphs in decreasing order of βs (e.g., $\beta_1 > \beta_2 > ... > \beta_n$). We call this layered structure of BI graphs a blocking island hierarchy (BIH). For example, we have such a BIH in Figure 9.9.

Given a request, using routing existence property we immediately know whether the request can be satisfied or not. With the abstract technique, instead of studying the whole network topology we focus our attention only on a small part. For example, given a demand (V1, V4, 40), according to the BIH, it is in 40-BI N_1. In the N_1 BI, different routing heuristic can be employed. If the route is allocated, the BIH may need to be modified. In this case route V1V2V4 is assigned. The available link capacity is less than 40 so we should reconstruct the BIH. Notice that although lower levels may be affected, all the modification is actually carried out within the N_1 BI.

The most frequent operation in the process is to construct a blocking island graph (BIG) according to a certain β. It is obtained with a simple greedy algorithm. Starting with an arbitrary node x, we add all the nodes that can be reached by links with at least β available bandwidth to form a β-BI. Then starting with another arbitrary node that is not in the previous β-BIs, we add all the nodes that can be reached by links with at least β available bandwidth to form a new β-BI. Repeat the process until all the nodes in the network are included in one of β-BIs. The complexity of constructing BIG is O(m),[59] where m is the number of links in the network.

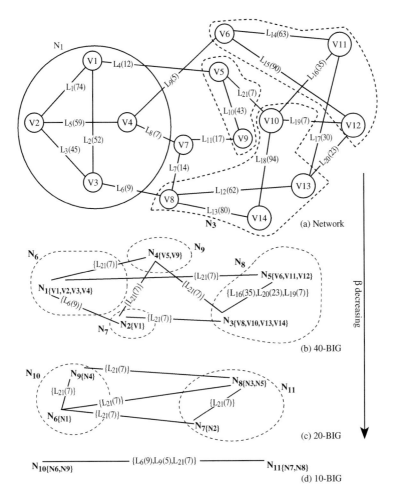

Figure 9.9 The blocking island hierarchy for bandwidth requirement {40, 20, 10}: (a) the network graph; (b) the 40-BIG; (c) the 20-BIG; (d) the 10-BIG.

BI is a natural abstraction of network resource. A β-BIG allows a clear picture about the load to be obtained because the nodes and links with enough resources are hidden behind an abstract node. Bottlenecks are identified by interlinks between BIs.

9.4 Summary

What should the future optical network look like? It will necessarily serve large, heterogeneous, geographically dispersed user populations and provide high quality of services. A mechanism to effectively control network bandwidth and virtual topology should be developed. It should be scalable and the protection and restoration scheme in optical layer should go well with

the high layer restoration such as layer-3 restorations. Architectures and protocols should be designed for the combination of user data and physical layer
transmission. Control and management planes need be developed to evaluate
and carry out different routing, scheduling, protection, and multiplexing
techniques. The objective of this chapter is to provide an overview of the
research development work in the area of optical network resource management and allocation. Some key issues on optical resource allocation and
management have been reviewed; namely RWA, traffic grooming, optical
network survivability, and optical opaque networks. We also discussed other
important techniques in handling optical network reconfiguration and QoS,
namely load balancing, blocking island, and optical burst switching. We
believe all these issues are very important in building the truly intelligent
optical network. Progress in these issues will help us manage and allocate
the optical network resource more efficiently and effectively.

References

1. B. Mukherjee, WDM optical communication networks: Progress and challeng-
    es, *IEEE J. Select. Areas. Commun.*, vol. 18, pp. 1810–1824, Oct. 2000.
2. I. Chlamtac, A. Ganz, and G. Karmi, Lightnet: lightpath-based solutions for
    wide bandwidth WANs, *INFOCOM '90*, vol. 3, pp. 1014–1021, 1990.
3. J.Y. Yoo and S. Banerjee, Design, analysis, and implementation of wave-
    length-routed all-optical networks: routing and wavelength assignment ap-
    proach, *IEEE Commun. Survey '98*, http://www.comsoc.org/pubs/surveys/
    Yoo/yoo.html.
4. H. Zang, J.P. Jue, and B. Mukherjee, A review of routing and wavelength
    assignment approaches for wavelength-routed optical WDM networks, *Op-
    tical Networks Mag.*, vol. 1, no. 1, Jan. 2000.
5. A. Mokhtar and M. Azizoglu, Adaptive wavelength routing in all-optical
    networks, *IEEE/ACM Trans. Networking*, vol. 6, pp. 197–206, Apr. 1998.
6. J.M. Yates, M.P. Rumsewicz, and J.P.R. Lacey, Wavelength converters in dy-
    namically reconfigurable WDM networks, *IEEE Commun. Surveys*, Second
    Quarter 1999, http://www.comsoc.org/pubs/surveys/
7. H. Harai, M. Murata, and H. Miyahara, Performance of alternate routing
    methods in all-optical switching networks, *Proc. IEEE INFOCOM '97*, pp.
    517–525, Apr. 1997.
8. S. Ramamurthy and B. Mukherjee, Fixed-alternate routing and wavelength
    conversion in wavelength-routed optical networks, *IEEE GLOBECOM*, pp.
    2295–2302, Nov. 1998.
9. L. Li and A.K. Somani, Dynamic wavelength routing using congestion and
    neighborhood information, *IEEE/ACM Trans. on Networking*, vol. 7, no. 5, pp.
    779–786, 1999.
10. S. Subramaniam and R. Barry, Wavelength assignment in fixed routing WDM
    networks, *IEEE Int. Conf. Commun.*, pp. 406–410, June 1997.
11. G. Li and R. Simha, On the wavelength assignment problem in multifiber
    WDM star and ring networks, *Proc. of IEEE INFOCOM '00*, pp. 1771–1780,
    Mar. 2000.

12. Y. Zhu, G.N. Rouskas, and H.G. Perros, A comparison of allocation policies in wavelength routing networks, *Photonic Networks Commun. J.*, vol. 2, no. 3, pp. 264–293, Aug. 2000.

13. E. Karasan and E. Ayanoglu, Effects of wavelength routing and selecting algorithms of wavelength conversion gain in WDM optical networks, *IEEE/ACM Trans. on Networking,* vol. 6, no. 2, pp. 186–196, Apr. 1998.

14. K. Chan and T.P. Yum, Analysis of least congested path routing in WDM lightwave networks, *Proc., IEEE INFOCOM '94,* Toronto, Canada, vol. 2, pp. 962–969, Apr. 1994.

15. E. Hyytia and J. Virtamo, Wavelength assignment and routing in WDM networks, *Nordic Teletraffic Seminar (NTS) 14,* Copenhagen, Denmark, Aug. 18–20, 1998.

16. J. Wang and B. Mukherjee, Interconnected WDM ring networks: Strategies for interconnection and traffic grooming, *Optical Networks Magazine,* Vol. 3, no. 5, pp. 10–20, Sept/Oct, 2002.

17. R. Berry and E. Modiano, Reducing electronic multiplexing costs in SONET/WDM rings with dynamically changing traffic, *J. Select. Areas. Commun.,* vol. 18, pp. 1961–1971, Oct. 2000.

18. A. Chiu and E. Modiano, Reducing electronic multiplexing costs in unidirectional SONET/WDM ring networks, *Proc. CISS,* Feb. 1998.

19. O. Gerstel, P. Lin, and G. Sasaki, Wavelength assignment in a WDM ring to minimize the cost of embedded SONET rings, *Proc. INFOCOM,* San Francisco, CA, Apr. 1998, pp. 662–672.

20. O. Gerstel, R. Ramaswami, and G.H. Sasaki, Cost-effective traffic grooming in WDM rings, *IEEE/ACM Trans. on Networking,* vol. 8, no. 5, pp. 618–630, Oct. 2000.

21. X. Zhang and C. Qiao, An effective and comprehensive approach for traffic grooming and wavelength assignment in SONET/WDM rings, *IEEE/ACM Trans. on Networking,* vol. 8, no. 5, pp. 608–617, Oct. 2000.

22. J. Wang, W. Cho, V.R. Vermuri, and B. Mukherjee, Improved Approaches for cost-effective traffic grooming in WDM ring networks: Nonuniform traffic and bidirectional ring, *Proc. IEEE ICC 2000 Conf.,* New Orleans, LA, June 2000.

23. E.H. Modiano and A.L. Chiu, Traffic grooming algorithms for minimizing electronic multiplexing costs in unidirectional SONET/WDM ring networks, *Proc, CISS '98,* Mar. 1998.

24. ITU-T Rec.G.872, Architecture of Optical Transport Networks, Feb. 1999.

25. T.H. Wu, *Fiber Network Service Survivability,* Artech House, Norwood, MA, 1992.

26. S. Ramamurthy and B. Mukherjee, Survivable WDM mesh networks, Part II — Restoration, *Proc. ICC '99,* 1999.

27. B.T. Doshi et al., Optical network design and restoration, *Bell Labs Technical. J.,* pp. 58–54, Jan.–Mar. 1999.

28. S. Ramamurthy and B. Mukherjee, Survivable WDM mesh networks, Part I — Protection, *Proc. IEEE INFOCOM '99,* pp. 744–751, 1999.

29. F. Bentivoglio and E. Iannone, The opaque optical network, *Optical Networks Mag.,* pp. 24–32, Oct. 2000.

30. H. Gysel, What we have to learn from the migration from PDH to SDH, *OADM Workshop,* Scheveningen/The Hague, The Netherlands, Apr. 23–24, 1998.

31. A. Narula-Tam and E. Modiano, Dynamic load balancing for WDM-based packet networks, *INFCOM 2000,* vol. 2, pp. 1010–1019, 2000.

32. D. Banerjee and B. Mukherjee, Wavelength-routed optical networks: linear formulation, resource budgeting tradeoffs, and a reconfiguration study, *IEEE/ ACM Trans. on Networking,* vol. 8., no. 5, pp. 598–607, Oct. 2000.

33. C. Chen and S. Banerjee, Optical switch configuration and lightpath assignment in wavelength routing multihop lightwave networks, *Proc. IEEE INFO-COM,* Boston, MA, pp. 1300–1307, June 1995.

34. J.F.P. Labourdette, G.W. Hart, and A.S. Acampora, Branch-exchange sequences for reconfiguration of lightwave networks, *IEEE Trans. Commun.,* vol. 42, pp. 2822–2832, Oct. 1994.

35. G.N. Rouskas and M.H. Ammar, Dynamic reconfiguration in multihop WDM networks, *J. High-Speed Networks,* vol., 4, no. 3, pp. 221–238, 1995.

36. D. Bienstorck and O. Gunluk, Computational experience with a difficult mixed-integer multicommodity flow problem, *Mathematical Programming,* vol. 68, pp. 213–237, 1995.

37. I. Baldine and G.N. Rouskas, Dynamic load balancing in broadcast WDM networks with tuning latencies, *IEEE INFOCOM,* vol. 1, pp. 78–85, 1998.

38. I. Baldine and G.N. Rouskas, Dynamic Reconfiguration policies for WDM networks, *IEEE INFOCOM,* vol. 1, pp. 313–320, 1999.

39. A. Narula-Tam and E. Modiano, Dynamic load balancing in WDM packet networks with and without wavelength constraints, *IEEE JSAC,* vol. 18, no. 10, pp. 1972–1959, Oct. 2000.

40. S. Keshav and R. Sharma, Issues and trends in router design, *IEEE Commun. Mag.,* pp. 144–151, May 1998.

41. D.K. Hunter, M.C. Chia, and I.L. Andonovic: buffering in optical packet switches, *IEEE/OSA J. Lightwave Technol.,* vol. 16, no. 12, pp. 2081–2094, Dec. 1998.

42. S. Yao, S. Dixit, and B. Mukherjee, Advances in photonic packet switching: an overview, *IEEE Commun. Mag.,* vol. 38, pp. 84–94, Feb. 2000.

43. M. Yoo and C. Qiao, Just-enough-time (JET): a high-speed protocol for bursty traffic in optical networks, *Dig. IEEE/LEOS Summer Topical Meetings — Technologies for a Global Info. Infrastructure,* pp. 26–27, Aug. 1997.

44. S. Verma, H. Chaskar, and R. Ravikanth, Optical burst switching: a viable solution for terabit IP backbone, *IEEE Network,* pp. 48–53, Nov./Dec. 2000.

45. C. Qiao and M. Yoo Choices, features and issues in optical burst switching (OBS), *Optical Network Mag.,* vol. 1, no. 2, pp. 36–44, Apr. 2000.

46. R. Ramaswami, Optical networking, *IEEE INFOCOM (tutorial),* Mar. 1996.

47. K.C. Lee and V. Li, A wavelength-convertible optical network, *IEEE/OSA J. Lightwave Technol.,* vol. 11, no. 5/6, pp. 962–970, May/June 1993.

48. W. Parys, B.C. Caenegem, B. Vandenberghe and P. Demeester, Meshed wavelength-divison multiplexed networks partially equipped with wavelength converters, *Optical Fiber Commun. Conf.,* pp. 359–360, Feb. 1998.

49. C. Brackett, Dense wavelength divison multiplexing networks: principles and applications, *IEEE JSAC,* vol. 8, no. 6, pp. 948–964, Aug. 1990.

50. X. Zhang, J.Y. Wei, and C. Qiao, Constrained multicast routing in WDM networks with sparse light splitting, *IEEE/OSA J. Lightwave Technol.,* vol. 18, no. 12, pp.1917–1927, Dec. 2000.

51. R. Karp, Reducibility among combinatorial problems, in *Complexity of Computer Computations,* pp. 85–103, Plenum Press, New York, 1972.

52. L.H. Sahasrabuddhe and B. Mukherjee, Light-trees: optical multicasting for improved performance in wavelength-routed networks, *IEEE Commun. Mag.,* pp. 67–73, Feb. 1999

53. N. Sreenath, K. Satheesh, G. Mohan, and C.S.R. Murthy, Virtual source-based multicast routing in WDM optical networks, *Proc. ICON 2000,* pp. 385–389, 2000.

54. M. Ali and J.S. Deogun, Cost-effective implementation of multicasting in wavelength-routed networks, *IEEE/OSA J. Lightwave Technol.,* vol. 18, no. 12, pp. 1628–1638, Dec. 2000.

55. R. Dutta., Virtual topology design for traffic grooming in WDM networks, Ph.D. thesis, North Carolina State University, Raleigh, Aug. 2001.

56. G. Sasaki and P. Lin, A minimal cost WDM network for incremental traffic, *Proc. SPIE,* vol. 3843, 1999.

57. V.R. Konda and T.Y. Chow, Algorithm for traffic grooming in optical networks to minimize the number of transceivers, *IEEE 2001 Workshop for High-Performance Switching and Routing,* pp. 218–221, 2001.

58. K. Zhu and B. Mukherjee, Traffic grooming in an optical WDM mesh network, *IEEE J. Select. Areas. Commun.,* pp. 122–133, Jan. 2002.

59. C.R. Frei, Abstraction techniques for resource allocation in communication networks, Ph.D. thesis, Lausanne, EPFL, 2000.

chapter ten

Real-time provisioning of optical communication networks

Chadi Assi
City College of the City University of New York
Abdallah Shami
Lakehead University
Mohamed A. Ali
City College of the City University of New York

Contents

10.1 Introduction

This chapter considers the problem of real-time provisioning of optical channels in hybrid Internet provider (IP)-centric dense wavelength division multiplexing (DWDM)-based networks. First, an overview of the emerging architectural alternatives for IP over optical networks is presented; namely the overlay, the peer, and the augmented models. Subsequently, lightpath provisioning issues are detailed for route selection, with a particular focus on the routing and wavelength assignment (RWA) problem. In particular, a broad overview is presented, with methodologies and associated algorithms for dynamic lightpath computation being outlined. Additionally, two broad constraint-based RWA algorithms for dynamic provisioning of the optical channels are presented. Finally, the implications of implementing the proposed RWA schemes for the lightpath provisioning aspects for each of the three emerging IP-over-optical network interconnection models are examined.

Recently there has been a dramatic increase in data traffic, driven primarily by the explosive growth of the Internet as well as the proliferation of virtual private networks (VPNs). At the same time, the rise of optical networking, first with wavelength-division multiplexing (WDM) transmission technology and, more recently, with optical multiplexers as well as optical cross-connects (OXCs) devices, is moving us toward the vision of creating an "all-optical" Internet. In particular, these technologies yield the ability to add, drop, and, in effect, construct wavelength-routed networks, heralding a new era in which bandwidth is relatively abundant and inexpensive. To some, a key realization of this vision will occur when lightpaths (wavelengths) can be provisioned automatically to create bandwidth between end users, with timescales on the order of minutes or seconds. Dynamic wavelength provisioning is the main focus of this chapter, and it will help open up a whole new world of responsive, customer-driven bandwidth services.[1]

To better understand and appreciate the provisioning issue, we need to look into how circuits are provisioned in a typical network today. Provisioning a cross-country SONET service today requires several steps. First, connectivity from the customer premise to the carrier's point of presence (POP) must be established for each end of the circuit. Second, a physical path must be mapped out between the many physical hubs in a carrier's network between the two points. Each path must be checked for fiber/ring bandwidth availability. Terminating equipment must be ordered and installed on each end of each fiber path, and each interconnect point must have capacity on the optical cross-connect system. Then, all the cross-connects and physical interconnects must be made and each segment must be documented and tested. This process is extremely manual and generally takes several months to accomplish. DWDM complicates this process even further because tens, and soon hundreds, of wavelengths are supported on individual fiber strands. Clearly, an automated optical routing layer will facilitate much faster provisioning.

Before this vision can be realized, however, networks need to slim down. Today's core network architecture model has four layers: IP and other content-bearing traffic, over-ATM for traffic engineering, over-SONET for transport, and over-WDM for fiber capacity. This approach has significant functional overlap among its layers and typically suffers from the lowest common denominator effect, where any one layer can limit the scalability of the entire network. When first conceived this layering made sense, but as IP and DWDM evolve, a more efficient interworking is needed (i.e., one that exploits the complementary features of each domain). In effect, high-performance routers plus a smart optical transport layer equipped with a new breed of photonic networking components and subsystems together are setting the foundation for the next-generation networking paradigm.

The solution, many believe, is to layer IP directly over the optical substrate.[2] If IP can be mapped directly onto the WDM layer, some of the functional overlap can be eliminated, potentially collapsing today's vertically layered network architecture into a horizontal model where all network elements work as peers to dynamically establish optical paths. To bring the IP and WDM layers together, however, new capabilities must be added to both layers. A framing standard is needed for carrying packets directly over lambdas. Signaling standards are needed so that IP devices can control optical resources.[3,4,5] More importantly, with the conventional multilayered architecture out of the way, automated provisioning systems will gain direct access to WDM resources, and dynamic lightpath provisioning will become easier and more practical to implement.

Once the view about network topology has changed, one will have to rethink routing as well. Initially, fixed routing over fixed circuits (public switched telephone network [PSTN]) was used, and next came dynamic routing over fixed circuits (IP). Subsequently, there was a move toward dynamic routing over virtual circuits (i.e., IP-over-ATM). Now, with recent advances in multi-protocol label switching (MPLS), we have label swapping over virtual circuits.[3] Furthermore, industry organizations, such as the Optical Internetworking Forum (OIF) and the Internet Engineering Task Force (IETF), are now extending the MPLS framework (generalized MPLS, also referred to as multi-protocol lambda switching [MPλS]) to support not only devices that perform packet switching (routers), but also those that perform switching in time (SONET), wavelength (OXCs), and space. Most likely, the next evolution will be label swapping over dynamic circuits or lightpaths.[4-11]

The Internet drafts cited previously describe several dynamic routing possibilities.[7-11] The simplest is to treat the optical layer as completely separate from the IP layer. In this "overlay" model, optical transport offers only higher capacity and higher reliability. A more ambitious "integrated" model links the routing decisions at the IP layer with the dynamic reconfiguration capabilities of optical cross-connects (MPλS).[8] The main goal of these initiatives is to provide a framework for real-time provisioning of optical channels, through combining recent advances in MPLS traffic-engineering control planes with emerging optical switching technologies in a hybrid IP-centric optical network.[2]

This chapter considers the problem of real-time provisioning of optical channels in a hybrid IP-centric DWDM-based networking model. Provisioning in this work implies that an optical channel is successfully routed if both an active path (working) and another alternate node/link-disjoint path (backup) are set up at the same time. Provisioning of connections requires algorithms for route selection, and signaling mechanisms to request and establish connectivity within the network along a chosen route. In particular, the problem of route selection in such wavelength-routed networks is referred to as the RWA problem.[12,13] Here, we present a review of RWA schemes and also compare the performance of two different constraint-based routing/RWA algorithms for dynamic provisioning of the optical channels. Specifically, the RWA schemes are used to compute end-to-end dedicated and shared backup paths to protect against single link/node failures. These algorithms are examples of approaches that might be used to simplify the complex problem of dynamic lightpath computation. Methodologies and associated algorithms for dynamic lightpath computation are outlined. We present an overview of the emerging architectural alternatives of the two-layer model, referred to in the literature as "the interconnection models," for IP over optical networks, namely the overlay, the peer, and the augmented models.[8] Finally, we examine the implications of implementing the proposed RWA schemes for the lightpath provisioning aspects for each of the three emerging interconnection models.

The remainder of this chapter is organized as follows. Section 10.2 presents an overview of the emerging architectural alternatives of the two-layer model. The proposed dynamic RWA algorithm is presented in Section 10.3. In Section 10.4, we present an overview of fault-tolerant routing. The implementation of real-time provisioning at the optical layer is presented in Section 10.5. Finally, Section 10.6 offers a summary and conclusion.

10.2   IP over optical network architectural alternatives

In the network model considered here, clients (e.g., IP/MPLS routers) are attached to an optical core network and connected to their peers over dynamically switched optical paths (lightpaths) spanning potentially multiple OXCs. The interaction between the client and the optical core is over a well-defined signaling and routing interface, referred to as the user–network interface (UNI). Meanwhile, the optical core network consists of multiple OXCs interconnected by optical links in a general mesh topology. This network may be multi-vendor, where individual vendor OXCs constitutes sub-networks. Each sub-network itself is assumed to be mesh-connected. The interaction between the sub-networks is over a well-defined signaling and routing interface, referred to as the network–network interface (NNI) (see Figure 10.1).

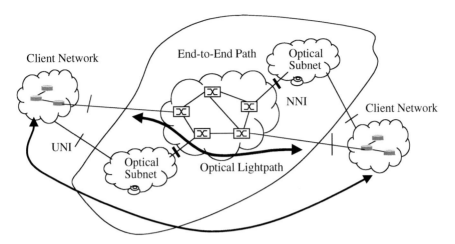

Figure 10.1 Optical networking model.

Each OXC is assumed to be capable of switching a data stream from a given input port to a given output port. This switching function is controlled by appropriately configuring a cross-connect table. A lightpath is a fixed bandwidth connection between two network elements such as IP/MPLS routers established via the OXCs. A single-hop channel logically connects two IP/MPLS routers to each other. This logical channel is the so-called lightpath. A continuous lightpath is a path that uses the same wavelength on all links along the entire route from source-to-destination.

10.2.1 Interconnection models

One approach for transporting IP traffic over WDM networks is to use a multi-layered architecture comprised of IP/MPLS layer over ATM over SONET over WDM.[2] If an appropriate interface is designed to provide access to the optical network, multiple higher layer protocols can request lightpaths to peers connected across the optical network. This architecture has four management layers. Another approach is to use a packet over SONET approach, doing away with the ATM layer, by putting IP/PPP/HDLC into SONET framing. This architecture has three management layers. The fact that both approaches support multiple protocols increases complexity for IP-WDM integration because of various edge inter-workings required to route, map, and protect client signals across WDM sub-networks.[7]

The two-layer model, which aims at a tighter integration between IP and optical layers, offers a series of advantages over the current multi-layer architecture model. MPLS[3] and its extension G-MPLS[9] have been proposed as the integrating structure between IP and optical layers. Nevertheless, routing in non-optical and optical parts of hybrid IP networks needs to be coordinated. To examine the architectural alternatives for the two-layer model (IP-over-optical network), it is important to distinguish between the data plane and control planes over the UNI. The IP-over-optical-network architecture is classified according to the organization of the control plane (i.e., whether a single integrated or separate independent monolithic routing and signaling protocol exists, which spans the IP and the optical domains). Several models have been proposed including overlay, augmented, and peer-to-peer models.[8]

10.2.1.1 The overlay model

Under the overlay model, the IP domain is more or less independent of the optical domain; that is, the IP domain acts as a client to the optical domain. The IP/MPLS routing and signaling protocols are independent of the routing and signaling protocols of the optical layer. Thus, the topology distribution, path computation, and signaling protocols would have to be defined for the optical domain. In this model, the client routers request high-bandwidth connections (lightpaths) from the optical network through the UNI. The client routers are provided with no knowledge of the optical network topology or resources. In this scenario, the optical network provides point-to-point connection to the IP domain. The overlay model may be statically provisioned using a network management system or may be dynamically provisioned.

10.2.1.2 The peer model

In the peer model, the two layers are collapsed into a single integrated layer managed and traffic engineered in a unified manner. In this regard, the OXCs are treated just like any other router (IP/MPLS routers and OXCs act as peers), and there is only a single instance of a routing protocol spanning an administrative domain consisting of the core optical network and the surrounding edge devices (IP/MPLS routers, ATM switches). Thus, from a routing and signaling point of view, no distinction is made among the UNI, the NNI, and any other router-to-router interface. This allows the SP edge devices to have a full access to the topology of the core network. A common IGP such as OSPF or IS-IS may be used to exchange topology information. The assumption in this model is that all the optical switches and the routers have a common addressing scheme.

10.2.1.3 The augmented model

In the augmented model, the IP and optical domains can be functionally separated, each running its own routing protocol, but exchanging full reachability information across the UNI using a standard protocol. For example, IP addresses could be assigned to optical network elements and carried by optical routing protocols to allow reachability information to be shared with the IP domain to support some degree of automated discovery. This model combines the best of the peer and overlay interconnection models; it is relatively easy to deploy compared to the peer model in the near term. Also, this is a convenient solution because it allows implementation of both provisioning and restoration procedures for optical sub-networks independent of the client network routing. In addition, this approach supports the common scenario where the optical network and client networks are administered by different entities.

The central issue in this model is the way in which the routing information is exchanged at the IP-optical UNI. Two possibilities exist for this. The first is to consider the inter-domain IP routing protocol, BGP, which may be adopted for exchanging routing information between IP and optical domains. The second is to consider the use OSPF areas (OSPF supports a two-level hierarchical routing scheme through the use of OSPF areas) to exchange routing information across the two domains. On the other hand, running a protocol such as BGP across the UNI may be considered too involved, at least for initial implementations of the UNI. A simpler approach would be to limit the reachability information passed through the optical network.

10.3 Lightpath provisioning

Provisioning of connections requires algorithms for route selection and signaling mechanisms to request and establish connectivity within the network along a chosen route. The problem of route selection in such wavelength-routed networks is referred to as the RWA problem, and consists of two sub-problems. The first is the routing problem, which determines the path along which the connection can be established. The second problem is to assign a wavelength (or a set of wavelengths) on each link along the selected path (wavelength assignment problem). Real-time provisioning implies that both the path and wavelength should be chosen/assigned dynamically (dynamic RWA), depending on the network state. In general, all networking models described previously, regardless, require route/wavelength computation/assignment to provision a lightpath (i.e., dynamic RWA engine).

10.3.1 Overview of the RWA problem

Given a set of connections, the problem of setting up lightpaths by routing and assigning a wavelength to each connection is called the RWA problem.[12] Typically, connection requests may be of three types: static, incremental, and dynamic.[13] With static traffic, the entire set of connections is known in advance, and the problem is then to set up lightpaths in a global fashion while minimizing network resources such as the number of wavelengths or the number of fibers in the network. Here, the RWA problem for static traffic is known as static lightpath establishment (SLE) and can be formulated as mixed-integer linear program.[12, 13] In the incremental-traffic case, connection requests arrive sequentially, a lightpath is established for each connection, and the lightpath remains in the network indefinitely.

For the case of dynamic traffic, a lightpath is set up for each connection request as it arrives, and the lightpath is released after some finite amount of time. The objective in the incremental and dynamic traffic cases is to set up lightpaths and assign wavelengths in a manner that minimizes the amount of connection blocking.[13] This problem is referred to as the dynamic lightpath establishment (DLE). Generally, the DLE is more difficult to solve, and, therefore, heuristics methods are generally employed. Heuristics exist for both the routing sub-problem and the wavelength assignment sub-problem.

For the routing sub-problem, three basic approaches can be found in the literature: fixed routing, fixed-alternate routing, and adaptive routing.[13] Fixed routing is one variant of the "static routing" in which routing decisions do not vary with time. Moreover, in fixed routing, the same fixed route for a given source-destination pair is always selected. The fixed alternate routing approach considers multiple routes between a source-destination pair and each node in the network maintains an ordered list of a number of fixed routes to each destination node. When a connection request arrives, the source node attempts to establish the connection on each of the routes from the list in sequence, until a route with a valid wavelength assignment is found. Conversely, in adaptive routing,[14–17] the route from a source node to a destination node is chosen dynamically, depending on the network state. Adaptive routing requires extensive support from the control and management protocols to continuously update the routing table at the node. An advantage of adaptive routing is that it results in lower connection blocking than fixed and fixed-alternate routing.

Meanwhile, for the wavelength assignment sub-problem, a number of heuristics have been proposed.[13,15,17,18] These heuristics are Random Wavelength Assignment, First-Fit, Least-Used, Most-Used, Min-Product, Least-Loaded, MAX-SUM, Relative Capacity Loss, Wavelength Reservation, and Protecting Threshold. In Mokhtar and Azizoglu, the authors propose an adaptive unconstrained routing (AUR), which incorporates network state information into route computation and channel allocation.[15]

Currently, the algorithms that offer the best performance are Relative Capacity Loss (RCL)[19] and Distributed Relative Capacity Loss (DRCL).[13] The RCL algorithm calculates the relative capacity loss for each path on each available wavelength and then chooses the wavelength that minimizes the sum of the relative capacity loss on all the paths. DRCL is proposed in Zang et al. and is based on RCL, but it is more efficient in a distributed environment.[13] For a tutorial review on the RWA problem, we refer the reader to Zang et al.[13]

Optical networks can also pose added wavelength continuity constraints,[12] and these may require the use of wavelength conversion (also referred to as wavelength translation or wavelength changing). A wavelength converter is a device, which takes at its input a data channel modulated onto an optical carrier with a wavelength λ_{in}, and produces at its output the same data channel modulated onto an optical carrier with a different wavelength λ_{out}. If wavelength converters are included in the OXCs in WDM networks, connections can be established without the need to find an unoccupied wavelength, which is the same on all the links making the route. This means that networks with wavelength converters are equivalent to traditional circuit switched networks. Wavelength converters thus result in improvements in network performance. On the other hand, it has been demonstrated that a careful wavelength assignment in wavelength-continuous network can lead to improved performance; thus, reducing the benefits of wavelength converters.[20] In Yates et al., the authors investigate the benefits of limited wavelength conversion for ring and mesh-torus topologies with fixed shortest path routing.[20] Harai et al. used a hypercube network to study limited conversion with fixed shortest path routing and a first-fit wavelength selection algorithm.[21] It is demonstrated that limited wavelength conversion (25%) achieves the same performance improvement as full wavelength conversion.[20,21,23]

Many other constraints can also serve to complicate the RWA process, especially in all-optical networks. Specifically, besides wavelength continuity requirements, these include analog attenuation effects and power limitations. For example, adequate signal-to-noise ratios (SNRs), cross talk, and dispersion effects caused by subsystem components and fiber links can be computed along candidate paths. This information can be incorporated into route resolution strategies by defining new cost functions.[22] In Ali et al., the authors have extended the routing and wavelength assignment problem to account for the power degradation of a routed signal due to nonideal behavior of optical components such as multiplexers, demultiplexers, taps, and fiber links.[22]

10.3.2 *The proposed dynamic RWA algorithms*

Some combined RWA algorithms are now presented. Specifically, these algorithms integrate and collapse both the routing and wavelength assignment sub-problems into a single dynamic constraint-based routing problem.[24] Thus, the emphasis here is on the adaptive routing problem instead of focusing on the wavelength-assignment problem. It has been demonstrated that the routing scheme has much more of an impact on the overall network performance than the wavelength-assignment scheme. Moreover, both algorithms are also capable of supporting fault-tolerant adaptive routing and are amenable to fully distributed implementations.

The network is viewed as multi-layered graphs, each corresponding to a specific wavelength. For a connection request and on a given wavelength, Dijkstra's shortest path algorithm that is suitably modified for WDM networks is used for computing a constraint path. This is achieved by associating each link in the network with a specific weight function that incorporates WDM specific information such as the number of available wavelengths and the total wavelengths.[16] This means that the algorithm might compute (online) W paths, each corresponding to one of the W wavelengths. Then, one of these paths is selected according to a global selection criterion. Thus, the problem of wavelength-assignment is totally mitigated, and both the routing and wavelength assignment sub-problems are now integrated and collapsed into a single dynamic constraint-based routing problem. This contrasts the work reported in Fabry-Asztlos et al., where a single path is first calculated, then a wavelength is assigned to the path by propagating a wavelength request to all the routers along the path.[16] This type of algorithm avoids the overhead associated with such a wavelength request (probe message).

The algorithm is implemented as per the following:

1. First, consider a multi-fiber WDM-based network whose physical topology consists of multiple OXC interconnected via point-to-point WDM links in an arbitrary mesh topology.
2. Assume that none of the OXCs has wavelength conversion capability. Hence, to meet a connection request, a lightpath, which uses the same wavelength on all the links along the entire route from source-to-destination, has to be set up.
3. Both algorithms are based on a fully distributed implementation in which all nodes maintain a synchronized and identical topology as well as link-state information (traffic-engineering database [TED]).
4. Assuming that W is the number of wavelengths per fiber, the network is represented by W identical graphs, each conforming to the physical topology and a particular wavelength. Hence, the network can be viewed as W identical wavelength graphs, each representing a wavelength. In view of this multi-graph model, each physical link is now represented by W virtual links (channels), each corresponding to one of the wavelength graphs. Figure 10.2 illustrates the concept of the multi-graph approach for a simple network with four nodes, four physical links, and $W = 2$.
5. For a given connection request, a constraint route is calculated, for each of the wavelength graphs, throughout the entire network from source to destination, typically using a shortest path algorithm, but with the link weights adjusted to attain some sort of local resource

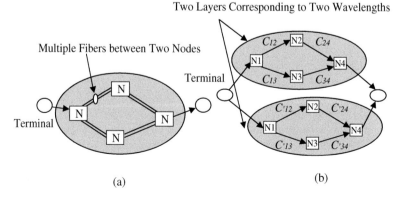

Figure 10.2 The sample network model (a) and its multi-layered graph approach (b)..

optimization. Clearly, at most, W paths can be calculated, each corresponding to a given wavelength, provided that each path can meet the given routing constraint. As a result, we get the vector V = <Path$_i$, Wavelength$_i$>, i = 1, ..., W, where the number of entries stored in V may vary from no entries at all (request is blocked), to a possible maximum of W entries. Finally, to globally optimize the network resources, provided that the number of entries stored in V is more than one, an entry (Path$_i$, Wavelength$_i$) out of all the other possible entries, has to be selected. Thus, by virtually separating wavelengths, both the routing and wavelength assignment sub-problems are now reduced into a single dynamic constraint-based routing problem.

10.3.2.1 Algorithm I — "full adaptive routing"
The implementation of this algorithm is as follows:

1. For a given wavelength graph λ_i, each virtual link, in each of the W wavelength graphs, is assigned a cost. Basically, the cost of a link at a given wavelength λ_i, is defined here as the inverse of the number of available channels over that particular link. Hence, initially the cost of a given link throughout the entire network is set = (1/F), where F is the number of fibers (per link) connecting two adjacent OXCs. In general, the cost of link L_j at wavelength λ_i, $C(L_j^{\lambda I})$, is given by:

$$C(L_j^{\lambda_i}) = \begin{cases} \dfrac{1}{F - N(L_j^{\lambda_i})} & \text{if } N(L_j^{\lambda_i}) < |F| \\ \infty & \text{if } N(L_j^{\lambda_i}) = |F| \end{cases} \tag{10.1}$$

where $N(L_j^{\lambda_i})$ is the number of occupied (unavailable) λ_i's on link L_j.

2. For a given wavelength λ_i, we associate each path throughout the entire network with a total cost, $C_{sd}^{\lambda_i}$, which is defined here as the summation of the costs of all individual links spanning the entire path from source to destination.

$$C_{sd}^{\lambda_i} = \sum_{j=1}^{n} C(L_j^{\lambda_i}) \tag{10.2}$$

where $< L_1, L_2 ......, L_n >$ is the set of n links that comprise the path.

3. For a given connection request, run the Dijkstra's algorithm on the first wavelength graph λ_1 to find the shortest path (the path with minimum $C_{sd}^{\lambda_i}$). Store the calculated path along with its corresponding wavelength λ_1 as the first entry of the vector V. Note that the

calculated local path, for a given wavelength graph λ_i, is not necessarily the path with the minimum number of hops.
4. Repeat step 3 for each of the remaining W-1 wavelength graphs. Note that the vector V might now have up to W entries.
5. Examine the contents of the vector V and perform one of the following instructions:
   a. If the vector V has no entries at all, reject the connection request; otherwise go to step b.
   b. If the vector V has only one entry, select this entry as the combination (Path$_i$, Wavelength$_i$) that satisfies the connection request. After assigning the path, update the weights associated with all links along the entire path (just on the corresponding wavelength graph λ_i) by basically decrementing the number of the available λ_i's (channels), on every link along the selected path, by one; otherwise go to step c.
   c. If the vector V has more than one entry, select an entry that satisfies one of the following global path selection schemes.

Total cost-based selection. In this scheme, a total cost, $C_{sd}{}^{\lambda_i}$, is associated with each computed path within the vector V (<Path$_i$, Wavelength$_i$>), given by:

$$C_{sd}{}^{\lambda_i} = \sum_{j=1}^{n} C(L_j^{\lambda_i}) \tag{10.3}$$

The path with the minimum total cost $C_{sd}{}^{\lambda_i}$ is selected and assigned to the connection. Note that this selection criterion skews the conventional shortest path search to favor less utilized network resources. This is illustrated in Figure 10.3a where the path with fewer hops is not selected.

Balanced cost-based selection. In this case, each path is assigned a balanced cost, $C^B{}_{sd}{}^{\lambda_i}$, defined by:

$$C^B{}_{sd}{}^{\lambda_i} = n \times \left(\sum_{j=1}^{n} C_{Lj}{}^{\lambda_i} \right)$$

where n is the total number of links along the path.
The path with the least balanced cost $C^B{}_{sd}{}^{\lambda_i}$ is selected and assigned to the connection. Note that this selection criterion strikes a balance between the minimum cost and the minimum number of hops. The main objective of this selection criterion is to avoid assigning long paths to a connection and route connections over the healthy part of the network. This is illustrated in Figure 10.3b where the connection is routed on the path with wavelength λ_j that has a balanced cost (2*1) less than that with wavelength λ_i [3*(11/

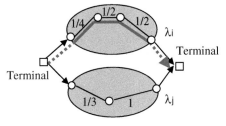

(a) Total Cost Path Selection, λ_i

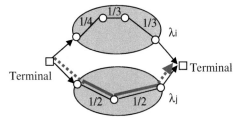

(b) Balanced Cost Path Selection, λ_j

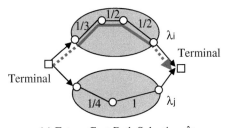

(c) Future Cost Path Selection, λ_i

Figure 10.3 Illustration examples.

12)], thus avoiding assigning the connection to the longer path (λ_j). Note, however, that the total cost of the path with wavelength λ_j (1) is higher than that with λ_i (11/12).

  Future cost-based selection.   This scheme uses the same total cost $C_{sd}^{\lambda_i}$ of Equation (10.3), but with the individual link cost $C_{Lj}^{\lambda_i}$ of Equation (10.1) redefined as:

$$C(L_j^{\lambda_i,Future}) = \frac{1}{F-1-N(L_j^{\lambda_i})} \qquad \text{if } N(L_j^{\lambda_i}) \; < \; |F|$$

Thus, the total future cost of this scheme is given by:

$$C_{sd}^{\lambda_i,Future} = \sum_{j=1}^{n} C_{L_j}^{\lambda_i,Future}$$

The path with the minimum future cost is selected and assigned to the connection. If all paths have the same future cost of infinity (i.e., along a path at least one link has one last channel available), select the path with the least cost using the original definition of individual link cost $C(L_j^{\lambda_i})$ of Equation (10.1). Note that this selection criterion strongly favors assigning paths with the least current/future-utilized resources. This is illustrated in Figure 10.3c, where the path with infinity future cost is not selected.

Two comments are in order here. First, for any of the three selection schemes described previously, once a path is assigned, the weights associated with all links along the selected path should be updated as indicated in step 5b. Second, all link-state updates must be propagated and advertised to all other core nodes throughout the network. Note that the frequency of the link-state update per unit time is proportional to the number of accepted/released calls. In other words, the link-state update is triggered once for every accepted/released call. As a result, the signaling overhead associated with a high volume of calls may be excessive.

10.3.2.2 Algorithm II — "semi-adaptive routing"

This algorithm adopts the same implementation procedures developed for the full adaptive algorithm described previously, except for the following fundamental differences:

1. A shortest path algorithm (Dijkstra's algorithm) is initially run offline to calculate the shortest path (just minimum number of hops) between every source-destination node pair (routing tables) throughout the entire network. These offline computed routing tables are stored at each node in every wavelength graph. Thus, the initial routing tables are identical for all W wavelength graphs.
2. For an initial connection request, the ingress node at every wavelength graph consults its own routing table for the shortest path. As a result, similar to algorithm I, we may get as many as W paths, where one of them can then be selected according to the selection schemes described previously.
3. For all the consecutive connection requests, the routing tables remain unchanged; so that step 2 is repeated until the cost of a link L_j in a given wavelength graph λ_i goes to infinity (no more available λ_i's). In this case, link L_j is removed from wavelength graph λ_i and the routing tables are calculated for each node again. Note that the Dijkstra's algorithm is run this time online to find the shortest path (the path with minimum $C_{sd}^{\lambda_i}$).

Note that the signaling overhead associated with the link-state updates for this algorithm is considerably less than that of algorithm I (link-state updates is only triggered when the cost of the link goes to infinity). In addition, the time associated with computing a path for the semi-adaptive algorithm is less than that of the full adaptive one because the path is directly read from the routing table.

Finally, in the case where the lightpath is wavelength continuous as this work has assumed, optical nonlinearities, chromatic dispersion, amplifier spontaneous emission, and other factors together limit the scalability of an all-optical network. Routing in such networks will then have to take into account noise accumulation and dispersion to ensure that lightpaths are established with adequate signal qualities. This work assumes that the all-optical (sub-)network considered is geographically constrained so that all routes will have adequate signal quality, and physical layer attributes can be ignored during routing and wavelength assignment; however, the policies and mechanisms proposed here can be extended to account for physical layer characteristics and require future work.

10.4   Fault-tolerant routing

Given the wide range of services envisioned for future IP networks, network survivability is a crucial concern. Survivability schemes can be classified into two forms, protection and restoration, where the former refers to preprovisioned failure recovery and the latter refers to more dynamic signaled recovery.[2] A common approach to protection is to set up two physical link-disjoint paths for every connection request. One path, called the primary, is used to transmit data, while the other path is reserved as a backup in the event that a link in the primary path fails. To further protect against node failures, the primary and backup paths may also be node-disjoint.[25,26]

Fixed-alternate routing provides a straightforward approach to handling protection.[15] On the other hand, in adaptive routing, a protection scheme may be implemented in which the backup path is set up immediately after the primary path has been established.[12] The same routing protocol may be used to determine the backup path with the exception that a link cost is set to infinity if that link is being used by the primary path. The resulting route will then be link-disjoint from the primary path. Because these schemes require backup path routing at setup time, they must be more closely incorporated with the primary lightpath RWA algorithms. On the other hand, channel restoration does not rely on precomputed backup routes, and instead dynamically recomputes a new path for a broken channel.[26] This has the advantage of low overhead in the absence of failures. This does not guarantee successful recovery, however, because the attempt to establish a new path may fail due to resource shortage at the time of failure recovery. Additionally, recovery timescales are usually longer.[2]

By making use of the WDM channel routing capabilities, a variety of lightpath protection schemes can be designed. For example, dedicated backup channels can be provisioned for users requiring high availability. Here, a pre-computed link-disjoint backup channel is reserved for each primary channel at setup time, and in case of a fault condition on the primary path, a channel switchover is performed. The dedicated backup reservation method has the advantage of a shorter restoration time because the resources are reserved for the backup path when establishing the primary path itself; however, this method reserves excessive resources. For better resource utilization, multiplexing techniques[27] can be employed. Here, if two primary lightpaths do not fail simultaneously, their backup lightpaths can share the same wavelength channel. This technique, known as shared path protection, improves the network resource utilization by sharing the reserved backup capacity. However, it experiences longer restoration times[31] when compared to dedicated path protection.

In addition, recently the concept of a shared risk link group (SRLG) definition has also been proposed to help identify risk associations between various entities (see Rajagopalan and Saha).[28] This concept is used to ensure that the primary and the backup path are not affected by the same failure. By using this concept, adequate resource "disjointness" can be introduced into the constraint-based path computation phase, thereby reducing the probability of simultaneous lightpath failures (e.g., between working and protection paths). Further details are out of the scope herein and interested readers are referred to Rajagopalan et al., and Rajagopalan and Saha.[8,28]

Overall, the proposed adaptive RWA scheme can be extended to ensure diversity in routes. This can be achieved by coordinating each diversely routed lightpath group by a single network entity. To create a diversely routed lightpath group, a user registers with a coordinator and receives the group identifier. For groups originating through the same client router, this router would typically act as the coordinator. To ensure diversity in routes, N SRLG and node disjoint routes through the network are selected, where N represents the number of diverse routes required.

10.5   Real-time provisioning at the optical layer

Provisioning end-to-end circuits is an endless source of struggle for service providers and a frustration for end users. Provisioning of connections requires algorithms for route selection and signaling mechanisms to request and establish connectivity within the network along a chosen route. This section examines the problem of route selection in the context of applying/adapting the RWA algorithm presented previously to each of the three interconnection models described in Section 10.2. The implications on both the route selection and signaling mechanism components will also be outlined for each of the three interconnection models.

10.5.1   Dynamic lightpath computation

Dynamic computation of a lightpath involves the implementation of two traffic engineering components: an information distribution mechanism that provides knowledge of the relevant attributes of available network resources, and a path selection process that uses the information distributed by the dynamic link-state advertisement algorithm to select a path that meets the specific requirements of the traffic flow. For a fully distributed network implementation, these requirements are described next.

10.5.1.1 An information distribution mechanism

The information distribution mechanism provides knowledge of network's topology and the available resources. This component is implemented by defining relatively simple extensions to the interior gateway protocol (IGP) (e.g., open shortest-path first [OSPF]) so that link attributes are included as part of each router's link-state advertisement. Some of the traffic-engineering extensions that need to be added to the IGP link-state advertisement include maximum link bandwidth, maximum reservable link bandwidth, current bandwidth reservation, current bandwidth usage, and link coloring. These extensions capture optical link parameters and any constraints specific to optical networks. Such topology and link-state information is then flooded to all nodes via update messages.[29] Another important component is to define naming and addressing convention for different elements of the physical plant hierarchy.[11] Here, we have defined and assigned a naming and addressing convention for different elements of the physical plant hierarchy (i.e., by implementing a simplified link-state advertisement algorithm to model extended OSPF). This algorithm is capable of periodically updating and advertising all of the preceding link attributes. The link-state updates can be triggered, for instance, based on a given threshold of the number of available wavelengths per fiber, below which the updates can be triggered. Once each node has a representation of the full physical network topology and the available resources on each link, a path selection algorithm is required (i.e., dynamic RWA).

10.5.1.2 A path selection process

This process uses the information distributed by the dynamic link-state advertisement algorithm to select an explicit route that meets the specific requirements of the traffic flow. This process can be performed either offline or online using a constraint based routing calculation. The source router (peer model), or the border OXC/central management node (augmented/overlay models) are basically responsible for computing the complete path all the way to the destination through the optical domain, and then initiating path setup using the signaling protocol (e.g., CR-LDP or RSVP). The route may be specified either as a series of nodes (routers/OXCs), or in terms of the specific links used (as long as IP addresses are associated with these links).

10.5.2 Route selection using the proposed RWA

Numerous policies can be used to route lightpaths through the network, such as the constraint-based routing algorithms proposed here. This scheme can be used directly for computing the route and assigning the wavelength for both the overlay and the augmented models. In this case, a connection request is initiated by a client IP/MPLS router (border router, that is, a router directly connected to the optical network) and sent to an ingress optical node (border OXC, that is, the OXC connected to the border router) using UNI signaling. Such provisioning request may specify the desired destination client router. Note that the source end point is implicit in this case. The ingress optical node processes the request and computes an appropriate route along with a wavelength through the network (using topology and state information that has been propagated using OSPF link-state advertisements[29]). Note that the request may also be received by an ingress OXC from a central management node, specifying the source and destination end points.

The routing within the optical and IP domains in the case of the augmented model may be separated, with a standard routing protocol running between domains.[7,8] This is similar to the IP inter-domain routing model, where the central issue is how the routing information is exchanged at the IP-optical UNI. Two possibilities exist for this. The first is to consider the inter-domain IP routing protocol, BGP, which may be adopted for exchanging routing information between IP and optical domains. The second is to consider the use OSPF areas (OSPF supports a two-level hierarchical routing scheme through the use of OSPF areas) to exchange routing information across the two domains.[7,8]

In the case of the peer model, however, additional extensions need to be added to the routing protocol (OSPF) so that the segment of the entire route that crosses the optical core (between the ingress and egress OXCs) must be treated as a virtual link of fixed capacity and advertised as such in further OSPF updates. The routing in this case is referred to as a "flat" routing organization.[78] Under this approach, there is only one instance of the routing protocol running in the IP and optical domains. An IGP, such as OSPF or IS-IS with suitable optical extensions, is used to exchange topology information. These optical extensions will capture the unique optical link parameters. The OXCs and the routers maintain the same link-state database. The routers can then compute end-to-end paths to other routers across the OXCs; these lightpaths are considered a tunnel across the optical network between edge routers. Once created, such lightpaths are treated as virtual links and are used in traffic engineering and route computation. Here, the notion of forwarding adjacency (FA) is essential in propating existing lightpath information to other routers. Finally, once an FA is advertised in a link state protocol, its usage is defined by the route computation and traffic engineering algorithms implemented.

10.5.3   RWA implications on signaling mechanisms

Once the ingress node receives a lightpath request from a source, it computes the complete path all the way to the destination through the optical domain using the proposed RWA algorithm. The output of this calculation is an explicit route consisting of a sequence of hops that provides the shortest path through the network that meets the constraints. This explicit route is then passed to the signaling component that initiates path setup (to reserve resources) using the signaling protocol (e.g., CR-LDP or RSVP-TE).[9] Note that the implications of using the proposed RWA scheme on the signaling is that the conventional overhead associated with the wavelength (i.e., probe message)[30] is no longer needed because the RWA scheme selects and assigns both the route and the wavelength simultaneously.

10.6 Conclusion

This chapter has considered the problem of real-time provisioning of optical channels in a hybrid IP-centric DWDM-based networking model. Provisioning implies that an optical channel is successfully routed if both an active path (working) and another alternate link-disjoint path (backup) are set up at the same time. Specifically, the work presented here has addressed the implementation issues of the path selection component of the traffic-engineering problem in such a network. Methodologies and associated algorithms for dynamic lightpath computation were outlined.

References

1. J.M. Elmirghani, and H. Mouftah, Technologies and architectures for scalable dynamic dense WDM networks, *IEEE Commun. Mag.*, vol. 38, no. 2, pp. 58–66, Feb. 2000.
2. N. Ghani et al., on IP-over-WDM integration, *IEEE Commun. Mag.*, March 2000.
3. R. Callon, A framework for multi-protocol label switching work in progress, Internet Draft, Sept 1999.
4. D. Awduche et al., Multiprotocol lambda switching, combining MPLS traffic engineering with optical cross-connects, work in progress, Internet Draft, draft-awduche-mpls-te-optical-01.txt
5. N. Ghani, Lambda-labeling: a framework for IP-over-WDM using MPLS, *Optical Networks Mag.*, vol. 1, pp. 45–58, April 2000.
6. Y. Ye, S. Dixit, and M.A. Ali, On joint protection/restoration in IP-centric DWDM-based optical transport networks, *IEEE Commun. Mag.*, vol. 38, no. 6, pp. 174–183, June 2000.
7. N. Chandhok et al., IP over optical networks: a summary of issues, work in progress, Internet Draft, November, 2000.
8. B. Rajagopalan et al., IP over optical networks: a framework, work in progress, Internet Draft, December, 2002.
9. P. Ashwood-Smith et al., Generalized MPLS, signaling functional description, work in progress, Internet Draft, August, 2002.
10. B. Rajagopalan et al., IP over optical networks: architecture aspects, *IEEE Commun. Mag.*, pp. 94–102, Sept. 2000.
11. S. Chaudhuri, G. Hjálmtýsson, and J. Yates, Control of lightpaths in an optical network, OIF submission OIF2000.04, Jan. 2000 and IETF draft, Feb. 2000.
12. B. Mukherjee, *Optical Communications Networks*, McGraw-Hill, New York, 1997.
13. H. Zang, J.P. Jue, and B. Mukherjee, A review of routing and wavelength assignment approaches for wavelength-routed optical WDM networks, *Optical Networks Mag.*, vol. 1, pp. 47–60, Jan. 2000.

14. S. Ramamurthy and B. Mukherjee, Fixed-alternate routing and wavelength assignment in wavelength routed optical networks, *Proc. IEEE Globecom '98,* vol. 4, pp. 2295–2302, Nov. 1998.
15. A. Mokhtar and M. Azizoglu, Adaptive wavelength routing in all optical networks, *IEEE/ACM Trans. on Networking,* vol. 6, pp. 197–206, April 1998.
16. T. Fabry-Asztlos, N. Bhide, and K. M. Sivalingam, Adaptive weight functions for shortest-path routing algorithms for multi-wavelength optical WDM networks, *Proc. IEEE ICC 2000,* New Orleans, LA, June 2000.
17. I. Chlamtac, A. Ganz, and G. Karmi, Purely optical networks for terabit communication, *Proc. IEEE INFOCOM '89,* Washington, DC, vol. 3, pp. 887–896, April 1989.
18. E. Karasan and E. Ayanoglu, Effects of wavelength routing and selection algorithms on wavelength conversion gain in WDM optical networks, *IEEE/ACM Trans. on Networking,* vol. 6, no. 2, pp. 186–196, April 1998.
19. X. Zhang and C. Qiao, Wavelength assignment for dynamic traffic in multi-fiber WDM networks, *Proc. 7th ICCCN,* pp. 479–485, Oct. 1998.
20. J. Yates, P. Rumsewicz, and J. Lacey, Wavelength converters in dynamically reconfigurable WDM networks, *IEEE Commun. Survey,* Q2, 1999.
21. H. Harai, M. Murata, and H. Miyahara, Performance of all-optical networks with limited range wavelength conversion, *Proc. IEEE ICC,* Montreal, Canada, pp. 416–421, April 1997.
22. M. Ali et al., Routing and wavelength assignment (RWA) with power considerations in all-optical wavelength-routed networks, *Proc. IEEE GLOBECOM '99,* Rio De Janeiro, Brazil, pp. 1443–1437, Dec. 1999.
23. C. Assi, A. Shami, and M. Ali, Impact of wavelength converters on the performance of optical networks, *Optical Network Mag.,* vol. 3, issue 2, March/April 2002.
24. C. Assi, A. Shami, and M. Ali, Optical networking and real-time provisioning: an integrated vision for the next-generation Internet, *IEEE Networks,* pp. 36–45, July/August 2001, vol. 15, no. 4.
25. R. Bahandary Survivable networks: algorithms for diverse routing, Kluwer Academic Publishers, Dordrecht, 1999.
26. B. Doshi et al., Optical network design and restoration, *Bell Labs Tech. J.,* vol. 4, no. 1, Jan.–Mar. 1999, pp. 58–84.
27. G. Mohan and C.S.R. Murthy, Lightpath restoration in WDM optical networks, *IEEE Network,* Nov./Dec. 2000.
28. B. Rajagopalan and D. Saha, Link bundling in optical networks, work in progress, Internet Draft.
29. A. Shami, C. Assi, and M. Ali, Performance evaluation of two GMPLS-based distributed control and management protocols for dynamic lightpath provisioning in future IP networks, IEEE ICC'2, NY, April, 2002.
30. A. Shami, C. Assi, and M. Ali, On the merits of flooding/parallel probing-based signaling algorithms for fast automatic setup and teardown of paths in IP/MPLS-over-optical-networks, *Proc. IEEE GLOBECOM,* San Antonio, Texas, Nov. 2001.
31. C. Assi, Y. Ye, S. Dixit, M. Au, A hybrid distributed fault management protocol for combating single fiber failures in mesh-based DWDM optical networks, *IEEE GLOBECOM,* Taipei, Taiwan, 2002.

chapter eleven

Routing and wavelength assignment with multi-granularity traffic in optical networks

Pin-Han Ho
University of Waterloo
Hussein T. Mouftah
University of Ottawa

Contents

0-8493-1333-3/03/$0.00+$1.50
© 2003 by CRC Press LLC

11.1 Introduction

We propose a novel switching architecture of multi-granularity optical cross-connects (MG-OXCs) for dealing with multi-granularity traffic in the optical domain. MG-OXCs can cooperate with the generalized multiprotocol label switching (MPLS) control plane, which provides the advantages of cost reduction, better scalability in physical size, and unified traffic management. Detailed discussions are provided on the characteristics and implementation issues for the switching architecture. Based on the proposed MG-OXCs, two routing and wavelength assignment (RWA) with tunnel allocation algorithms are presented: dynamic tunnel allocation (DTA) and capacity-balanced static tunnel allocation (CB-STA). In the former, we use fixed alternate routing with k-shortest paths to inspect network resources along each alternate path for dynamically setting up lightpaths. For the latter, fiber- and waveband-tunnels are allocated into networks at the planning stage (or offline) according to weighted network link-state (W-NLS). We will show that, with the proposed algorithms, the RWA problem with tunnel allocation in the optical networks containing MG-OXCs can be solved effectively. Simulation is conducted on networks with different percentages of switching capacity and traffic load. The simulation results demonstrate that DTA is outperformed by CB-STA in the same network environment due to a well-disciplined approach for allocating tunnels with CB-STA. We also find that the mix of the two approaches yields the best performance given the same network environment apparatus.

The emergence of generalized multiprotocol label switching (GMPLS)[1,2] has opened a new era for control and management of the optical Internet with a core technology of dense wavelength division multiplexing (DWDM). GMPLS was extended from MPLS[3] by the IETF (Internet Engineering Task Force), which has collected numerous, valuable ideas from both industry and academia on how to migrate an MPLS-based control plane to the optical domain. Although a label switched path (LSP) and a lightpath are intrinsically different, GMPLS is equipped with the ability of bundling lambda LSPs (L-LSPs), which provision multi-granularity optical flows and provide a suite of mechanisms that assign a generalized label to bundled consecutive wavelengths, such as a waveband LSP (WB-LSP) or a fiber LSP (F-LSP). With this, a hierarchical traffic structure can be achieved, and the pop/push mechanisms defined in the MPLS-based control plane can possibly be migrated to the optical domain, which is conceptually an "aggregation" of traffic with respect to the phase of control and management.

In addition to the functionality expanded in the control plane, the bundling of lightpaths can be implemented in the transport layer to reduce the network cost and node size at the expense of higher control complexity and loss of throughput, which is conceptually a "bypassing" mechanism with respect to some switching functions in nodes. With MG-OXCs, the switching types and traffic granularities are no longer limited to L-LSP in which a single lightpath is switched according its wavelength, but they are also WB-LSP or F-LSP in which a waveband or a fiber of lightpaths are switched as a whole in the optical domain. Therefore, the *tunneling* (i.e., allocation of *tunnels* into networks) is not only limited to a logical programming in the control phase, but is also limited to a physical configuration upon the associated network resources. Here, the *tunnel* is defined as a group of consecutive wavelength channels bundled and switched together, which is either a fiber- or waveband-tunnel containing a fixed number of wavelength channels.

Although the idea of multi-granularity traffic in the optical layer has generated the concept of GMPLS since early 2000, and the Internet draft focusing on this topic first emerged in mid-2001,[7] the associated research and relative simulation-based study can only be seen in research by Gerstel et al. and Noirie et al.,[4,5] which, however, did not conduct a performance study (in Gerstel et al.[4]) or did not consider the grooming of traffic from different source nodes to different destination nodes (in Noirie et al.[5]). This chapter proposes a switching architecture for MG-OXCs along with discussions on the operational principles and various aspects of implementation issues. In addition to a reduction of cost and node size, the architecture of MG-OXCs is also devised to cooperate with the control plane of GMPLS in the optical domain that supports the same routing and signaling instances with IP/MPLS domain. For achieving an efficient use of network resources, a routing algorithm that can perform intermediate grooming to all types of traffic is desired (i.e., the end-to-end traffic, the traffic from a source node to different destination nodes, the traffic from different sources to a destination, and the traffic from different sources to different destinations).

The design objective of the proposed RWA schemes is to achieve load balancing among resources of different switching types, so that the throughput of the whole system can be as close as possible to the case in which only traditional lambda-switched OXCs exist in the network. With this, we propose a suite of algorithms for performing RWA and tunnel allocation, which include DTA with fixed alternate routing (FAR) for allocating tunnels according to dynamically arrived connection requests, and a capacity-balanced static tunnel allocation (CB-STA) scheme with the W-NLS for allocating tunnels offline according to the capacity-balanced characteristic of the network. A simulation-based study is conducted to examine the proposed schemes.

This chapter is organized as follows. Section 11.2 introduces the proposed four-tier (i.e., electronic-switching, lambda-switching, waveband-switching, and fiber-switching) architecture. Section 11.3 presents the network planning algorithms to facilitate RWA with tunnel allocation. Section 11.4 presents the algorithms of DTA and CB-STA. Section 11.5 conducts a simulation on the performance of the proposed algorithms in terms of blocking probability, and Section 11.6 concludes this chapter.

11.2 Multi-granularity optical cross-connects

11.2.1 Traffic hierarchy defined in generalized MPLS

GMPLS can be described in brief as an optical extension to the MPLS-based control plane, which is in its ongoing progress of standardization for supporting optical traffic with multi-granularity and provides generalized labels to traffic flows with different switching types.[1,6,7] In the GMPLS architecture, the labels in the forwarding plane of label switch routers (LSRs) are not only limited to the packet headers and cell boundaries, but also time slots, wavelengths, or physical ports. The arrangement of labels vs. different types of interfaces are summarized as follows:[7]

> For the packet-switched capable (PSC) interface, the label is a 20-bit number.
> For the layer-2 switch capable (L2SC) interface, the label is frame/cell headers.
> For the time-division-multiplexed capable (TDM) interface, the label could be the serial number of the time slot.
> For the wavelength-switched capable (LSC) interface, the label is the wavelength number of the lightpath.
> For the waveband-switched capable (WBSC) interface, the label is the ID of waveband in a local port.
> For the fiber-switched capable (FSC) interface, the label is the local port ID of the fiber.

A connection can only be established between interfaces of the same type, which is termed a G-LSP in the context of GMPLS. Nodes in the optical networks equipped with GMPLS signaling are defined as G-LSRs. A hierarchy can be built if an interface is capable of multiplexing several G-LSPs with the same technology, which is called a nesting of traffic flows. With the same manner, the nesting can also occur between G-LSRs with different interfaces. A list of interface hierarchy from the highest to the lowest level is as follows: FSC, WBSC, LSC, TDM, L2SC, and PSC, as illustrated in Figure 11.1. G-LSP

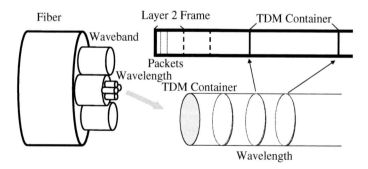

Figure 11.1 Illustration of the traffic hierarchy defined in GMPLS. (P.H. Ho and H.T. Mouftah, Routing and wavelength assignment with multi-granularity traffic in optical networks, IEEE, Aug. 2002. With permission.)

nesting is realized in such a way that a G-LSP that starts and ends on a lower level of interface is nested (together with the other G-LSPs of the same level) into a G-LSP that starts and ends on a higher level of interface.[7] In the optical domain, no difference exists between the processing of LSC, PSC, L2SC, and TDM traffic because they all have to be de-multiplexed into wavelength-level flows in order to be further processed. Thus, we must have at least three classes of granularities, lambda- (or wavelength), waveband-, and fiber-switching, in which several lambda-switching flows can be nested into a waveband-switching flow. In turn, several waveband-switching flows can be nested into a fiber-switching traffic flow.

11.2.2 Functional architecture

The previous description of the control and management of multi-granularity traffic has been documented in some of the most recently reported Internet drafts.[1,2,6,7] The realization of each hierarchy of traffic in the physical layer is well defined, except for the waveband switching. Three alternatives to implement waveband switching in the optical domain have been reported: *inverse multiplexing, wavelength concatenation,* and *physical switching of a waveband.*[7] The first two methods are for bundling L-LSPs into a logical waveband without a realization in the transport plane, and do not consider the fact that a waveband has a physical significance as well as a specific interface. We need a physical waveband that directly switches a spectrum of the wavelengths and is transparent to its inner wavelength individuals.

Figure 11.2 demonstrates the switch architecture of an MG-OXC for manipulating multi-granularity traffic, in which a physical waveband

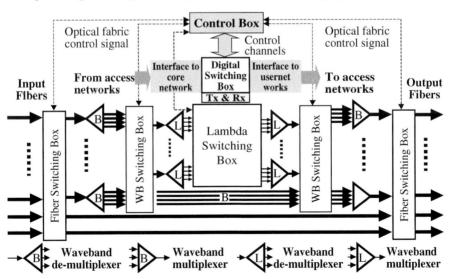

Figure 11.2 Architecture of a node that implements multi-granularity traffic tunneling. (P.H. Ho and H.T. Mouftah, Routing and wavelength assignment with multi-granularity traffic in optical networks, IEEE, Aug. 2002. With permission.)

switching is realized. The fiber- and waveband-switching boxes at the left-hand side play the role of selectors on the input fibers and wavebands. The waveband- and fiber-switching boxes on the right-hand side perform space switching on the traffic that is leaving the node. The input traffic takes one of the three switching types in the optical domain, according to how the switching boxes of selection are configured.

The upper control box is where signaling mechanisms are generated, in which the control packets can be transmitted either out-of-band or in-band. For the out-of-band case, an IP network is built either on a single, dedicated lightpath in one of the fibers or on a single dedicated fiber bundled with the other fibers in the data plane. For the in-band case, the control packets are embedded in the data frame. The functional architecture of MG-OXC we proposed can be applied to each of the three transmission types for signaling mentioned previously.

11.2.3 Implementation issues

The basic operation principle for the control and management of MG-OXCs is such that the lambda-switching channels are used mainly for packet switching, for layer-2 switched and TDM data (which may be the accessing traffic), or for protection or transition between two tunnels, instead of forwarding traffic. On the other hand, traffic bypassing through a node should be bundled into the tunnel as much as possible. The operation principles are listed as follows:

1. A certain percentage of fibers along a direction link are assigned as fiber-switched, waveband-switched, or lambda-switched fibers.
2. In a waveband or fiber-switched fiber, all the wavelength channels in a waveband or a fiber have to be switched together.
3. A connection can take a tunnel (either a waveband or a fiber tunnel) if and only if it takes a path that traverses the *ingress* and *egress nodes* of the tunnel.
4. The connection must be lambda-switched at the source and the destination node of a lightpath.
5. A lightpath may take more than one tunnel. A transition (i.e., a lambda-switching) is required when a lightpath is traversing between any two tunnels so that the other traffic can be grouped or de-grouped as well.

With the MG-OXC architecture, Figure 11.3 illustrates an example of the switching process during the transportation of a lightpath. Two tunnels are depicted in the figure. The first one is a fiber tunnel from node 0 to node 4; and the second one is a waveband tunnel from node 4 to node 6. Note that the selectors determine the switching type of a fiber into a node. The FS selector can determine whether the input fiber is fiber-switched or waveband-/lambda-switched. The WBS selector determines whether the input

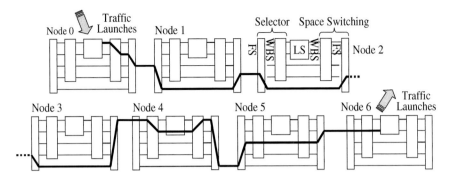

Figure 11.3 Switching process of a lightpath that is sourced at node 0 and destined at node 6. (P.H. Ho and H.T. Mouftah, Routing and wavelength assignment with multi-granularity traffic in optical networks, IEEE, Aug. 2002. With permission.)

fiber is waveband-switched or lambda-switched. On the other hand, the WBS space switch exchanges wavebands between fibers, and FS space switch exchanges a whole fiber of data flow between two fibers. It is clear that the resource of lambda-switching boxes along these nodes is consumed at only node 0, node 4, and node 6.

11.2.4 Advantages/disadvantages of using MG-OXCs

The most notable advantage of using the four-tier MG-OXCs is the cost reduction, which is due to the reduced amount of lambda-switched devices including demultiplexers and switching fabrics. Another important characteristic of adopting the MG-OXCs is that the traffic hierarchy in the optical domain can be achieved, which facilitates the migration of an MPLS-based control plane to the optical domain. This is, however, limited by wavelength continuity constraint because of the intrinsic difference between LSPs and L-LSPs. We have to consider whether an L-LSP is on a right wavelength plane before knowing whether it can be physically switched together with the other L-LSPs in the fiber- or waveband-tunnel. To guarantee the availability of transition between different switching types (e.g., F-LSP $\leftrightarrow$ L-LSP) at ingress and egress nodes of a tunnel, a partial wavelength conversion with a percentage of

$$\frac{\min[F_3, (F_2 + F_1)]}{\Delta} \times 100\%$$

is necessary, where F_1, F_2, and F_3 are treated as the switching capability for L-LSP, WB-LSP, and F-LSP in the node, respectively. On the other hand, if the conversion capability is insufficient, an L-LSP may not be able to be bundled into a tunnel even if the tunnel has residual bandwidth. A similar situation happens when an L-LSP, originally tunneled in an F-LSP/WB-LSP, may not be able to de-multiplex at the egress node of the tunnel to a specific wavelength plane.

The preceding two characteristics are gained at the expense of a decrease in throughput and an increase in the number of network states that are required to track. The constraints on bundling lightpaths in the waveband- and fiber-switched fibers reduce the utilization of these fibers and impair the performance. The computation complexity in path selection is increased as more network states need to be considered.

11.3 Network planning

This section describes the assumptions of this study as well as some planning efforts required by the proposed tunnel allocation and path selection approaches.

11.3.1 Basic assumptions

The study is based on the following assumptions:

Networks are densely meshed with DWDM technology.
Nodes are partially wavelength convertible only for transition between different switching types.
Each link is directional with a fixed number of fibers, each of which contains a fixed number of wavelength planes.

To simplify the analysis, we assume a Poisson model and uniform traffic in this study. For the *kth* S–D pair, each connection request is for a lightpath that occupies a concatenation of wavelength channels along a simple path with an arrival rate λ_m. Once a connection between the *mth* S–D pair is established, it holds the wavelength channels along the selected route for a time period defined by an exponential distribution function with rate 1. We define the load of the *mth* S–D pair $\rho_m \equiv \lambda_m$, which can be derived from historical data. With the traffic model assumed previously, every S–D pair in the network follows the behavior of a multi-server loss system[9] in which the number of servers equals to the number of all possible concatenations of wavelength channels connecting the S–D pair. Without loss of generality, every connection request is for a single wavelength channel.

11.3.2 Weighted network link-state (W-NLS)

We propose a link metric called W-NLS, which is designed to reflect the network topology, location, and potential traffic load of each S–D pair. The design motivation is that, in general, a connection may be built between an S–D pair upon one of a limited number of paths, most of which are the shortest or second shortest paths. In case only the shortest paths are considered, each shortest path connecting between an S–D pair has an equal tendency to be adopted when a connection request arrives at the S–D pair. Due to the fact that the load of a link is the summation of all the S–D pairs that launch traffic onto paths that traverse the link, a parameter (or a weighting) on each link can be defined to show how much traffic the link is likely to afford under a specific load condition during a period of time (e.g., an hour or two). Because we have assumed that the load of the *mth* S–D pair ρ_m can be derived from historical record, the W-NLS for the link *i-j* is defined as:

$$w_{i,j} = \sum_{m=0}^{M} \sum_{r=0}^{J_m} \frac{\rho_m \cdot \alpha_{m,r}}{J_m} \cdot (\delta_{m,r}^{i,j})$$

where ρ_m is the load of *mth* S–D pair, J_m is the number of the prearranged paths between the *mth* S–D pair, $\alpha_{m,r}$ is the weighting on the *rth* route of the *mth* S–D pair, and M is the total number of S–D pairs in the network. The binary parameter $\delta_{m,r}^{i,j}$ is defined as:

$$\delta_{m,r}^{i,j} = \begin{cases} 1 & \text{if route } r \text{ of } S\text{-}D \text{ pair } m \text{ passes link } i\text{-}j \\ 0 & \text{otherwise} \end{cases}$$

The prearranged paths of each S–D pair for calculating $w_{i,j}$ can be all the shortest paths, or all the shortest paths as well as some of the suboptimal paths. $\alpha_{m,r}$ is a weighting on how likely the *rth* prearranged path be taken. In general, the shortest paths between an S–D pair are used much more often than the longer paths; as a result, $\alpha_{m,r}$ is suggested to be a function of hop count of the *rth* path. In this study, only the shortest paths are considered, that is,

$$\alpha_{m,r} = \begin{cases} 1 & \text{if route } r \text{ of } S\text{-}D \text{ pair } m \text{ passes link } i\text{-}j \\ 0 & \text{otherwise} \end{cases}$$

As an example illustrated in Figure 11.4, four shortest paths (assume all the links have equal cost) run between node 3 and node 13. The traffic in the

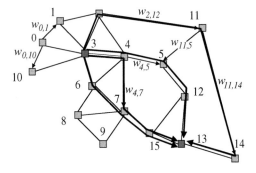

Figure 11.4 An example for deriving the weighted network link state (W-NLS) for the link *i-j*. (P.H. Ho and H.T. Mouftah, Routing and wavelength assignment with multi-granularity traffic in optical networks, IEEE, Aug. 2002. With permission.)

network contributed by the S–D pair (3, 13) is highly likely to load onto the links along the four shortest paths. If no preference occurs on the path selection, each of the four paths could be equally weighted (i.e., $\alpha_{m,r} = 1$ and $J_m = 4$). Applying this process to all S–D pairs in the network, $w_{i,j}$ is derived by a linear combination of the traffic load from all the S–D pairs that are likely to traverse it.

11.4 Solving RWA with tunnel allocation

11.4.1 Introduction

This section presents two approaches for performing dynamic RWA with multi-granularity traffic: DTA and CB-STA. DTA is used to allocate tunnels dynamically according to the arrival of connection requests. The dynamic selection of a lightpath needs to consider both the existing tunnels that can be possibly taken by the lightpath as well as any possibility of further establishing a tunnel for use. The design originality of DTA emphasizes the ability of swiftly adapting to traffic changes by setting up and tearing down tunnels according to real-time traffic.

The second scheme, CB-STA, is used to allocate tunnels in an offline manner according to W-NLS for balancing the potential traffic load. Dijkstra's algorithm is performed for each connection request on the network topology, taking the prescheduled tunnels into consideration. With CB-STA, tunnels are well disciplined with their location and length at the expense of less dynamicity. Because it is a pure waste in the tunnel layers without being covered by any tunnel, CB-STA needs to reduce the network resources uncovered by a tunnel as much as possible.

We will also investigate a combination of the previous two approaches (termed "mixed" in the following context), in which the network is deployed with tunnels at the planning stage that will never be torn town, while dynamic allocation of tunnels is allowed and performed while the network is running.

11.4.2 Dynamic tunnel allocation with path selection

The algorithm for the implementation of DTA with path selection is described as follows:

1. Derive a combination of feasible tunnels along an alternate path under inspection (including the existing tunnels and establishable tunnels), such that a lightpath can be built from the source to the destination.
2. Try all the feasible alternate paths and derive an optimal combination of tunnels along a path. The first-fit wavelength assignment scheme is adopted.

The grooming strategy for dynamic allocation is listed as follows:

1. Priority: use of an existing fiber tunnel > use of an existing band tunnel > newly creating a fiber tunnel > newly creating a waveband tunnel > use of a lambda-switched channel.
2. If a tunnel is registered as "long life," it will never be torn down no matter if any lightpath is using the tunnel; otherwise, it is torn down whenever a lightpath is not traversing through it.

To increase the utilization of fibers dedicated for tunnels, the link cost with different switching types should be such that $c(lambda) \gg c(waveband) > c(fiber)$. The cost of the t-th tunnel along the ith alternate path between the mth S–D pair is:

$$CT^i_{t,m} = h^i_{t,m} \cdot c(S)$$

where $h^i_{t,m}$ is the hop count of the t-th tunnel along the ith alternate path, and S is the switching type, which could be either *lambda*, *waveband*, or *fiber*. Note that a tunnel is assumed to be nonwavelength-specific by giving enough wavelength conversion capability at the ingress and egress nodes of the tunnel. Therefore, the cost of the ith path is:

$$CP^i_m = \sum_t CT^i_{t,m} + (H^i_m - \sum_t h^i_{t,m}) \cdot c(lambda)$$

where H^i_m is the hop count of the ith alternate path between the mth S–D pair. The final decision is made by taking a lightpath on the first available wavelength plane along the ith alternate path with the smallest CP^i_m, where $i = 0 \sim I_m$, which is the number of alternate paths for the mth S–D pair.

Figure 11.5 is an example of allocating tunnels along an alternate path with the source at node 0 and the destination at node 6. Five fibers exist along a directional link where $F_1 = 1$, $F_2 = 2$, and $F_3 = 2$. The optical traffic

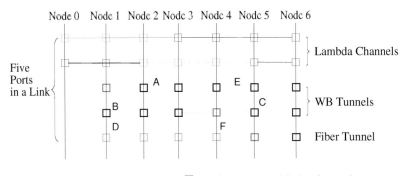

Figure 11.5 An example of dynamic tunnel allocation. (P.H. Ho and H.T. Mouftah, Routing and wavelength assignment with multi-granularity traffic in optical networks, IEEE, Aug. 2002. With permission.)

launches on the transport layer via lambda switching at node 0. The only fiber for it is the second one. For allocating a feasible combination of tunnels, consider the four existing tunnels (A, B, C, and D) and 2 establishable tunnels (E and F). The optimal tunnel allocation for this particular path can be one out of the $2^6 = 64$ combinations. Fortunately, some of the combinations may not be feasible. For example, the tunnel A and B should not be considered at the same time because they overlap each other. Another example is that the establishable tunnel F is exclusive to the fiber tunnel D. After trying all the possible combinations of tunnels and wavelength channels, the optimal solution can be derived. In this case, the optimal lightpath goes from the second fiber at node 0, joins fiber tunnel D, then goes out of D at node 4, and joins waveband tunnel C.

The dynamic tunnel allocation is exponential in complexity because the number of possible combinations of tunnel allocation is proportional to 2^n, where n is the number of feasible tunnels along an alternate path. If the number of fibers in a link and the size of the network are small, derivation of the optimal combination of tunnels can be conducted by selecting the best among all possible solutions. If the network is large, however, the number of feasible tunnels residing along an alternate path can also be large, which may result in a huge amount of computation time. In this case, a heuristic is needed to get a suboptimal solution so that the computation efficiency is improved. For this purpose, we propose a greedy heuristic that keeps the longest and most valuable tunnels first and discards the other tunnels that conflict. The selection of the *t-th* tunnel along the *ith* alternate path between the *mth* S–D pair is based on another cost function:

$$Ct^i_{t,m} = (H^i_m - h^i_{t,m}) \cdot c(S)$$

The flowchart describing the algorithm is depicted in Figure 11.6.

11.4.3 *Capacity-balanced static tunnel allocation (CB-STA)*

CB-STA is designed to allocate tunnels into networks at the planning stage (or offline). CB-STA is different from DTA because the tunnels built in *a priori* will be marked as "long life" and will never be torn down while the network is running. The task of allocating tunnels into tunnel layers can be subdivided into two subtasks:

1. Determine a tunnel ingress–egress (I–E) pair for each tunnel based on W-NLS.
2. Route for each tunnel I–E pair according to the information in task 1.

For task 1, we define the *potential* and *sink* of a node, P_n, and SI_n, for quantifying how likely the *nth* node initiates and terminates lightpaths:

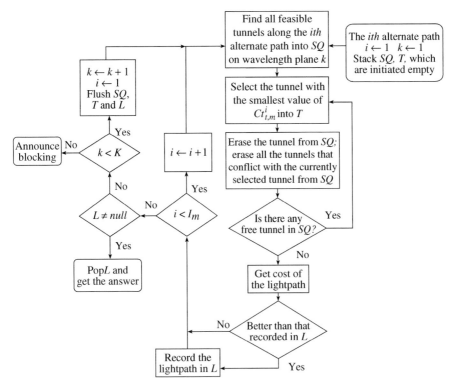

Figure 11.6 The flowchart for describing the dynamic tunnel allocation algorithm. In each iteration, the wavelength has to be specified because the concatenation of lambda-switched channels subjects to the wavelength continuity constraint. (P.H. Ho and H.T. Mouftah, Routing and wavelength assignment with multi-granularity traffic in optical networks, IEEE, Aug. 2002. With permission.)

$$P_n = \sum_{r \in \tilde{n}} w_{n,r} \qquad \text{for all } n$$

$$SI_n = \sum_{r \in \tilde{n}} w_{r,n} \qquad \text{for all } n$$

where $w_{n,r}$ is the W-NLS between node n and node r; $\tilde{n}$ is defined as the node set containing all the nodes adjacent to node n. The node with the largest *potential* has the highest priority to be ingress of a tunnel, and the node with the largest *sink* has the highest priority to be egress of a tunnel. After a tunnel is set up between an I–E pair, an adjustment is made in such a way that the *potential* of the ingress and its neighbor nodes are decreased by $\delta(S)$ and $(\delta(S)) / (nd)$, and the *sink* of the egress and its neighbor nodes, on the other hand, are decreased by $\delta(S)$ and $(\delta(S)) / (nd)$. In the preceding description, the deviation $\delta(S)$ is a function of the switching type:

$$\delta(fiber) = \delta(waveband) \times B$$

where B is the number of waveband in a fiber, and nd is the nodal degree of the ingress or egress node of a tunnel.

Another constraint is made on the hop count of the tunnels, D, which is set to the minimum integer that is larger than the average distance of paths between each S–D pair in the network. The number of the tunnel I–E pairs generated at the beginning is the maximum number of tunnels that can be accommodated. For example, for a network with $|L|$ directional links (L is the set of all directional links), the upper limit on the number of tunnels in switching layer S are

$$NO_{upper}(S) = \frac{|L| \cdot SL(S) \cdot NF(S)}{D}$$

where $SL(S)$ is the number of *sub-layers* for a layer of switching type S, and $NF(S)$ is the number of fibers dedicated to the switching layer S. In case F_1 fibers for fiber-switching and F_2 fibers for waveband-switching exist, we get $SL(fiber) = 1$ and $SL(waveband) = B$; $NF(fiber) = F_1$ and $NF(waveband) = F_2$. After the algorithm generates all fiber tunnels, the values of *potential* and *sink* of each node (i.e., P_n and SI_n) are recorded and used for the generation of waveband tunnels. With this knowledge, the deviation $\delta(S)$ is customized as

$$\delta(S) = \frac{\overline{P} + \overline{SI}}{NO_{upper}(fiber) + NO_{upper}(waveband)/B}$$

where $\overline{P}$ and $\overline{SI}$ are the average values of *potential* and *sink* at the beginning of tunnel allocation. Because each sublayer is independent from the others for waveband tunnels, the upper bound on the number of tunnels to be inspected is

$$\frac{|L|}{D} \cdot (F_1 + B \cdot F_2)$$

The completion of task 1 generates a series of I–E pairs sorted according to their importance to network load balancing (we term the I–E pairs as "series" in the following context). The series of I–E pairs may overload the network because the number of tunnels in the series is an upper limit by estimation of NO_{upper}. The problem can be formulated into Integer Programming (InP) in which the algorithm allocates as many tunnels as possible into the network. We assume that every node pair is prepared with all the shortest paths to every other node.

Target: maximize K, subject to the following constraints:

$$\sum_{s=1}^{NS_k} x_{k,s} = 1 \quad for \quad k = 1 \sim K \tag{11.1}$$

$$\sum_{s=1}^{NS_k} x_{k,s} = 0 \quad for \quad k = (K+1) \sim NO_{upper}(S) \tag{11.2}$$

$$NF(S) \geq \sum_{k=1}^{K} \sum_{s=1}^{NS_k} x_{k,s} \cdot \zeta_{k,s}^l \quad for \quad l \in L \tag{11.3}$$

where K is the number of tunnels from the top of the series to be allocated; NS_k is the number of shortest paths from which the kth I–E pair can choose; $x_{k,s}$ is a binary decision index, which yields a "1" if the sth shortest path of the kth I–E pair is taken and a "0" otherwise; $\zeta_{k,s}^l$ is a binary decision index, which yields a "1" if the sth shortest path of the kth I–E pair traverses span l and a "0" otherwise.

In the preceding formulation, Equation (11.1) limits the number of shortest paths taken by an I–E pair in the series to one. Note that more than one tunnel may be established between a node pair, therefore, the number of I–E pairs dedicated to the same node pair may be larger than one. Equation (11.2) indicates that the I–E pairs in the series, with a sequence number larger than K, will not be allocated with a tunnel. Equation (11.3) puts a constraint on the capacity along each directional link.

The preceding formulation is subject to an NP-hard computation complexity. Heuristics must be used to trade performance with computation complexity. I–E pairs are simply allocated with a tunnel sequentially instead of being optimized as a whole. The first-fit scheme[8] is adopted by assigning the first available sub-layer of the switching layer to a tunnel. Different from the InP formulation, the algorithm skips any blocked I–E pair, and routes for the next I–E pair in the series. The algorithm continues until all the I–E pairs are tried. The computation complexity becomes

$$O(\frac{|L|}{D} \cdot (F_1 + B \cdot F_2) \cdot N^2)$$

if no special heap data structure is adopted in performing Dijkstra's algorithm.

After the network is allocated with tunnels, a *makeup process* is conducted to fill each fiber-switching and waveband-switching layer, with tunnels as full as possible. An inspection is made at a node along all the shortest paths to all the other nodes to see if any possibility exists to set up more tunnels. The inspection has to be based on the database of keeping the residual resources in a switching layer and following the sequence according to the

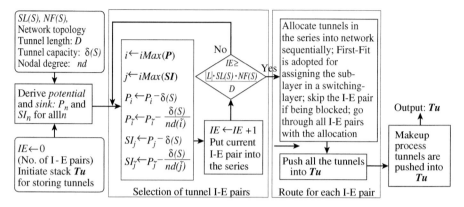

Figure 11.7 Flowchart for summarizing the algorithm of the static tunnel allocation as a strategy of network planning. Assume that the network has N nodes. We define $iMax(Q_i)$, which returns the indexes I, such that $Q_I = \max_{i=1\sim N}(Q_i)$. IE is an index to count the number of tunnels in the series. (P.H. Ho and H.T. Mouftah, Routing and wavelength assignment with multi-granularity traffic in optical networks, IEEE, Aug. 2002. With permission.)

value of P_n and SI_n (node-pairs with larger values of $P_n + SI_n$ are prioritized). The database is updated whenever a makeup tunnel is allocated. The makeup mechanism aims to increase the link utilization in the fiber- and waveband-switching layers. Note that the links that are not traversed by any tunnel in the two layers are totally wasted. The makeup process can be iterated through all the network nodes of each layer until no tunnel can be built, with which 100% of tunnel coverage can be achieved. The flowchart of the whole algorithm is presented in Figure 11.7.

The cost matrix M that takes tunnels into consideration can be derived with the following pseudo code:

Input: W: the cost matrix without considering tunnels; Tu: all the fiber- and waveband-switched tunnels. $|Tu|$: number of elements in Tu. CF, CB, and CL: the unit cost for a fiber-, waveband-, and lambda-switched channel, respectively, which are defined as 1, B, and K in this study.

$M = W$;

For $i = 1$ to N **Do**

For $t = 1$ to $|Tu|$ **Do**

If (node i is the ingress of the tunnel $Tu[t]$ with an egress node e) **Then**

If $(M[i][e] == \infty)$ **Then** /* if node i and e are not adjacent; */

If $(Tu[t]$ **is a fiber tunnel**)

Then

$$M[i][e] = \frac{CF \cdot (length(i,e) - 1) + CL}{B \cdot Num(Tu[t], e, fiber) \cdot f_{i,e} + Num(Tu[t], e, waveband) \cdot b_{i,e}}$$

/* for a fiber tunnel */

Else

$$M[i][e] = \frac{CB \cdot (length(i,e) - 1) + CL}{B \cdot Num(Tu[t], e, fiber) \cdot f_{i,e} + Num(Tu[t], e, waveband) \cdot b_{i,e}}$$

/* for a waveband tunnel */
End If;
End If;
End For;
End For;
Return *M*;
Output: The reconfigured cost matrix *M* that considers tunnels.

In the preceding pseudocode the function *length(i,e)* and *Num(Tu[t], i, e, S)* returns the minimum distance and the number of tunnels with a switching type *S* from node *i* to node *e*, respectively; $f_{i,e}$ and $b_{i,e}$ are the number of free wavelength channels in the fiber and waveband tunnels connecting between *i* and *e*, respectively, which are dynamic link metrics and need to be updated whenever there is any lightpath built up or torn down. Figure 11.8 illustrates the approach for deriving the cost defined between two non-adjacent nodes in the network. The cost between the two nodes, *M[A][B]*, is a parallel effect of all the fiber and waveband tunnels between them, which is dynamically changed with traffic distribution.

A tunnel is brought up if and only if both of the following resources are available: the lambda-switching capability between the tunnel ingress to the second node, and any free channel along the tunnel from the second node to the tunnel egress. An example is presented in Figure 11.9a. A connection request for a lightpath from G to A arrives. A fiber tunnel from F to A is available to the connection request if the lambda-switched capacity from F to C and any one of the free channels along the fiber tunnel are available. The virtual topology is presented in Figure 11.9b, where the tunnel is expressed as an extra link from F to A at a much cheaper cost. Whenever the tunnel is taken, a lambda-switched channel from F and C is also taken

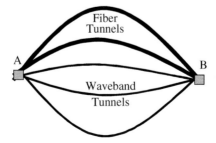

Figure 11.8 An example illustrating the cost between two nodes connected by tunnels. (P.H. Ho and H.T. Mouftah, Routing and wavelength assignment with multi-granularity traffic in optical networks, IEEE, Aug. 2002. With permission.)

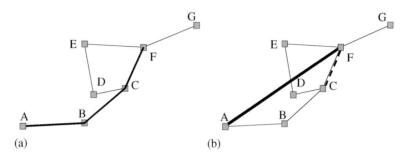

Figure 11.9 An example for the policy of availability announcement during the static tunnel allocation process; (a) physical connection of a tunnel; (b) virtual connection recorded in the cost matrix *M*. (P.H. Ho and H.T. Mouftah, Routing and wavelength assignment with multi-granularity traffic in optical networks, IEEE, Aug. 2002. With permission.)

and updated in the link-state database. If there is only a single lambda-switched channel available along the link F–C, both the tunnel from F to A and the lambda-switching capacity between F and C can be announced available without a conflict because the lightpath can never be built in such a way as (G,F,C,D,E,F,C,B,A), which traverses link F–C twice.

    After deriving the cost matrix *M*, Dijkstra's algorithm is performed on each wavelength plane for each connection request, based on the modified network topology and link-state that takes the tunnels into account. The lightpath with the least cost among all the wavelength planes will be taken. If two or more optimal lightpaths are present after inspecting all the wavelength planes, the one with smaller sequence number of wavelength planes is taken.

11.5 Simulations

Simulation is conducted for an evaluation of the performance for the proposed algorithms on the 16-node network of Figure 11.4 and a 22-node network in Figure 11.10. We assume that the number of wavelengths in a fiber is 64, and the number of fibers along a directional link is 5. A waveband-switched fiber has 8 wavebands: the 1st to 8th lambdas are in the first waveband; the 9th to 16th lambdas are in the second waveband; and, subsequently, the 57th to 64th lambdas are in the 8th waveband. We assume that every node has the same percentage of capacity for each switching type (i.e., the switching capacity dedicated to a specific switching type is the same from node to node).

    For node pair (i,j), we define the traffic load $\rho_{i,j} = \bar{\rho} \cdot \varsigma$, where $\bar{\rho}$ is the *average load*, and ς is a random number in the range of (0.5, 1.5). In this simulation, with different $\bar{\rho}$, the blocking probability of each network with different percentages of switching type is observed. Each node is assumed to have sufficient wavelength conversion capability for transition between different switching types. Except for the transition purpose, the wavelength continuity constraint is held.

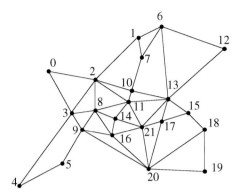

Figure 11.10 The 22-node EUPAN network for this simulation. (P.H. Ho and H.T. Mouftah, Routing and wavelength assignment with multi-granularity traffic in optical networks, IEEE, Aug. 2002. With permission.)

In addition to using purely DTA and CB-STA for performing RWA, a mix of the two schemes is also examined, which is abbreviated as *mixed*. In this case, the networks are planed with "long life" tunnels scheduled by a modified CB-STA algorithm, in which the *makeup tunnels* are not adopted. Instead, the DTA algorithm is performed on the network link-state with "long life" tunnels. All the shortest, second shortest, and third shortest paths between an S–D pair are inspected for a connection request.

Figure 11.11 presents the simulation results. We abbreviate the experiment with F_1 fibers for fiber-switching, F_2 fibers for waveband-switching, and F_3 fibers for lambda-switching as $(F_1)L(F_2)B(F_3)F$. The following four combinations of switching type are examined: 3L1B1F, 2L2B1F, 1L1B3F, and 1L3B1F. *CB-STA w/o* is an abbreviation for the CB-STA case where the makeup tunnels are *not* adopted. Different from *mixed*, *CB-STA w/o* uses Dijkstra's algorithm to find a lightpath and does not allow a dynamic allocation of tunnels. *ADR-5L* stands for the case using adaptive dynamic routing (ADR) with lambda-switched OXC's in the same network topology. Instead of using a threshold to determine the availability of a link adopted in Spath,[10] this simulation uses the cost function for link a–b on wavelength k:

$$C_{a,b}^k = \frac{TB_{a,b}^k}{RB_{a,b}^k}$$

where $TB_{a,b}^k$ and $RB_{a,b}^k$ are the total and residual bandwidth along link a–b on wavelength k. After calculating all the wavelength planes, the lightpath with least cost is selected. With ADR, as a connection request arrives, Dijkstra's shortest path first algorithm is invoked with the link-state derived by the preceding cost function. The number of connection requests for each data is 60,000.

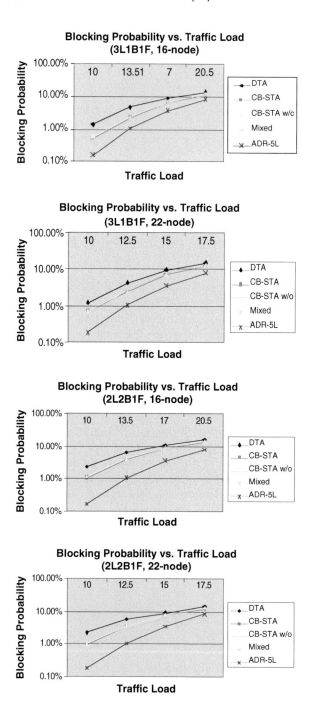

Figure 11.11 Blocking probability vs. traffic load with different percentages of switching types in the 16-node and 22-node networks. CB-STA w/o is a CB-STA tunnel allocation without the makeup process.   (continued)

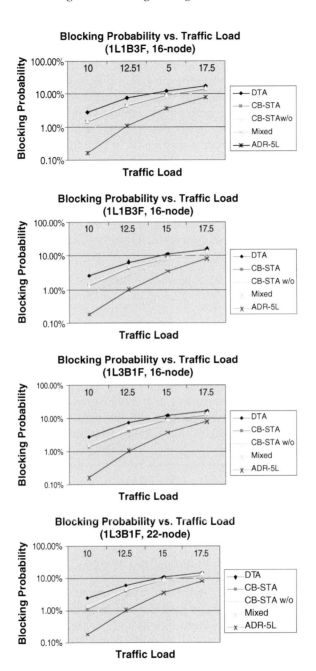

Figure 11.11 (continued) Blocking probability vs. traffic load with different percentages of switching types in the 16-node and 22-node networks. CB-STA w/o is a CB-STA tunnel allocation without the makeup process. (P.H. Ho and H.T. Mouftah, Routing and wavelength assignment with multi-granularity traffic in optical networks, IEEE, Aug. 2002. With permission.)

From the simulation results, it is clear that ADR-5L yields the best performance because the constraint on the resource utilization in the tunnel layers is relaxed. It can also be observed that with the percentage of lambda switching increased, the blocking probability is reduced; with the makeup tunnels, CB-STA has gained only a small amount of performance improvement; and with the same amount of lambda-switching capability, performance is impaired with more fiber-switching capability. In addition, *mixed* yields the better performance, especially when the percentage of tunnel layers is getting larger.

Table 11.1 lists some statistics of resource utilization in each scheme. *UT* (link utilization) is defined as the average percentage of wavelength channels in a specific switching layer to be occupied by a lightpath. *PC* (percentage of coverage) is the average percentage of network resources in a tunnel layer to be covered by tunnels so that the links are utilizable. We also define

$$EL = \sum_{v}^{V} \frac{hop_v - hop_{v,min}}{hop_{v,min} \cdot V}$$

(ratio of extra length), where V is the total number of connections in the network, hop_v is the length of the vth connection, $hop_{v,min}$ is the minimum distance of the S–D pair for the vth connection. *EL* measures the extra percentage of network resource consumed by a single connection request. A larger *EL* impairs performance because more network resources (i.e., hops) are consumed by a single connection. All the data are measured and averaged at the arrival of the 20,000th, 30,000th, 40,000th, 50,000th, and 60,000th connection requests.

The DTA scheme is outperformed by all the other schemes at all traffic load conditions, percentages of switching types, and network topology adopted in this simulation. The reasons for the inferiority of DTA are in the following two aspects. First, the tunnels in the dynamic scheme are established according to the random arrival of connection requests. As a result, the tunnels may be inappropriately long or short such that fragmentation is easier to occur. The short tunnels (e.g., tunnels with hop count of one) generated in DTA are hard to be erased because the shorter a tunnel is, the easier it can be fitted into a lightpath. The short tunnels may fragment the network resources and prevent long tunnels from being established. Therefore, the link utilization in the tunnel layers is reduced, and, as a result, impairs the performance. Table 11.1 demonstrates that, with the DTA scheme, a smaller percentage of network resources are covered with tunnels than when other schemes are used. Around 93% of capacity in the tunnel layers is covered with tunnels when using CB-STA scheme, while the DTA yields an average of around 80% of coverage. In other words, an average of 20% of network resources in the tunnel layer are totally wasted.

Second, the static scheme allows a path with loops to be set up while the dynamic one does not. An example is presented in Figure 11.9. With

Table 11.1 Statistical Data in the Simulation

Switching type	DTA			CB-STA			Mixed			ADR-5L
	Fiber	WB	Lambda	Fiber	WB	Lambda	Fiber	WB	Lambda	Lambda
UT (heavy load)	63.5%	65.3%	71.1%	68.3%	69.2%	70.3%	69.8%	70.1%	70.8%	73.4%
UT (light load)	54.5%	56.8%	59.3%	57.1%	59.6%	61.8%	63.0%	64.6%	63.9%	61.3%
EL (heavy load)	11.5%	11.5%	11.5%	12.8%	12.8%	12.8%	11.3%	11.3%	11.3%	6.1%
EL (light load)	8.2%	8.2%	8.2%	9.1%	9.1%	9.1%	8.5%	8.5%	8.5%	3.9%
PC (average)	78.6%	80.8%	N/A	100%[a] 85.9%[b]	100%[a] 87.3%[b]	N/A	92.7%	93.4%	N/A	N/A

[a] With the tunnel makeup process.

[b] Without the tunnel makeup process.

Note: "heavy load" is defined as the data with $\bar{\rho}$ = 17 or 20.5 in the 16-node network and the data with $\bar{\rho}$ = 20 or 24 in the 22-node network; "light load" is defined as the data with $\bar{\rho}$ = 10 or 13.5 in the 16-node network and the data with $\bar{\rho}$ = 12 or 16 in the 22-node network.

Source: P.H. Ho and H.T. Mouftah, Routing and wavelength assignment with multi-granularity traffic in optical networks, IEEE, Aug. 2002. With permission.

DTA, the connection request that arises at B or C can never take the fiber tunnel. On the other hand, with CB-STA or *mixed*, a lightpath may take a lambda-switched channel and a tunneled channel along the same physical edge. In Figure 11.9, a connection request arriving at node B to node C can take a lambda-switched channel on link (B–A), then get on the fiber tunnel (A,B,C,F) to node F, and then take a lambda-switching channel on link (F–C) to node C. Therefore, the use of tunneled network link-state in CB-STA and *mixed* makes the utilization of tunnels more flexible.

Mixed outperforms CB-STA by having smaller *EL* and larger value of *UT.* The dynamical establishment of tunnels in *mixed* instead of the randomly making up the tunnel layers facilitates the adaptation to traffic variation. *Mixed* keeps some tunnels with well-disciplined characteristics (i.e., load balancing and length limitation) as "long life" while allowing dynamically establishing/tearing down some others, which yields the best performance. We also observe in Figure 11.11 that the randomly generated makeup tunnels can improve the performance only to a very limited extent.

11.6 Conclusions

This chapter proposed a switching architecture of four-tier multi-granularity optical cross-connects (MG-OXCs) in dealing with multi-granularity traffic, with which fiber-, waveband-, and lambda-switching are supported in the optical domain. The use of MG-OXCs can bring us the advantage of cost reduction along with the disadvantage of more control complexity. We proposed two algorithms for solving the RWA problem with tunnel allocation; namely dynamic tunnel allocation (DTA) with path selection, and capacity-balanced static tunnel allocation (CB-STA). DTA allocates tunnels into networks according to dynamically arrived connection requests, which can adapt to traffic variation swiftly but may not guarantee that each tunnel is at a proper location and with a proper length. On the other hand, CB-STA achieves capacity balancing by allocating "long life" tunnels into networks at the planning stage (or offline), based on weighted network link-state (W-NLS), which allocates tunnels with fixed length at proper locations in the network. Furthermore, we devised a strategy that is a combination of the previous two (or the scheme *mixed*), in which networks are deployed with "long-life" tunnels determined by CB-STA, while a dynamic setup of tunnels is allowed using DTA. Extensive simulation was conducted for the proposed schemes in networks with different percentages of switching type and traffic load. The simulation results show that, with the same experimental apparatus and assumptions, DTA is outperformed by the other schemes due to its unscheduled way of allocation that may yield short or improperly positioned tunnels. Statistics are provided with detailed discussions for investigating the characteristics of each scheme. We conclude in this simulation that the approach of the mix of DTA and CB-STA yields the best performance because it can meet both requirements of capacity balancing and adaptation to traffic variation.

References

1. E. Mannie et al., Generalized multi-protocol label switching (GMPLS) archi-tecture, Internet draft, <draft-ietf-ccamp-gmpls-architecture-03.txt>, work in progress, Aug. 2002.
2. D. Cheng, GMPLS extensions for dynamic trunking, Internet draft, <draft-cheng-gmpls-dynamic-trunking-00.txt>, work in progress, Dec. 2001.
3. E. Rosen, A. Viswanathan, and R. Callon, Multi-protocol label switching architecture, *Request for Comment 3031*, Jan. 2001.
4. O. Gerstel, R. Ramaswami, and W.-K. Wang, Making use of a two-stage multiplexing scheme in a WDM network, *Proc. Optical Fiber Commun. (OFC.'00)*, Baltimore, MD, March 2000, ThD1.1-ThD1.3.
5. L. Noirie, M. Vigoureus, and E. Dotaro, Impact of intermediate traffic group-ing on the dimensioning of multi-granularity optical networks, *Proc. Optical Fiber Commun. (OFC '01)*, Anaheim, CA, March 2001, TuG3.1-TuG3.3.
6. R. Douville, D. Papadimitriou, L. Ciavaglia, M. Vigoureux, and E. Dotaro, Extensions to GMPLS for waveband switching, Internet draft, <draft-dou-ville-ccamp-gmpls-waveband-extensions-02.txt>, work in progress, Oct., 2002.
7. E. Dotaro, D. Papadimitriou, L. Noirie, M. Vigoureux, and L. Ciavaglia, Op-tical multi-granularity architecture framework, Internet draft, <draft-dota-ro-ipo-multi-granularity-00.txt>, work in progress, July 2001.
8. H. Zang, J.P. Jue, and B. Mukherjee, A review of routing and wavelength assignment approaches for wavelength-routed optical WDM networks, *Op-tical Networks Mag., SPIE*, vol. 1, no. 1, pp. 47–60, Jan. 2000.
9. S.M. Ross, *Introduction to Probability Models*, 7th ed., Academic Press, New York, 2000.
10. J. Spath, Resource allocation for dynamic routing in WDM networks, *Proc. All-Optical Networking 1999: Architecture, Control, and Management Issues*, pp. 235–246, Boston, MA, Sept. 1999.

chapter twelve

Adaptive routing and wavelength assignment in all-optical networks: the role of wavelength conversion and virtual circuit deflection

Emmanuel A. Varvarigos
University of Patras
Theodora Varvarigou
National Technical University of Athens
Evangelos Verentziotis
National Technical University of Athens

Contents

Session establishment in an all-optical network involves two kinds of decisions:

1. The selection of a route, or sequence of hops, that the session must traverse
2. For each hop along the route and according to the wavelength conversion capability of the corresponding switching node, the selection of a wavelength on which the session will be carried for that hop

This chapter describes schemes for adaptive routing and for adaptive wavelength assignment in an all-optical network, and examines the improvements these two kinds of adaptivity can offer on performance. For the specific adaptive routing scheme that we examine, we demonstrate that, for the hypercube and torus topologies considered, providing (at most) one alternate link at every hop gives a per-wavelength throughput that is close to that achieved by oblivious routing with twice the number of wavelengths per link. Also, we examine the effect of limited wavelength conversion in network performance and find that limited conversion to only one or two adjacent wavelengths can provide a considerable fraction of the improvement that full-wavelength conversion provides over no-wavelength conversion. These results clearly emphasize the need for network designers to investigate the tradeoffs between wavelength conversion, routing flexibility, and hardware cost when designing future optical networks.

12.1 Introduction

The recent advances in fiber-optic technology are strongly affecting our everyday life and habits leading the way to the "information technology society." The ever-increasing demand for bandwidth dictates imperatively the use of techniques and protocols that can optimally exploit the fiber's potential capabilities. One of the most promising technologies in this direction is *wavelength division multiplexing* (WDM), which is the current favorite multiplexing technology for long-haul communications in optical communication networks.[1] WDM divides the huge fiber's bandwidth (about 50Tbps) into many non-overlapping WDM channels, each corresponding to a different wavelength. With each WDM channel assigned to a different communication channel operating at (potentially) peak electronic rate (e.g., 40Gbps), WDM manages to accommodate multiple communication channels from different users (with dissimilar data formats) in parallel using the same fiber.

Emerging wide-area-networks that employ WDM are capable of switching data entirely in the optical domain by means of optical wavelength routing switches (WRSs) or wavelength selective cross-connects (WSXCs). In this way, all-optical connections, or *lightpaths*, which may span multiple fiber links, can be established across a network without undergoing any intermediate optical-to-electronic-to-optical (O-E-O) conversion. Figure 12.1 illustrates a wavelength-routed optical WDM network consisting of a set of

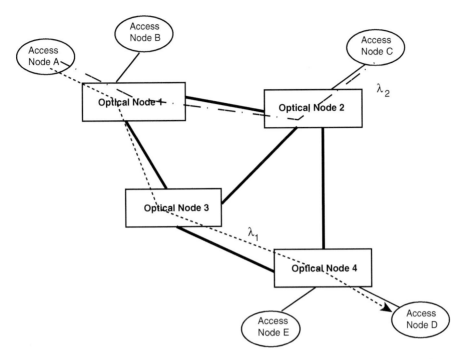

Figure 12.1 A wavelength-routed optical WDM network including optical switching nodes and access nodes. Lightpaths are established between pairs of access nodes on different wavelengths.

nodes and a set of links. The network nodes can be either optical switching nodes or access nodes, while the links that join the nodes can be either optical backbone links (connecting optical switching nodes, set in bold in Figure 12.1) or access links (connecting the access nodes to the optical network switches). The access nodes provide the necessary electronic-to-optical (and vice versa) conversion to interface the optical network to the conventional networks; they are connected to the optical network switches at a specific input fiber and wavelength. If a different wavelength is required, then wavelength conversion has to be used; the same is true for the traffic absorbed by an access station. Lightpaths can be established among pairs of access nodes on different wavelengths.

A *lightlink* between two switches is a specific wavelength within a specific link. A lightpath in an optical network is considered as a sequence of unidirectional *lightlinks* from a starting node s to a terminating node d. A lightpath consists of a specific wavelength of the input fiber in switch s, a specific wavelength of the output fiber in d, and the intermediate lightlinks that are used to connect the intermediate switches. The *physical length* of a lightlink is the length of the corresponding fiber. The *amplifier length* of a lightlink is the number of amplifiers on the lightlink. The physical and the amplifier length of a lightpath are defined as the sums of the corresponding physical and the amplifier lengths of the lightlinks that comprise it. Dispersion and other factors pose an upper limit on the maximum allowable physical length of a lightpath. An upper limit also exists on the amplifier length of a lightpath (before undergoing O-E-O conversion). Since the amplifier length of a lightpath is usually proportional to its physical length (amplifiers are usually placed at equal distance), the two types of constraints can often be treated as one.

12.1.1 Wavelength conversion

An optical switching device can be equipped, apart from amplifiers, with wavelength-conversion facilities, which enable switching data from a wavelength λ_i on an incoming link to a wavelength λ_j on an outgoing link. This conversion may be required for a connection arriving from some optical node that is switched to some other optical node (continuing connection), for a connection arriving from an access node (originating connection), or for a connection switched to an access node (terminating connection). In this context, we distinguish three classes of wavelength routing switches:

1. Switches with full-wavelength conversion capability, which can switch an incoming wavelength to *any* outgoing wavelength
2. Switches with limited-wavelength conversion capability, which can switch an incoming wavelength to a *subset* of the outgoing wavelengths
3. Switches with no-wavelength conversion capability, which can switch an incoming wavelength *only* to the same outgoing wavelength

In the last case, a lightpath is required to occupy the same wavelength on each fiber link along its path in the network, a restriction known as the *wavelength-continuity constraint*, which increases the probability of call blocking. At each node, there is an upper limit on the number of O-E-O conversions that can be performed at that node. If wavelength conversion is also performed using O-E-O conversion (as opposed to using all-optical wavelength converters), then there is a single constraint for the total number of O-E-O conversions and wavelength conversions that can be performed at that node.

Full-wavelength conversion has been extensively studied and, although it has been shown to dramatically decrease the blocking probability and improve network performance,[7,8,11,12,13] it is not often feasible to provide such a capability to all the optical switches of a network, due to technological and financial limitations; wavelength conversion technology is still immature and expensive. These restrictions led researchers to some more practical alternatives: *sparse* and *limited* wavelength conversion. In sparse conversion, only *a few* of the switches have (full) conversion capabilities and, in this case, the objective is to minimize the number of such switches. Subramaniam et al. demonstrated that no significant degradation occurs in network performance when sparse — instead of full — wavelength conversion is employed.[19] In limited conversion, on the other hand, each switch provides wavelength conversion but with limited capabilities. For example, in *k-adjacent wavelength switching*, an incoming wavelength can be translated only to a subset (called the *feasible wavelength set*) consisting of k of the W outgoing wavelengths (i.e., to k –1 wavelengths in addition to itself). Limited conversion has received particular attention: Yates et al. presented a simple, approximate probabilistic analysis for single paths in isolation;[20] Ramaswami and Sasaki provided a non-probabilistic analysis for ring networks and, under certain restrictions, for tree networks and networks of arbitrary topology;[21] while Wauters and Demeester provided new upper bounds on the wavelength requirements for a WDM network under a static model of the network load.[22]

12.1.2 *Adaptive routing*

In an all-optical WDM network, connection establishment for a session involves *routing* and *wavelength assignment*, both of which may be either *oblivious* or *adaptive*. In oblivious (or static) routing, the route is selected at the source and is independent of the state (loading or congestion) of the network, while in adaptive (or dynamic) routing, the route is selected either at the source or on a hop-by-hop basis, based on the state of the network at the time of connection establishment.

Two well-known examples of static routing algorithms are *fixed routing* and *fixed-alternate routing*. In fixed routing, a single fixed path is predetermined for each source-destination pair, and a connection request is blocked if the associated path is not available. In fixed-alternate routing, a fixed set of predetermined paths is assigned to each source-destination pair. When a connection request arrives at a node, the set of predetermined paths is searched according to some policy (e.g., in fixed or adaptive order) to select an available path. Though static algorithms are much simpler to implement than adaptive routing schemes, they can lead to high blocking probabilities.

In adaptive routing schemes, on the other hand, the path for a source-destination pair is selected dynamically by taking into account the network state, thus resulting in lower blocking probabilities. One form of adaptive routing is adaptive least-cost-path routing based on global network state information, in which every link is assigned a cost (based, for example, on wavelength availability) and, upon arrival of a connection request, a least-cost routing algorithm determines the path for the given source-destination pair. This scheme may be implemented in either a centralized or a distributed manner.[2] In the former version, a centralized entity maintains global network state information and establishes lightpaths in response to connection requests. In the distributed version, two common approaches are used: the *link-state* approach, where each node in the network maintains global network state information and, therefore, can find suitable routes in a distributed manner;[3] and the *distance-vector* approach, where each node maintains a routing table that only indicates, for each destination and on each wavelength, the next hop to the destination and the distance to the destination.[4] A disadvantage of all the preceding adaptive routing schemes is the need for continuous network updates of the routing tables at each node, whenever network changes take place, which results in a significant increase in control overhead and a requirement for elaborate control and management protocols. Therefore, global knowledge-based adaptive routing schemes are mostly used in networks where lightpaths are quite static and do not change much with time.[2]

An important type of adaptive routing is *deflection routing*. Deflection routing protocols have been analyzed by several researchers under a variety of assumptions on the underlying network topology, and have been shown to perform outstandingly in many cases due to their low overhead and high adaptivity.[23-33] The deflection routing schemes proposed to date, however, are based on packet-by-packet (datagram) deflections, and may be inappropriate for high-speed networks due to their excessive per-packet processing requirements, the loss of packet order, etc. Thus, in this chapter we will concentrate on a form of deflection routing called the *virtual circuit deflection* (VCD) protocol, first proposed by Varvarigos and Lang, which performs deflections on a per-session (virtual circuit) basis.[17] The VCD protocol is a hybrid of virtual circuit switching and deflection routing, combining some of their individual advantages. It alleviates to a large extent many of the problems of datagram deflections schemes, while its small buffer requirements make it particularly appropriate for high-speed networks that use optical switching.

The VCD scheme is a virtual circuit switching protocol of the tell-and-go variety, where data starts being transmitted shortly after the setup packet of the session is sent. In this scheme, the intermediate links (and wavelengths) of a path are determined dynamically on a hop-by-hop (instead of end-to-end) basis, depending on link (and wavelength) utilization. At each node, an outgoing link is selected from among the subset of outgoing links that lie on a shortest route to the destination. If wavelength resources are unavailable on the chosen link, an alternate link lying on the shortest route to the destination is tried; we then say that the session is deflected. This process continues until either an available link is found or all the alternate links have been examined. Hence, routing-table updates in the network are not needed, and control overhead is greatly reduced.

12.1.3 Routing and wavelength assignment

Wavelength assignment usually does not take place in parallel with the selection of the links of the path, as we assume it happens with the VCD scheme; instead, once a route has been selected for a source-destination pair, a wavelength assignment algorithm assigns suitable wavelengths to each link of the route, so that any two lightpaths sharing the same physical link are assigned different wavelengths. In the static case (i.e., when the lightpaths that are to be set up are known in advance), and under the wavelength continuity constraint discussed above, wavelength assignment reduces to the graph-coloring problem, which is known to be NP-complete. Heuristic methods (such as random assignment, first-fit, least-used assignment, etc.) are usually employed to assign wavelengths to lightpaths. For a review of these methods and for performance comparison in terms of connection blocking, see Zang et al.[5]

The design of efficient routing and wavelength assignment (RWA) algorithms in all-optical networks has been the objective of many research initiatives. Karasan and Ayanoglu analyzed the first-fit wavelength assignment strategy in a network with no-wavelength conversion and fixed shortest-hop routing.[14] They also proposed an adaptive RWA algorithm and evaluated its performance via simulations. Mokhtar and Azizoglu also proposed several adaptive RWA algorithms for networks with no-wavelength conversion; they also analyzed oblivious alternate routing using a fixed-order wavelength search.[15] Harai et al. analyzed oblivious alternate routing with fixed wavelength assignment and no-wavelength conversion.[16] Harai et al. also analyzed oblivious alternate routing with various wavelength assignment schemes for networks with limited wavelength conversion.[10] All these algorithms, however, require information on global wavelength utilization, assuming either a periodic exchange of such information[15] or a centralized network controller.[14,16]

The remainder of this chapter is organized as follows. Section 12.2 describes the virtual circuit deflection scheme and shows how it can be combined with other techniques to improve network performance. Section 12.3 presents analytical results for oblivious and adaptive routing and wavelength assignment in the torus and hypercube networks with full-wavelength conversion. Section 12.4 focuses on the performance results obtained for the VCD scheme and examines the effects of wavelength conversion in network performance. Also, some interesting design options when building an all-optical network are discussed. Finally, Section 12.5 concludes the chapter.

12.2 *Virtual circuit deflection (VCD): an adaptive routing scheme*

This section describes a specific adaptive routing scheme — VCD. We assume that connection requests are generated at the source nodes with a specified destination and bandwidth requirement (number of wavelengths required). A source node tries to accommodate each request by choosing one of its outgoing links that lies on a shortest route to the destination (according to a static routing table, which is held at each node* and is based on [possibly outdated] network topology information). If the chosen link does not have the wavelength resources required for the connection, an alternate link laying on the shortest route to the destination is tried until either an available link is found or all the alternate links have been examined. After determining an outgoing link, a setup packet is transmitted to the next node of the path to set the routing tables and reserve resources at intermediate nodes. At each hop, the setup packet randomly selects a wavelength from among the available wavelengths of a link and, if it is successful in establishing a connection, the wavelengths required by the session are reserved for the session duration; otherwise, the session is randomly assigned a new time at which to try. The setup packet is thus forwarded hop-to-hop and is followed after a short delay by the data packets. If the setup packet is successful in reserving resources on all the links on the path to the destination without deflection, the VCD scheme looks like the usual (forward) reservation protocols, with the difference that the reservation (setup) phase and the transmission phase overlap in time.

In a large 2-dimensional mesh, most intermediate nodes have two outgoing links lying on a shortest route. In a hypercube, i outgoing links are lying on a shortest route when the packet is at a distance i from its destination. The number of alternate links that lie on a shortest route to the destination may change as the setup packet progresses toward its destination. Furthermore, a limit on the number of alternate links that are examined could be used to reduce the processing overhead at intermediate nodes. We let l be the number of outgoing links that a session may try at each hop, which we refer to as the *routing flexibility*. The number of *feasible* outgoing links at a node t is given by $\min(l, n_{t,d})$, where $n_{t,d}$ is the number of outgoing links at node t that lie on a shortest route to the destination d. Therefore, if the capacity of each link is divided in k wavelengths, a session currently at node t will be blocked and scheduled to retry only if all of the k wavelengths on each of its feasible outgoing links are unavailable.

It is possible for sessions to be deflected such that the paths contain loops. This may arise after a series of deflections or if a setup packet is deflected immediately to the previously visited node. In either case, the resources reserved in the loop are inefficiently used and it is desirable to remove the loop; however, unless the setup packet visits the intermediate node for the second time prior to the arrival of the first data packet, it is unclear whether the added protocol complexity associated with removing the loop outweighs the efficiency benefits.

Allowing sessions to follow very long paths can waste network resources, increasing the probability that future sessions will be blocked or forced to take even longer paths. To avoid the waste that occurs when a session follows a very long path due to deflections, we may request that a session is dropped when the setup packet has traveled more than H hops without reaching its destination. The parameter H can be chosen to be equal to a multiple (e.g., two or three times) of the shortest distance between the source and the desti-

---

* Nodes do no need to have any global information about the utilization of the network links other than for their outgoing links. The performance results that will be described in the following sections assume that only this minimal information is available at the nodes.

nation of the session, and it may also be dependent on the current congestion in the network. A session that has undergone too many deflections is dropped by transmitting a control packet to the source, requesting it to cease transmitting new packets. Data packets sent prior to the arrival of the control packet at the source can either be dropped or allowed to remain in the network until they reach their destination (possibly over a very long path), while the remaining data is sent later, over a different path, by the source.

The time gap between the transmission of the setup packet and the transmission of the first data packet from a source is chosen to be equal to the maximum number of hops H allowed for the particular session times the time required to process a setup packet at a node. In other words, the gap must be at least as large as the minimum time by which the connection setup phase and the data transmission phase should be separated in order to ensure that data packets do not overpass the setup packet.

12.3 Performance analysis: flexibility in routing vs. flexibility in wavelength assignment

This section presents analytical results for oblivious and adaptive routing and wavelength assignment in the *torus* and *hypercube* networks with full-wavelength conversion. Our choice of the torus and hypercube topologies reflects our interest in analyzing two popular topologies with very different characteristics. The torus is a sparse topology with a small (fixed) node degree and rather large diameter, while the hypercube is a dense topology, with a node degree and diameter that increase logarithmically with the number of nodes. The results are based on the analysis found in Lang et al., where the reader is referred for a more in-depth study.[6] The analytical results apply to regular, all-optical networks with full-wavelength conversion. These results hold for any vertex and edge-symmetric topology and, with modifications, to any vertex symmetric (but not edge-symmetric) topology.

We assume a distributed network model where the routing decision is made locally at each node, using information only about the state of each node's outgoing links and wavelengths. Also, we do not require that the alternate paths between a source-destination pair be link disjoint,[15,16] instead allowing links (and wavelengths) to overlap between alternate paths. In the network model considered, new sessions with uniformly distributed over all nodes destinations arrive independently at each node of the network according to a Poisson process. The capacity of each link is divided into k wavelengths, and each node has full-wavelength conversion capability. An outgoing link of a node with k wavelengths per link is modeled by an auxiliary $M/M/k/k$ queuing system. Using the occupancy distribution of this system, a closed-form expression for the probability P_{succ} of successfully establishing a circuit can be produced without the need to use the link independence blocking assumption, but instead by taking into account partially the dependence between the acquisition of successive wavelengths on the path followed by a session. The analysis is general, computationally inexpensive, and scales easily for larger network sizes and arbitrary k. It applies to both oblivious and adaptive routing, and applies equally well to multi-fiber networks with no-wavelength conversion.[6]

In Section 12.4 we will examine how the extent of improvement in achievable throughput for a fixed P_{succ} depends on the *number of wavelengths k per link* and on the number of links l that may be tried at each hop. This is important because it impacts on the cost and the complexity of the switch. We will see that increasing the *routing flexibility l* increases the switch complexity and delay. Similarly, with full-wavelength conversion, increasing the number of wavelengths k per link increases hardware complexity and may be difficult to realize with current technology. We find that although the throughput per wavelength increases superlinearly with k, the incremental gain in throughput per wavelength (for a fixed P_{succ}) saturates rather quickly to a linear increase. We also see that when the routing flexibility l is varied, the largest incremental gain in throughput per wavelength occurs when l is increased from one to two. We also compare the performance obtainable with a certain number of wavelengths k with that obtainable with a certain routing flexibility l. For the torus and hypercube topologies, we find that for a fixed P_{succ}, a system with k wavelengths per link and only one alternate choice of an outgoing link (i.e., $l = 2$) gives a per-wavelength throughput that is close to that achieved by a system using oblivious routing with $2k$ wavelengths per link, with only a small additional improvement as l is increased further. The preceding observations imply several interesting alternatives for the provisioning and expansion of all-optical networks, some of which we discuss in Section 12.4.

12.3.1 Torus networks

We consider the $p \times p$ torus network, which consists of $N = p^2$ nodes arranged along the points of a two-dimensional grid with integer coordinates, with p nodes along each dimension. Two nodes (x_1, x_2) and (y_1, y_2) are connected by a bidirectional link if and only if, for some $i = 1, 2$, we have $(x_i - y_i)$ mod $p = 1$ and $x_j = y_j$ for $j = i$. In addition to these links, wraparound links connecting node $(x_1, 1)$ with node (x_1, p), and node $(1, x_2)$ with node (p, x_2), are also present.

In oblivious routing with full-wavelength conversion, the route followed by a session is chosen at the source and is independent of the state of the links. In this case, a session is blocked and scheduled to retry only if all wavelengths on the desired outgoing link are unavailable, where we assume that a setup packet selects the outgoing wavelength from among the available wavelengths on the link with equal probability. We consider an XY routing scheme where a session follows a shortest route to its destination, first traversing all the links in one dimension (horizontal or vertical) and then traversing all the links in the other dimension (vertical or horizontal); the first dimension is selected as random at the source. For uniformly distributed destinations, the average probability of success for a new session can be calculated to be:

$$
P_{succ} = \begin{cases} \dfrac{\delta}{\alpha_1}\left[\left(1+2\alpha_1\left(\dfrac{1-\alpha_2^{(p-1)/2}}{1-a_2}\right)\right)^2-1\right]_1 & p \text{ odd} \\[4ex] \dfrac{\delta}{\alpha_1}\left[\left(1+2\alpha_1\left(\dfrac{1-\alpha_2^{p/2}}{1-a_2}\right)+a_1\left(\dfrac{1-\alpha_2^{(p/2)-1}}{1-a_2}\right)\right)^2-1\right]_1 & p \text{ even} \end{cases}
\tag{12.1}
$$

where $\delta = \alpha_0/(p^2 - 1)$, α_0 expresses the probability that a wavelength is available on the outgoing link of the originating node of the session, and α_1 and α_2 express the probability that a wavelength is available on an outgoing link at a transit node of the path, given that a wavelength was available on an incoming link of that node.

In adaptive VCD routing, a link is selected at random, at each hop, from among all the outgoing links that lie on a shortest route to the destination; if all the wavelengths on the chosen link are unavailable, an alternate link lying on the shortest route to the destination is tried. This process continues until either an available link is found or all the alternate links have been examined. For the torus network, there are at most two outgoing links at a node that lie on a shortest route to the destination. For uniformly distributed destinations, the average probability of success for a new session P_{succ} can be calculated to be:

$$
P_{succ} = \begin{cases} \dfrac{1}{p^2-1}\left[\displaystyle\sum_{i-1}^{(p-1)/2}4iP_{succ}(i)+\sum_{i=(p+1)/2}^{p-1}4(p-i)P_{succ}(i)\right]_1 & p \text{ odd} \\[5ex] \dfrac{1}{p^2-1}\left[\begin{array}{l}\displaystyle\sum_{i-1}^{(p/2)-1}4iP_{succ}(i)\\[3ex]+\displaystyle\sum_{i=(p/2)+1}^{p-1}4(p-i)P_{succ}(i)\\[3ex]+2(p-1)P_{succ}\left(\dfrac{p}{2}\right)+P_{succ}(p)\end{array}\right]_1 & p \text{ even} \end{cases}
\tag{12.2}
$$

where $P_{succ(i)}$ is defined as:

$$P_{succ}(i) = \begin{cases} \left(a_0(1) + (i-1)a_0(2)\right)\left(\dfrac{1}{i!}\right) \\ \cdot \displaystyle\prod_{b=1}^{i-1} (a_1(1) + (b-1)a_1(2)), \qquad i \le \left[\dfrac{p-1}{2}\right] \\ a_0(2)a_1(2)^{i-[(p+1)/2]}\left(\dfrac{1}{([(p-1)/2])!}\right) \\ \cdot \displaystyle\prod_{b=1}^{[(p-1)/2]} (a_1(1) + (b-1)a_1(2)), \qquad i > \left[\dfrac{p-1}{2}\right] \end{cases} \tag{12.3}$$

where $\alpha_0(i)$, $i = 1, 2$ expresses the probability that at least one wavelength on $i = \min(l, n_{s,d})$ outgoing links at the origin is available, and $\alpha_1(i)$, $i = 1, 2$ expresses the probability that, at each transit node t, a wavelength is available on one of $i = \min(l, n_{t,d})$ alternate outgoing links, given that a wavelength was available on an incoming link of that node.

12.3.2 Hypercube networks

This section considers the 2^r-node hypercube network, where each node can be represented by a binary string $(x_1, x_2, \ldots, x_r)$, and two nodes are connected via a bidirectional link if their binary representations differ in only one bit.

In oblivious routing, where a shortest route is chosen at random at the source, and for uniformly distributed destinations, the average probability of success for a new session can be calculated to be:

$$P_{succ} = \frac{a_0}{a_1(2^r - 1)}\left[(1 + a_1)^r - 1\right] \tag{12.4}$$

where α_0 and α_1 are as defined in Equation (12.1).

In adaptive VCD routing we note that in the hypercube network, a node that is i hops away from the destination has i outgoing links lying along a shortest route to the destination. We let l, $l = r$, be the maximum number of outgoing links that may be tried at any hop. Assuming the source is at a distance i hops from its destination, the probability of successfully establishing a connection is given by:

$$P_{succ}(i) = a_0\left(\min(l, i)\right)\prod_{j=1}^{i-1} a_1\left(\min(l, j)\right) \tag{12.5}$$

where $\alpha_0(\min(l, i))$, $i = 1, \ldots, r$ and $\alpha_1(\min(l, j))$, $j = 1, \ldots, r$ are as defined in Equation (12.3).

12.4 Analytical and simulation results

Before we examine the simulation results for the VCD scheme and compare them with the analytical results presented in Section 12.3, it is worth turning our attention first to the effects of a number of parameters on the performance of the network under study.

First, we examine the effect of wavelength conversion in network performance, considering the torus network with oblivious routing and in three different cases:

1. No-wavelength conversion (or 1-adjacent wavelength switching)
2. Limited-wavelength conversion using k–adjacent wavelength switching, where $k = 2, 3$
3. Full-wavelength conversion (or W-adjacent wavelength switching) in a WDM network with W wavelengths per link.

We note that full-wavelength conversion provides the best achievable performance (in terms of the realizable probability of success for a given arrival rate per wavelength or in terms of the realizable throughput per wavelength for a given probability of success) for a given number of wavelengths W per link. When no-wavelength conversion is used, the different wavelengths on a link do not interact with one another. Thus, an all-optical network with W wavelengths per fiber is essentially equivalent to W disjoint single-wavelength networks operating in parallel. To obtain the probability of success in this special case, it is therefore enough to focus attention on any one of the W independent parallel networks, for which the analysis given in Sharma applies.[34]

We define the degree of conversion δ of a k-adjacent wavelength switching system with W wavelengths per fiber to be

$$\delta = \frac{k-1}{W-1} \times 100\%$$

Thus, $\delta = 100\%$ corresponds to the case of full-wavelength conversion (or W-adjacent wavelength switching), while $\delta = 0\%$ corresponds to the case of no-wavelength conversion (or 1-adjacent wavelength switching).

We define $P_{succ}(\lambda, k)$ to be the probability of success in a k-adjacent wavelength switching system when the arrival rate per node per wavelength is equal to λ; and we define $\lambda(P_{succ}, k)$ to be the throughput per node per wavelength of a k-adjacent wavelength switching system, when the probability of success is equal to P_{succ}. To quantify the performance of limited wavelength conversion vs. full- or no-wavelength conversion, we also define the *throughput efficiency* $\lambda(P_{succ}, k)$ of a k-adjacent wavelength-switching scheme, with W wavelengths per fiber, for a given probability of success P_{succ}, to be:

$$\Delta\lambda\left(P_{succ},k\right)=\frac{\lambda\left(P_{succ},k\right)-\lambda\left(P_{succ},1\right)}{\lambda\left(P_{succ},W\right)-\lambda\left(P_{succ},1\right)}\times100\%$$

and the *success efficiency* $P_{succ}(\lambda,\ k)$ of a *k*-adjacent wavelength switching system, for a given arrival rate per node per wavelength λ, to be:

$$\Delta P_{succ}\left(\lambda,k\right)=\frac{P_{succ}\left(\lambda,k\right)-P_{succ}\left(\lambda,1\right)}{P_{succ}\left(\lambda,W\right)-P_{succ}\left(\lambda,1\right)}\times100\%$$

The throughput and success efficiencies represent the degree of improvement (over no-wavelength conversion) in the throughput and in the probability of success respectively, which is obtained when limited wavelength conversion with *k*-adjacent wavelength switching is used, as a percentage of the improvement obtained when full-wavelength conversion is used. For $k = W$ (full-wavelength conversion), we get $\lambda(P_{succ},\ k) = 100\%$ and $P_{succ}(\lambda,\ k) = 100\%$, while for $k = 1$ (no-wavelength conversion), we get $\lambda(P_{succ},\ k) = 0\%$ and $P_{succ}(\lambda,\ k) = 0\%$ (no improvement).

In Figure 12.2, we present performance results for the probability of success P_{succ} plotted vs. the arrival rate per node per wavelength λ when limited wavelength conversion is permitted. The results depicted here were obtained using the analysis presented in Sharma and Varvarigos, and Sharma.[9,34] Observe that limited conversion to only one or two adja-

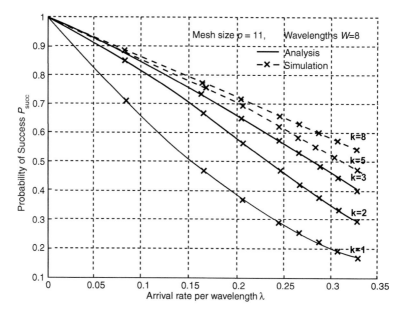

Figure 12.2 Success probability P_{succ} vs. the arrival rate per wavelength λ, for a $p \times p$ torus ($p = 11$), for $W = 8$ wavelengths per link.

Table 12.1 Quantifying the Benefits Obtained with Limited Wavelength
Conversion when $K = 2, 3$

p	W	$\Delta P_{succ}(0.25, 2)$	$\Delta P_{succ}(0.25, 3)$	$\Delta\lambda(0.7, 2)$	$\Delta\lambda(0.7, 3)$
11	5	51%	86%	61%	87%
11	20	47%	79%	49%	76%

cent wavelengths provides a considerable fraction of the improvement
that full-wavelength conversion provides over no-wavelength conversion. These benefits are summarized in Table 12.1, where we illustrate
the throughput and success efficiencies for a $p \times p$ torus ($p = 11$) for a few
selected points.

Also, the benefits of wavelength conversion diminish as the extent of
conversion k increases and, eventually, appear to saturate. We see, therefore, that limited conversion of small range (i.e., $k = 2$ or 3) gives most of
the benefits obtained by full-wavelength conversion, where $k = W$. For
instance, in Figure 12.3, which also illustrates the network performance
for $W = 20$ wavelengths, increasing the extent of conversion k beyond
some value leads to diminishing returns. Similar remarks regarding the
effects of the extent of wavelength conversion also apply in the case of
hypercube networks, using either descending dimensions switches or
crossbar switches.[9,34]

Next, we present performance results for the VCD protocol, focusing
mainly on the adaptivity that VCD can exhibit. The results are obtained from
Varvarigos and Lang, where a Manhattan street (MS) network topology is

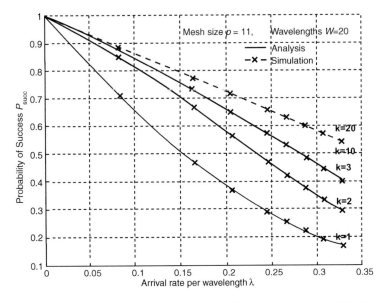

Figure 12.3 Success probability P_{succ} vs. the arrival rate per wavelength λ, for a $p \times p$
torus ($p = 11$), for $W = 20$ wavelengths per link.

considered.[17] The MS network is a two-connected regular mesh network with unidirectional communication links, which has been analyzed extensively in the literature for datagram deflection schemes due to its regularity and symmetry properties.[24,26,30,31] The MS $X \times Y$ -dimensional wraparound mesh consists of $N = XY$ processors arranged along the points of a 2-D space that have integer coordinates. X processors exist along the x-dimension, and Y processors exist along the y-dimension, where X and Y are even numbers. Each processor has two outgoing links, one horizontal and one vertical. The horizontal links are directed eastward on even rows and westward on odd rows, while the vertical links are directed northward on even columns and southward on odd columns. Each processor is represented by a pair (x, y) with $0 \leq x \leq X - 1$ and $0 \leq y \leq Y - 1$.

A natural measure of the performance of the VCD protocol is the *inefficiency ratio* $\eta(\lambda)$, defined as the ratio

$$\eta(\lambda) = \frac{D(\lambda)}{D(0)}$$

of the average path length $D(\lambda)$ taken by a session for a given arrival rate , over the average shortest-path length $D(0)$ of the MS network topology. The inefficiency ratio characterizes the effectiveness with which the VCD protocol uses the network bandwidth for a given network load. In the performance results that we present, we assume that each session requires 1 unit of capacity, and each link has capacity equal to m units. In WDM networks, 1 unit may correspond to the capacity of one wavelength, in which case m will correspond to the number of W of wavelengths per link.

In Figures 12.4 and 12.5, we illustrate the blocking probability B (the probability that a session attempting to establish a connection is blocked at its first hop) and the inefficiency ratio $\eta(\lambda)$, as a function of the normalized arrival rate per unit of link capacity $\lambda = \lambda/m$, for an 8×8 MS network. In fact, m in this case can be viewed as the number of sessions or channels that can simultaneously use a link and hence, m provides a measure of the adaptivity of VCD. The dashed lines in these figures highlight the stability boundary (points to the left of the boundary belong to the stable region), where the stable region is defined as the region where the connection request queue remains finite; stability is not directly related to B, and it is possible to have B considerably less than one and still be in the unstable region. From Figures 12.4 and 12.5, it is evident that when link capacity m is large (i.e., more channels can simultaneously use the link and, therefore VCD adaptivity is enhanced), the efficiency of the VCD protocol increases significantly. For example, for $m = 20$, the blocking probability B is always less than 0.4 and the lengths of the paths taken are, on the average, within 5% from the shortest path length for any value of the external arrival rate.

Now we will focus on the effect of full-wavelength conversion on oblivious and adaptive routing. In Figures 12.6 and 12.7, we present the success

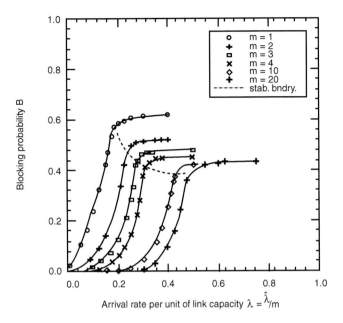

Figure 12.4 Performance results for the VCD protocol, illustrating the blocking probability B as a function of the normalized arrival rate per unit of link capacity $\hat{\lambda}/m$ for an 8×8 MS network and several values of m. The dashed lines correspond to the stability boundary.

probability P_{succ} predicted by the analytical results in Section 12.3 vs. the results obtained from simulations for oblivious routing in the hypercube and torus networks, respectively; Figures 12.8 and 12.9 present the respective results for adaptive VCD routing. We observe that in all the figures, close agreement exists between the simulations and the analytically predicted values over the entire range of applicable input rates.[21] Despite its accuracy, the presented analysis is considerably simpler than the analyses available in the literature and its computational requirements are modest, allowing it to scale easily for large k.

To compare the performance of systems with varying k and l, we define the *incremental per-wavelength throughput gain* $\lambda(k_1, l_1; k_2, l_2)$ of a system with k_2 wavelengths and a choice of l_2 links per hop, over a system with k_1 wavelengths and a choice of l_1 links per hop, for a given P_{succ}, to be:

$$\Delta\lambda\left(k_1, l_1; k_2, l_2\right) = \frac{\lambda\left(P_{succ}, k_2, l_2\right) - \lambda\left(P_{succ}, k_1, l_1\right)}{\lambda\left(P_{succ}, k_1, l_1\right)} \times 100\% \qquad (12.6)$$

where $\lambda(P_{succ}, k, l)$ is the throughput per node per wavelength in a system with k wavelengths and routing flexibility l, when the probability of success is equal to P_{succ}.

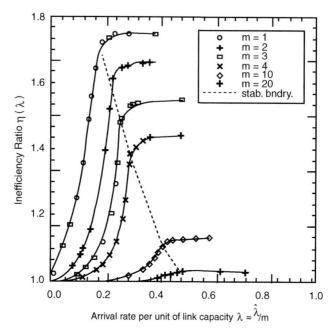

Figure 12.5 Performance results for the VCD protocol, illustrating the inefficiency ratio $\eta(\lambda)$ as a function of the normalized arrival rate per unit of link capacity $\lambda = \hat{\lambda}/m$, for an 8×8 MS network and several values of m. The dashed lines correspond to the stability boundary.

We also define the *incremental probability of success gain* $P_{succ}(k_1, l_1; k_2, l_2)$ of a system with k_2 wavelengths and a choice of l_2 links per hop, over a system with k_1 wavelengths and a choice of l_1 links per hop, for a given arrival rate per wavelength, to be:

$$\Delta P_{succ}\left(k_1, l_1; k_2, l_2\right) = \frac{P_{succ}\left(\lambda, k_2, l_2\right) - P_{succ}\left(\lambda, k_1, l_1\right)}{P_{succ}\left(\lambda, k_1, l_1\right)} \times 100\% \qquad (12.7)$$

where $P_{succ}(\lambda, k, l)$ is the probability of success in a system with k wavelengths and routing flexibility l, when the arrival rate per wavelength is equal to $\lambda = \hat{\lambda}/k$.

The throughput and probability of success gains measure the degree of improvement that a full-wavelength conversion system with k_2 wavelengths and a choice of l_2 outgoing links per hop provides over a similar system with k_1 wavelengths and a choice of l_1 links per hop.

In Figures 12.10 and 12.11, we illustrate the analytically predicted probability of success P_{succ} vs. the arrival rate per wavelength λ/k, for k ranging from 1 to 16, for the torus and hypercube networks, respectively.

As can be seen from Figures 12.10 and 12.11, for a given P_{succ} and fixed l, the throughput per wavelength increases with increasing k. In other words,

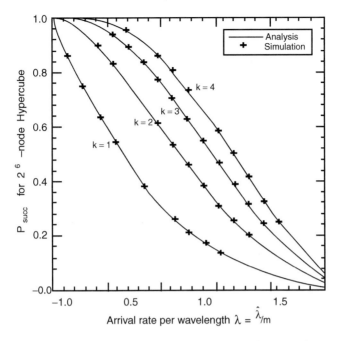

Figure 12.6 Analytical and simulation results for P_{succ} vs. the arrival rate per wavelength $\lambda = \hat{\lambda}/k$, for a 2^6-node hypercube network, using oblivious routing.

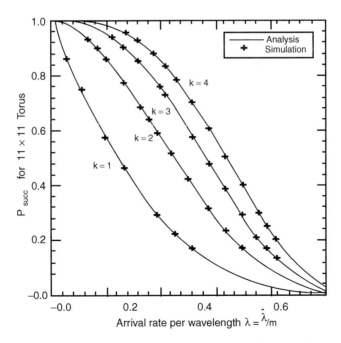

Figure 12.7 Analytical and simulation results for P_{succ} vs. the arrival rate per wavelength $\lambda = \hat{\lambda}/k$, for an 11×11 torus network, using oblivious XY routing.

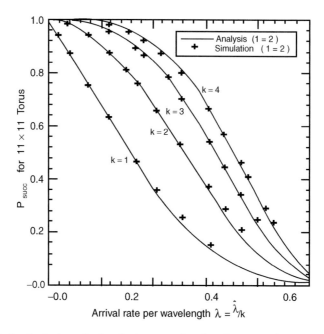

Figure 12.8 Analytical and simulation results for P_{succ} vs. the arrival rate per wavelength for an 11×11 torus hypercube network, using adaptive VCD routing.

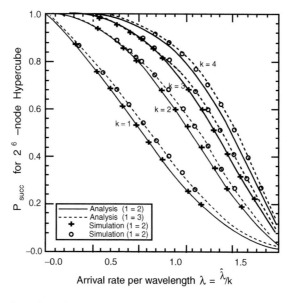

Figure 12.9 Analytical and simulation results for P_{succ} vs. the arrival rate per wavelength $\lambda = \hat{\lambda}/k$, for a 2^6-node hypercube network, using adaptive VCD routing.

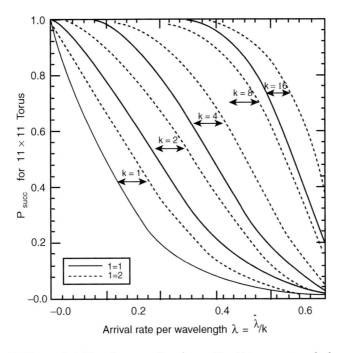

Figure 12.10 The probability of success P_{succ} for an 11×11 torus network, for k varying from 1 to 16.

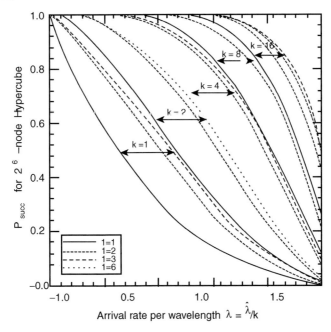

Figure 12.11 The probability of success P_{succ} for a 2^6-node hypercube network, for k varying from 1 to 16.

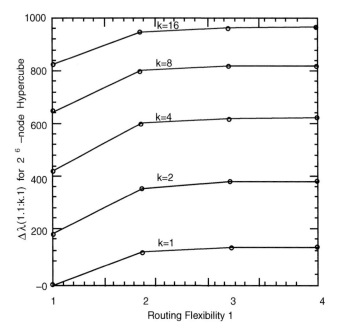

Figure 12.12 The incremental throughput gain $\lambda(1, 1; k, l)$, for $k = 1, 2, 4, 8, 16$ and l = 1, 2, 3, 4], for a 2^6-node hypercube network with $P_{succ} = 0.8$.

the throughput per link (and the network throughput) increases superlinearly with k. The linear part of the increase in throughput is because of the increase in capacity, while the superlinear part of the increase is due to more efficient use of that capacity because of the greater flexibility in establishing a circuit when a larger number of wavelengths is available. The incremental gain in achievable throughput per wavelength for a given l, $\lambda(k_1, l; k_2, l)$, however, decreases rapidly with increasing k. This result holds for both oblivious and adaptive VCD routing, and is in agreement with the results for oblivious routing presented in Sharma and Varvarigos, and Koch.[9,18] Similarly, the incremental throughput gain for a given k, $\lambda(k, l_1; k, l_2)$, decreases rapidly with increasing l. If we fix l_1 and l_2, and increase k, the incremental gain decreases, suggesting that the performance improvement for adaptive VCD routing is tightly coupled with the number of wavelengths, and that the benefits of alternate routing are not as significant when the number of wavelengths k is large.

Another interesting feature of adaptive VCD routing in networks with full-wavelength conversion is that the per-wavelength throughput for fixed number of wavelengths k and increasing the routing flexibility l, appears to saturate at or near the per-wavelength throughput of a system using oblivious routing with wavelength conversion over twice as many wavelengths.

In Figure 12.12, we plot the incremental throughput gain for the hypercube network when $P_{succ} = 0.8$ and the number of wavelengths k ranges from 1 to 16, and the routing flexibility l ranges from 1 to 4 (i.e., we plot $\lambda(1, 1; k$, *l*) for $k = 1, 2, 4, 8, 16$ and $l = 1, 2, 3, 4$). As depicted in Figure 12.12, the largest increase in incremental throughput gain occurs when the routing flexibility increases from $l = 1$ to $l = 2$, regardless of the number of wavelengths. Furthermore, this gain obtained by increasing the routing flexibility from $l = 1$ to $l = 2$, with fixed k, approaches the gain obtained by doubling the number of wavelengths to $2k$, with $l = 1$. For example, the incremental throughput gain for $k = 8$ and $l = 2$ is within 3% of the incremental throughput gain for $k = 16$ and $l = 1$.

The previous discussion leads to some interesting design options when building an all-optical network. For instance, because the per-wavelength throughput gain saturates quickly with increasing k, simply building a network in which every node can translate between k wavelengths may not be the most efficient option. Instead, it may be preferable to build a network in which every node consists of k/n simpler switching elements operating in parallel (each switching between a nonintersecting subset of n wavelengths) that achieves performance comparable to that of the k-wavelength system at a much lower cost. This suggests that a network designer may initially choose to build the network with nodes that have a small number of parallel channels, with n wavelengths per channel. As network traffic grows, the designer may expand the nodes by adding more parallel channels. Better yet, instead of increasing the number of channels per link at every network node, the designer may focus on the routing algorithms and may choose to increase the routing flexibility to obtain equivalent performance at no extra hardware cost. For instance, the designer may simply increase the number of outgoing links that may be tried at each hop. Observe, however, that the routing flexibility is limited by the network topology and is also a function of the switch architecture. Our results emphasize the need for network designers to investigate the tradeoffs between wavelength conversion, routing flexibility, and hardware cost when designing future optical networks.

12.5 Conclusion

We presented adaptive routing and wavelength assignment protocols which are suitable for all-optical networks with wavelength conversion and outperform oblivious routing schemes in the hypercube and torus topologies. We demonstrated that for the topologies considered, the performance of a system using the adaptive VCD routing scheme, with only one alternate link per hop, approaches that of a system using oblivious routing with twice as many wavelengths per link. We also presented performance results, focusing mainly on the adaptivity VCD can exhibit and noticed that when link capacity is large (i.e., more channels can simultaneously use the link and therefore VCD adaptivity is enhanced), the efficiency of the VCD protocol increases significantly. We also examined the effect of wavelength conversion in network performance, considering the cases of no-wavelength conversion, of limited-wavelength conversion and full-wavelength conversion and observed that limited conversion to only one or two adjacent wavelengths can provide a considerable fraction of the improvement that full-wavelength conversion provides over no-wavelength conversion. These results clearly emphasize the need for network designers to investigate the tradeoffs between wavelength conversion, routing flexibility, and hardware cost when designing future optical networks.

References

1. B. Mukherjee, WDM optical communications networks: progress and challenges, *IEEE J. Select. Areas Commun.*, vol. 18, Oct. 2000, pp. 1810–1823.
2. H. Zang et al., Dynamic lightpath establishment in wavelength-routed WDM networks, *IEEE Commun. Mag.*, Sep. 2001, pp.100–108.
3. R. Ramaswami and A. Segall, Distributed network control for optical networks, *IEEE/ACM Trans. Networking*, vol. 5, no. 6, Dec. 1997, pp. 936–943.
4. H. Zang et al., Connection management for wavelength-routed WDM networks, *Proc. IEEE GLOBECOM '99*, Dec. 1999, pp. 1428–1432.
5. H. Zang, J. Jue, and B. Mukherjee, A review of routing and wavelength assignment approaches for wavelength-routed optical WDM networks, *Opt. Networking Mag.*, vol. 1, no. 1, Jan. 2000, pp. 47–60.
6. J. Lang, V. Sharma, and E. Varvarigos, An analysis of oblivious and adaptive routing in optical networks with wavelength translation, *IEEE/ACM Trans. Networking*, vol. 9, no. 4, Aug. 2001, pp. 503–517.
7. R. Barry and P. Humblet, Models of blocking probability in all-optical networks with and without wavelength changers, *IEEE J. Selected Areas Commun.*, vol. 14, June 1996, pp. 858–867.
8. M. Kovacevic and A. Acampora, Benefits of wavelength translation in all-optical clear-channel networks, *IEEE J. Selected Areas Commun.*, vol. 14, June 1996, pp. 868–880.
9. V. Sharma and E.A. Varvarigos, An analysis of limited wavelength translation in regular all-optical WDM networks, *Proc. IEEE INFOCOM '98*, vol. 2, Mar.–Apr. 1998, pp. 893–901.
10. H. Harai, M. Murata, and H. Miyahara, Performance analysis of wavelength assignment policies in all-optical networks with limited-range wavelength conversion, *IEEE J. Selected Areas Commun.*, vol. 16, Sep. 1998, pp. 1051–1060.
11. B. Ramamurthy and B. Mukherjee, Wavelength conversion in WDM networking, *IEEE J. Selected Areas Commun.*, vol. 16, Sep. 1998, pp. 1061–1073.
12. K.N. Sivarajan and R. Ramaswami, Lightwave networks based on de Bruijn graphs, *IEEE/ACM Trans. Networking*, vol. 2, Feb. 1994, pp. 70–79.
13. Y. Zhu, G. Rouskas, and H. Perros, Blocking in wavelength routing networks. Part 1: The single path case, *Proc. IEEE INFOCOM '99*, vol. 1, Mar. 1999, pp. 321–328.
14. A. Karasan and E. Ayanoglu, Effects of wavelength routing and wavelength selection algorithms on wavelength conversion gain in WDM optical networks, *IEEE/ACM Trans. Networking*, vol. 6, Apr. 1998, pp. 186–196.
15. A. Mokhtar and M. Azizoglu, A new analysis of wavelength translation in regular WDM networks, *IEEE/ACM Trans. Networking*, vol. 6, Apr. 1998, pp. 197–206.

16. H. Harai, M. Murata, and H. Miyahara, Performance of alternate routing methods in all-optical switching networks, *Proc. IEEE INFOCOM '97*, vol. 2, Apr. 1997, pp. 516–524.

17. E.A. Varvarigos and J. Lang, A virtual circuit deflection protocol, *IEEE/ACM Trans. Networking*, vol. 7, no. 3, June 1999, pp. 335–349.

18. R. Koch, Increasing the size of a network by a constant factor can increase performance by more than a constant factor, *Symp. Foundations of Computer Science*, vol. 18, Oct. 1988, pp. 221–230.

19. S. Subramaniam, M. Azizoglu, and A.K. Somani, All-optical networks with sparse wavelength conversion, *IEEE/ACM Trans. Networking*, vol. 4, no. 4, Aug. 1996, pp. 544–557.

20. J. Yates, J. Lacey, D. Everitt, and M. Summerfield, Limited-range wavelength translation in all-optical networks, *Proc. IEEE INFOCOM '96*, vol. 3, Mar. 1996, pp. 954–961.

21. R. Ramaswami and G.H. Sasaki, Multi-wavelength optical networks with limited wavelength conversion, *Proc. IEEE INFOCOM '97*, Kobe, Japan, Apr. 1997.

22. N. Wauters and P. Demeester, Wavelength translation in optical multi-wavelength multi-fiber transport networks, *Int. J. Optoelectronics*, vol. 11, no. 1, Jan.–Feb. 1997, pp. 53–70.

23. N.F. Maxemchuk, Problems arising from deflection routing: live-lock, lock-out, congestion and message reassembly, in *High-Capacity Local and Metropolitan Area Networks*, G. Pujolle, Ed., Berlin, Germany: Springer-Verlag, June 1990.

24. F. Borgonovo, L. Fratta, and J. Bannister, Unslotted deflection routing in all-optical networks, *Proc. IEEE GLOBECOM '93*, vol. 1, pp. 119–125.

25. Z. Haas and D.R. Cheriton, Blazenet: a packet-switched wide-area network with photonic data path, *IEEE Trans. Commun.*, vol. 38, pp. 818–829, June 1990.

26. A.G. Greenberg and J. Goodman, Sharp approximate models of adaptive routing in mesh networks, *Proc. Teletraffic Analysis and Computer Performance Evaluation*, June 1986, pp. 255–270.

27. A. G. Greenberg and B. Hajek, Deflection routing in hypercube networks, *IEEE Trans. Commun.*, vol. 40, June 1992, pp. 1070–1081.

28. J. Bannister, F. Borgonovo, L. Fratta, and M. Gerla, A versatile model for predicting the performance of deflection-routing networks, *Perform. Eval.*, vol. 16, nos. 1–3, Nov. 1992, pp. 201–222.

29. F. Borgonovo, L. Fratta, and J. Bannister, On the design of optical deflection-routing networks, *Proc. IEEE INFOCOM '94*, vol. 1, pp. 120–129.

30. N.F. Maxemchuk, Comparison of deflection and store-and-forward techniques in the Manhattan Street and Shuffle-Exchange networks, *Proc. IEEE INFOCOM '89*, vol. 3, pp. 800–809.

31. J.T. Brassil, Deflection routing in certain regular networks, Ph.D. dissertation, University of California at San Diego, 1991.

32. A. Krishna and B. Hajek, Performance of shuffle-like switching networks with deflection, *Proc. IEEE INFOCOM '90*, vol. 2, pp. 473–480.

33. E.A. Varvarigos and D.P. Bertsekas, Performance of hypercube routing schemes with or without buffering, *IEEE/ACM Trans. Networking*, vol. 2, June 1994, pp. 299–311.

34. V. Sharma, Efficient communication protocols and performance analysis for gigabit networks, Ph.D. dissertation, Department of Electrical and Computer Engineering, University of California, Santa Barbara, Aug. 1997.

35. D. Bertsekas and R. Gallager, *Data Networks,* Englewood Cliffs, NJ: Prentice-Hall, 1992.

36. A. Butner and D. Skirmont, Architecture and design of a 40 gigabit per second ATM switch, *Proc. ICCD '95,* pp. 352–357.

37. E.A. Varvarigos and V. Sharma, The ready-to-go virtual circuit protocol: a loss-free protocol for multigigabit networks using FIFO buffers, *IEEE/ACM Trans. Networking,* vol. 5, Oct. 1997, pp. 705–718.

chapter thirteen

Connection management in wavelength-routed all-optical networks

Xiaohong Yuan
North Carolina Agricultural and Technical State University

Contents

0-8493-1333-3/03/$0.00+$1.50
© 2003 by CRC Press LLC

13.1 Introduction

Wavelength-division multiplexing (WDM) divides the tremendous bandwidth of a fiber into many nonoverlapping wavelengths (WDM channels) based on which multichannel lightwave networks can be built. A wavelength-routed (wide-area) all-optical WDM network consists of optical routing nodes interconnected by optical links. Signals on different wavelengths are coupled into the link using wavelength multiplexers. An optical routing node is capable of routing each wavelength on an incoming link to any outgoing link without having optoelectronic conversion.[1,2]

In a wavelength-routed WDM network, end users communicate with one another via lightpaths. A lightpath is an all-optical channel that may span multiple fiber links to provide a circuit-switched interconnection between two nodes. The end nodes of a lightpath access the lightpath with transmitters and receivers that are tuned to the wavelength on which the lightpath operates, while the intermediate nodes route the lightpath using their active switches. In the absence of wavelength converters, a lightpath must occupy the same wavelength on all the fiber links that it traverses. This property is known as the wavelength-continuity constraint. Two or more lightpaths traversing the same fiber link must be on different wavelengths so that they do not interfere with one another.[1,3,4]

To establish a circuit-oriented connection between two end nodes in a wavelength-routed WDM network, a lightpath needs to be set up between them. However, due to the limited number of wavelengths available in a fiber link and the limited number of transceivers equipped at each node (transceivers are very expensive), it is not possible to establish a lightpath between every pair of nodes.[1,3] Given a set of lightpaths that need to be established and a constraint on the number of wavelengths, determining the routes over which these lightpaths should be established and the wavelengths that should be assigned to them is called the routing and wavelength assignment (RWA) problem. Lightpaths that cannot be established due to constraints on routes and wavelengths are blocked. Therefore, minimizing the blocking probability is one objective of RWA algorithms.[1]

In a wavelength-routed WDM network, a lightpath is set up when a connection request occurs and is taken down when the connection terminates. To set up a lightpath, a network control and management protocol is employed to select a route, assign a wavelength to the lightpath, signal control information, reserve resources, and configure the appropriate optical switches in the network. It also provides updates to reflect the wavelengths currently being used on each fiber link so that the nodes may make correct routing decisions. The performance of various network control and management protocols could be evaluated by measuring the blocking probability of connection requests, the connection setup delays, the bandwidth used for control messages, or the stabilizing time (time required for nodes to update topology information after a connection has been established or taken down).[4,5]

RWA algorithms as well as network control management mechanisms play a key role in improving the performance of wavelength-routed WDM networks. Various RWA approaches and network control and management protocols are introduced in this chapter. The remainder of the chapter is organized as follows. Section 13.2 addresses the RWA approaches, Section 13.3 introduces network control and management mechanisms, and Section 13.4 summarizes the chapter.

13.2 Approaches for RWA problem

A wavelength-routed WDM network may deal with three types of traffic (connection requests): static, incremental, and dynamic. The traffic is static when the entire set of connections is known in advance, and the connections remain in the network for a long period of time. Incremental traffic refers to the situation when the connection requests arrive sequentially, and the connections remain in the network indefinitely. The traffic is dynamic when the connection requests arrive sequentially, and each connection lasts some finite amount of time. The RWA problem for static traffic is known as static RWA, or static lightpath establishment (SLE). The RWA problem for incremental and dynamic traffic is known as dynamic RWA problem, or dynamic lightpath establishment (DLE).

Existing approaches for RWA either consider the routing and wavelength assignment jointly or partition the RWA problem into two subproblems: routing and wavelength assignment. Each subproblem is solved separately.[6,7] The remainder of this section discusses the approaches for SLE and DLE, respectively.

Table 13.1 The routing and wavelength assignment method

Traffic Type	Routing and WA Considered Jointly	Routing and WA Considered Disjointly	
		Routing	WA
Static	ILP-formulation[1]	ILP-fomulation[1] Fixed routing Fixed-alternate routing[6-8]	Graph-coloring[1,6]
Dynamic	AUR[15] Dynamic routing[11] LLR[16]	Fixed routing Fixed-alternate routing[9,10,11] Adaptive routing[8,12,13]	RANDOM First-Fit Least-Used/SPREAD Most-Used PACK Min-Product Least-Loaded MAX-SUM Relative Capacity Loss Wavelength Reservation Protecting Threshold[6,9,15]

13.2.1 Approaches for SLE

13.2.1.1 ILP formulation of SLE

For SLE, a set of lightpaths is set up all at once for a set of connections known in advance. The typical objective is to minimize the number of wavelengths needed to set up a certain set of lightpaths for a given physical topology. Considering routing and wavelength assignment jointly, SLE with wavelength-continuity constraint is formulated as an integer linear program (ILP) in which the objective function is to minimize the flow in each link. This ILP problem is NP-complete.[1]

13.2.1.2 Routing methods for SLE

For the routing subproblem of SLE, the following methods can be used: ILP formulation, fixed routing, and fixed-alternate routing.[6]

ILP formulation of static routing. The routing subproblem of SLE without wavelength-continuity constraint can also be formulated as ILP with the objective function being to minimize the flow in each link. This problem is NP-complete but can be approximated by reducing the problem size and using randomized rounding.[1] The problem size is reduced by tracking only a limited number of alternative breadth-first paths between source-destination pairs. In randomized rounding, the formulated ILP is first relaxed to allow fractional flows in the interval [0, 1]. The nonlinear version of the problem can be solved by a suitable linear programming method. After that, path stripping is applied to find a set of possible routes for each lightpath and assign weights to each possible route. Next, one route is randomly selected for each lightpath according to the weights of the possible paths assigned in the path stripping.

Fixed routing and fixed-alternate routing. In fixed routing, each source-destination pair is assigned a fixed route. Any connection between the specified pair of nodes is established using the assigned (predetermined) route. A connection request will be blocked if no wavelength is available along the links in the fixed route. One possible fixed route for a source-destination pair is the shortest path between the pair.

Instead of using a single predetermined route, fixed-alternate routing considers multiple predetermined routes for each source-destination pair. These pre-computed routes are stored in an ordered list at the source node's routing table. When a connection request arrives, the source node searches the routing table to find a route with a valid wavelength. The first route in the routing table is usually referred to as the direct route. An alternate route is any route other than the first route in the list of routes in the routing table. The term "alternate route" sometimes also refers to all routes from a source to a destination.[6–8] Fixed routing and fixed-alternate routing are easy to implement and are also used for DLE, which will be discussed later in this chapter.

13.2.1.3 Wavelength assignment for SLE

After the routes have been determined for the lightpaths to be established for static traffic, a wavelength needs to be assigned to each lightpath such that no two lightpaths passing through the same fiber link share the same wavelength. This problem is also called static wavelength assignment. Assigning wavelengths to different lightpaths so as to minimize the number of wavelengths used under the wavelength-continuity constraint reduces to the graph-coloring problem:[1] Construct an auxiliary graph G in which each node represents a lightpath. If two lightpaths pass through a common fiber link, then the two nodes are connected by an undirected edge. The graph-coloring problem is to color the nodes of graph G such that no two adjacent nodes have the same color. It has been demonstrated that this problem is NP-complete, and the efficient sequential graph-coloring algorithm can be used.[1,6]

13.2.2 Approaches for DLE

This section discusses approaches for DLE from three aspects:

1. Methods for the routing subproblem of DLE
2. Methods for the wavelength assignment subproblem of DLE
3. Methods that consider routing and wavelength assignment jointly

13.2.2.1 Routing methods for DLE

Routing methods for DLE are fixed routing, fixed-alternate routing, and adaptive routing.

Fixed routing and fixed-alternate routing.   As for SLE, fixed routing and fixed-alternate routing can also be used for DLE. In Birman and Kershenbaum, fixed routing and fixed-alternate routing were combined with wavelength reservation (Rsv) and protecting threshold (Thr).[9] In the wavelength reservation method, a given number of wavelengths on each link is dedicated to connections with more hops. In the technique of protecting threshold, the single hop connection is assigned an idle wavelength only if the number of idle wavelengths on the link is at or above a given threshold. In Birman, an approximate method for calculating the blocking probability for fixed routing was presented.[10] In Harai et al., an alternate routing method with limited trunk reservation (a limited alternate routing method for short) was presented in which connections with more hops are provided with more alternate routes in proportion to the number of hop counts.[11] A connection is established on route *R* if the available number of wavelengths for links on route *R* is larger than the number of reserved wavelengths on route *R*. Otherwise, the next alternate route for the connection is examined. Compared with traditional fixed-alternate routing, the limited alternate routing method improves fairness among connections because the blocking probabilities of connections with more hops will decrease while the blocking probabilities of ones with shorter hops increase. The overall network performance was also improved. An approximate analytic approach for fixed-alternate routing including the limited alternate routing method was also given in Harai et al.[11]

Adaptive routing. Adaptive routing considers the network state information when selecting a route from a source to a destination. One form of adaptive routing is the adaptive shortest-cost path routing algorithm.[8] In this algorithm, a cost is assigned to each link based on wavelength availability. When a connection arrives, the shortest-cost path between the source and the destination is selected. This approach can be used in both wavelength-continuous and wavelength-converted networks. In a wavelength-converted network, the cost assigned to the links can be selected in such a way that wavelength-converted routes are chosen only when wavelength-continuous paths are not available.

Another form of adaptive routing is the least-congested-path (LCP) routing.[12] Similar to alternate routing, a set of routes is selected for each source-destination pair. A connection will be established on the route that is least congested (i.e. the route that has the largest number of available wavelengths). The least-loaded routing scheme described in Birman is similar to the LCP method.[10] In Banerjee and Mukherjee, the LCP algorithm was simulated and the congestion results were compared with the SLE case (the SLE was solved by employing multi-commodity flow formulation combined with randomized rounding and graph-coloring techniques).[13] It was noted that when using LCP routing, congestion results were very close to the optimal results obtained in the SLE case. The LCP algorithm tries to adaptively minimize the congestion as each connection arrives.

In Li and Somani, LCP performs much better than the fixed-alternate routing, and the LCP is extended to use neighborhood information to reduce setup delay and control overhead.[14] Instead of examining all the links on the preferred route for availability of free wavelengths as in LCP, only the first k links on the preferred route were searched, where k is a parameter to the algorithm. It has been demonstrated that, when $k = 2$, this algorithm can achieve similar performance to fixed-alternate routing.

13.2.2.2 Wavelength assignment for DLE

Once a route is established for a lightpath, it remains to select a wavelength for this lightpath if multiple feasible wavelengths are available. Heuristics are used for wavelength assignment for DLE. Ten heuristics for DLE found in the literature were reviewed in Zang et al.[6] These ten heuristics are summarized as follows:

1. Random Wavelength Assignment (RANDOM) — A wavelength is chosen randomly from the set of available wavelengths.
2. First-Fit (FF) — In this scheme, all wavelengths are searched sequentially, and the first available wavelength is selected. This scheme has

small computational overhead and low complexity, and performs well in terms of blocking probability and fairness. Harai et al. demonstrated that a routing method with FF wavelength assignment slightly reduces blocking probability, compared with the one with random wavelength assignment.

3. Least-Used (LU)/SPREAD — LU selects the wavelength that is the least utilized in the network in order to achieve a near-uniform distribution of the load over the wavelength set. The utilization of a wavelength is defined as the number of links on which the wavelength is currently being used.

4. Most-Used (MU)/PACK — MU is the opposite of LU in that it selects the wavelength that is the most utilized in the network. Mokhtar and Azizoglu examined the performance of adaptive routing in conjunction with the wavelength assignment schemes PACK, RANDOM, and SPREAD.[15] It was shown that the PACK scheme had the best performance, followed by RANDOM, and then by SPREAD.

5. Min-Product (MP) — MP is used in multi-fiber networks, and it attempts to minimize the number of fibers in the network. It chooses the wavelength that has the minimum

$$\prod_{l\in\pi(p)} D_{lj} ,$$

where D_{lj} is the number of assigned fibers on link l and wavelength j. $\pi(p)$ is the set of links comprising path P.

6. Least-Loaded (LL) — Designed for multi-fiber networks, the LL heuristic selects the wavelength that has the largest residual capacity on the most-loaded link along the path. The residual capacity of wavelength j on link l is the number of fibers on which wavelength j is available on link l.

7. MAX-SUM (MΣ) — MΣ was proposed for multi-fiber networks but it can also be applied to the single-fiber case. MΣ attempts to maximize the remaining path capacities after the lightpath establishment. The path capacity of path p is the sum of the path capacities on all wavelengths, and the path capacity on wavelength j is the number of fibers on which wavelength j is available on the most-loaded link along path p.

8. Relative Capacity Loss (RCL) — RCL is based on MΣ, which can be viewed as choosing the wavelength that minimizes the capacity loss on all lightpaths (i.e., to minimize the total capacity loss on this wavelength, because only the capacity on this wavelength will change after the lightpath is set up). RCL differs from MΣ in that it chooses wavelength to minimize the capacity loss relative to capacity on this wavelength before the lightpath is set up. This algorithm is extended to distributed relative capacity loss (DRCL) in Zang et al.

to simplify the computation and speed up the wavelength assignment procedure.[6] DRCL works well with adaptive routing for distributed-controlled networks.

9. Wavelength Reservation (Rsv) — Rsv heuristic reserves a given wavelength on a specified link for a traffic stream, usually a multi-hop stream to reduce the blocking for multi-hop traffic.[9]

10. Protecting Threshold (Thr) — In Thr, a single-hop connection is assigned a wavelength only if the number of idle wavelengths on the link is at or above a given threshold.[9]

Heuristics Rsv and Thr are different from the other heuristics in that they did not specify which wavelength to choose; therefore, they must be combined with other wavelength assignment methods. Instead of minimizing the overall blocking probability, they attempt to protect only the multi-hop connections.

The first eight wavelength assignment heuristics were simulated with fixed routing in Zang et al.[6] It was found that more complicated heuristics, such as MAX-SUM and RCL, provide smaller blocking probability than simpler heuristics; however, the difference in performance among the various heuristics is not very large.

13.2.2.3 Considering routing and wavelength assignment jointly

Rather than solving the routing and WA subproblems separately, sometimes methods that consider routing and WA jointly are used. Three such methods are discussed next.

The least-loaded routing (LLR) in Karasan and Ayanoglu selects the route-wavelength pair jointly.[16] It chooses the least congested path and wavelength among the available wavelengths over the predetermined k shortest paths. It considers networks with multiple fiber links. Let M_l denote the number of fibers on link l, and let A_{lj} denote the number of fibers for which wavelength j is utilized on link l. Let W be the number of wavelengths. For each connection request with wavelength-converted network, the LLR chooses the path p and wavelength j from the set of k shortest paths that achieves

$$\max_{p} \min_{l \in p} (WM_l - \sum_{j=1}^{W} A_{lj})$$

With wavelength-continuous network, the LLR chooses the path p and wavelength j that achieves

$$\max_{p,j} \min_{l \in p} M_l - A_{lj}$$

It was demonstrated that this LLR algorithm achieves much better blocking performance compared with the fixed shortest-path routing algorithm in networks both with and without wavelength converters.

The dynamic routing method in Harai et al. combines routing and wavelength selection.[11] A weighted function δ_{ij} is defined as:

$$\delta_{ij} = \alpha_1 \beta_i + (1 - \alpha_1)\theta_j(\gamma_j, h_j)$$

where α_1 is a constant ($0 \le \alpha_1 \le 1$), β_i denotes the number of links on which the wavelength λ_i is idle, θ_j is a function of hop counts h_j and the number of available wavelength γ_j on the route S_j. Then, the route S_j with the smallest δ_{ij} value is selected to establish a connection that uses wavelength λ_i. $\theta_j(\gamma_j, h_j)$, which could be selected as:

$$\theta_j(\gamma_j, h_j) = \alpha_2(W - \gamma_j) + (1 - \alpha_2)h_j$$

where α_2 is a constant, W is the total number of wavelengths. Simulation shows that this method is efficient when traffic load is low or many wavelengths are prepared on the link.

The adaptive unconstrained routing (AUR) algorithms in Mokhtar and Azizoglu select path and wavelength jointly.[15] An unconstrained routing scheme considers all paths between the source and the destination in the routing decision. Given a network with k links and W wavelengths, the state of the network at time t is given by the matrix

$$\sigma_t = (\sigma_t^{(0)}, ..., \sigma_t^{(k-1)})$$

where $\sigma_t^{(i)}$ ($0 \le i \le k - 1$) is a column vector

$$\sigma_t^{(i)} = (\sigma_t^{(i)}(0), \sigma_t^{(i)}(1), ..., \sigma_t^{(i)}(W - 1))^T$$

where $\sigma_t^{(i)}(j) = 1$ if wavelength λ_j is utilized by some connection at time t and $\sigma_t^{(i)}(j) = 0$ otherwise. The search for a route and a wavelength assignment can be viewed as a search over the rows and the columns of the network state matrix. Each row specifies the available topology at wavelength λ_j, where a standard shortest-path algorithm is used to find a path on the effective topology. The wavelength set is thus searched sequentially until an available path is found. Different adaptive RWA algorithms were proposed by considering different sorting mechanisms of the wavelength set:

- PACK — Wavelengths are searched in descending order of utilization.
- SPREAD — Wavelengths are searched in ascending order of utilization.
- RANDOM — The wavelength set is searched in a random order with a uniform distribution.
- EXHAUSTIVE — All the wavelengths are searched for the shortest available path.
- FIXED — The search order is fixed *a priori*.

It was demonstrated through simulation that unconstrained routing improves blocking performance compared with fixed routing and fixed-alternate routing. Incorporating network state information about wavelength utilization into the wavelength selection process results in marginal improvement in the blocking probability.

The RWA methods reviewed previously are summarized in Table 13.1.

13.3 Network control and management mechanisms

To set up and take down all-optical connections, a control mechanism is needed to select a route, assign a wavelength to the connection and configure the appropriate optical switches in the network. The control mechanism can either be centralized or distributed. Distributed mechanisms are preferred because they are more robust. The three distributed mechanisms examined in the literature[4,5,7,17,18] are described next.

13.3.1 The link-state approach

In the link-state approach,[17] each node maintains information of the network topology and the usage of wavelengths on the network links. A topology update protocol is employed through which each node periodically broadcasts pertinent topology information to all of the other nodes. Upon the arrival of a connection request, the originator node utilizes the topology information to select a route and a wavelength. The originator node then sends reservation requests for the selected wavelength to all the other nodes in the route in parallel. Each node on the route sends a positive or a negative response to the originator. If all the responses are positive, the originator node sends a SETUP message to all the other nodes in the path. The appropriate switches are then configured at each node, and the connection is established. If the reservation was not successful, the originator sends a TAKEDOWN message to each node on the route to release the reservation.

13.3.2 The distributed routing approach

In the distributed routing approach,[5] the knowledge of the overall network topology is not maintained at each node. Each node maintains information about the next hop and the cost associated with the shortest path to each destination on a given wavelength, in the form of a routing table. The cost may reflect hop counts or actual fiber link distances. When a connection request arrives, the source node will choose the wavelength, which results in the shortest distance to the destination. It will then reserve the wavelength to the next hop given by the routing table. Thus, each node selects the next hop and reserves the desired wavelength independently according to the routing information maintained in its routing table. This way the connection request is routed hop by hop toward the destination node. If a node is unable to reserve the designated wavelength on a link, it will send a negative acknowledgment (ACK) back to the source along the reverse path, and the nodes on the reverse path will release the reservation as they receive the negative ACK. Once the connection request reaches the destination node, the destination node sends a positive ACK back to the source node along the reverse path. Each node along the reverse path will configure its wavelength-routing switch as it receives the ACK. Whenever a connection is established or taken down, each node along the route sends to each of its neighbors an update message and the routing table at each node is updated subsequently.

In Zang et al., the two distributed control mechanisms previously described were compared through simulation based on five metrics:

1. Connection setup time
2. Blocking probability
3. Stabilizing time
4. Bandwidth requirement for control messages
5. Scalability

The link-state protocol has a lower stabilizing time than the distributed routing protocol, while the distributed routing protocol has a lower connection setup time, slightly lower blocking probability, and is more scalable than the link-state protocol.

13.3.3 The alternate link routing/deflection routing

The alternate link routing (or deflection routing) method[18] utilizes only local state information for routing. Each node maintains a routing table that indicates, for each destination, one or more predetermined alternate outgoing links to reach that destination as well as information regarding the status of wavelength usage on its own outgoing links. Upon the arrival of a connection request, the source node chooses one of its outgoing links based on either the "shortest-path" policy or the "least congested first" policy, and then forwards the connection request to the next-hop node on the link. The next-hop node will again choose one of its outgoing links and forward the connection request to the next node when the connection request moves from node to node toward the destination. It does not reserve any network capacity. Instead, it gathers the set of wavelengths that are available on all links in the route. When the connection request reaches the destination, the destination node may choose one wavelength according to some policy, such as first-fit or RANDOM selection. The destination node will then send a SETUP message along the reverse path back to the source node, reserving the selected wavelength on the way. A connection request is blocked when, upon arrival to a node, it finds that none of the alternate outgoing links has the desired wavelength available, or the alternate outgoing links lead to nodes that have already been visited. Because the alternate link routing method does not require the node to maintain any global information and only requires local state updates, the control bandwidth demand is greatly decreased. Its scalability is also high compared with the link-state approach and the distributed routing approach.

13.4 Summary

This chapter has discussed the various RWA algorithms in wavelength-routed WDM networks. The network control and management mechanisms examined in the literature were also summarized.

The RWA problem can either be solved by considering the routing and WA jointly or by partitioning the problem into two subproblems: (1) routing and (2) WA. For static traffic with wavelength continuity constraint, the RWA problem can be formulated as an ILP in which the objective function is to minimize the flow in each link. Fixed routing and fixed-alternate routing can be utilized for solving the routing subproblem for static traffic. Graph-coloring algorithms can be used for solving the WA subproblem for static traffic.

Fixed routing, fixed-alternate routing, and adaptive routing methods can be employed for solving the routing subproblem for the dynamic traffic. Various heuristics have been used for the WA subproblem for dynamic traffic. Several algorithms that consider routing and WA jointly for dynamic traffic were also examined in the literature.

The network control and management mechanism can either be centralized or distributed. The link-state approach, the distributed routing approach, and the alternate-link routing approach have been proposed as distributed control mechanisms.

References

1. Mukherjee, B., *Optical Communication Networks,* New York, McGraw-Hill, 1997.
2. Ramaswami, R. and Sivarajan, K.N., *Optical Networks: A Practical Perspective,* Morgan Kaufmann Publishers, San Francisco, CA, 1998.
3. Chlamtac, I., Ganz, A., and Karmi, G., Lightpath communications: an approach to high bandwidth optical WANs, *IEEE Trans. on Commun.,* vol. 40, no. 7, pp. 1171–1182, July 1992.
4. Mukherjee, B., WDM optical communication networks: progress and challenges, *IEEE J. on Selected Areas in Commun.,* vol.18, no.10, pp. 1810–1824, Oct. 2000.
5. Zang, H., Sahasrabuddhe, L., Jue, J.P., Ramamurthy, S., and Mukherjee, B., Connection management for wavelength-routed WDM networks, *Proc. Global Telecommun. Conf.,* pp. 1428–1432, 1999.
6. Zang, H., Jue, J.P., and Mukherjee, B., A review of routing and wavelength assignment approaches for wavelength-routed optical WDM networks, *Optical Networks Mag.,* vol. 1, no. 1, Jan. 2000.
7. Zang, H., Jue, J.P., Sahasrabuddhe, L., Ramamurthy, R., and Mukherjee, B., Dynamic lightpath establishment in wavelength-routed WDM networks, *IEEE Commun. Mag.,* vol. 39, no. 9, pp. 100–108, Sept. 2001.

8. Ramamurthy, S. and Mukherjee, B., Fixed-alternate routing and wavelength conversion in wavelength-routed optical networks, *Proc. IEEE Global Telecommun. Conf.*, pp. 2295–2302, 1998.

9. Birman, A. and Kershenbaum, A., Routing and wavelength assignment methods in single-hop all-optical networks with blocking, *Proc. 14th Annu. Joint Conf. IEEE Computer and Commun. Soc.*, pp. 431–438, 1995.

10. Birman, A., Computing approximate blocking probabilities for a class of all-optical networks, *IEEE J. on Selected Areas in Commun.*, vol. 14, no. 5, pp. 852–857, June 1996.

11. Harai, H., Murata, M., and Miyahara, H., Performance of alternate routing methods in all-optical switching networks, *Proc. 16th Annu. Joint Conf. IEEE Computer and Commun. Soc.*, pp. 516–524, 1997.

12. Chan, K. and Yum, T.P., Analysis of least congested path routing in WDM lightwave networks, *Proc. Networking for Global Commun.*, pp. 962–969, 1994.

13. Banerjee, D. and Mukherjee, B., A practical approach for routing and wavelength assignment in large wavelength-routed optical networks, *Proc. IEEE J. on Selected Areas in Commun.*, vol. 14, no. 5, pp. 903–908, June 1996.

14. Li, L. and Somani, A.K., Dynamic wavelength routing using congestion and neighborhood information, *IEEE/ACM Trans. on Networking*, vol. 7, no. 5, pp. 779–786, Oct. 1999.

15. Mokhtar, A. and Azizoglu, M., Adaptive wavelength routing in all-optical networks, *IEEE/ACM Trans. on Networking*, vol. 6, no. 2, pp. 197–206, April 1998.

16. Karasan, E. and Ayanoglu, E., Effects of wavelength routing and selection algorithms on wavelength conversion gain in WDM optical networks, *IEEE/ACM Trans. on Networking*, vol. 6, no. 2, pp. 186–196, April 1998.

17. Ramaswami, R. and Segall, A., Distributed network control for optical networks, *IEEE/ACM Trans. on Networking*, vol. 5, no. 6, pp. 936–943, Dec. 1997.

18. Jue, J.P. and Xiao, G., An adaptive routing algorithm with a distributed control scheme for wavelength-routed optical networks, *Proc. Int. Conf. on Computer Commun. Networks*, pp. 192–197, Las Vegas, NV, Oct. 2000.

chapter fourteen

A novel distributed protocol for path selection in dynamic wavelength-routed WDM networks

Pin-Han Ho
University of Waterloo
Hussein T. Mouftah
University of Ottawa

Contents

14.1 Introduction

This chapter solves the problem of path selection for WDM mesh networks with a special focus on the implementation in middle-sized networks, such as metropolitan-area networks (MANs). A novel routing and signaling protocol, called asynchronous criticality avoidance (ACA), is proposed to improve the network performance. With the ACA protocol, a specific set of wavelength channels are defined as *critical links* between a node pair according to dynamic link-state. Criticality information is defined as the critical links and the associated information, which is coordinated and disseminated by each source node to every other source node as an *inter-arrival critical coordination*. Routing and wavelength assignment is performed along with the criticality avoidance mechanism, in which path selection process is devised to take the criticality information into consideration. Simulation is conducted in 22- and 30-node networks to examine the proposed approach. The simulation results show that the ACA protocol significantly outperforms the fixed-path least-congested (FPLC) and adaptive dynamic routing (ADR) schemes under the fixed-alternate routing architecture.

In dynamic wavelength-routed WDM mesh networks without wavelength conversion, achieving load balancing in terms of physical topology while avoiding resource fragmentation in terms of independent wavelength channels is the most critical issue for improving the network throughput. The preceding two tasks are also referred to as routing and wavelength assignment, respectively. In the case that a network link in a direction contains multiple fibers (or a multi-fiber network), a single wavelength plane along a network edge contains multiple interchangeable wavelength channels. In such circumstance, a dilemma emerges: the "spread" of working capacity into the network can achieve better load-balancing on each wavelength plane; however, resource fragmentation may be caused between different wavelength planes. On the other hand, the "pack" of working capacity can eliminate resource fragmentation between different wavelength planes; however, the physical load-balancing characteristic is lost. The objective of this chapter is to solve this dilemma.

Extensive studies have been reported in solving the dynamic routing and wavelength assignment (RWA) problem in a multi-fiber WDM network mentioned previously. Because the provisioning latency is required to be less than several hundreds of milliseconds, the RWA algorithm needs to be time-efficient, in which a global optimization is totally impossible. Therefore, most of the reported schemes separate routing and wavelength assignment into two steps to simplify the RWA process (i.e., physical route is selected followed by a wavelength assignment process). For selecting the physical route, fixed-alternate routing (FAR), which is aimed at reducing the amount of dynamic link-state dissemination and computation efforts, is one of the most commonly used strategies in WDM mesh networks.[1–4] For implementing the FAR, basically each network node is equipped with a routing table, in which the *alternative paths* (or abbreviated as APs in the following context) to every other node in the network are determined at the network planning stage. As a connection request arrives at a node pair, one of the end nodes behaves as a *coordinator*, which looks up the routing table for deriving the APs to the other end node (called a *subordinator*) and sends probing packets along each of the APs for gathering dynamic link-state and wavelength availability of each span. We use *coordinator* and *subordinator* instead of source and destination because every connection request is assumed to be for a bidirectional lightpath, in which a source or a destination node is not explicitly defined. We also use *coordinator–subordinator pair* (or *C–S pair*) to replace the term *source-destination* pair in this study.

This study adopted the architecture of FAR due to the following reasons. The FAR is more suitable to a middle-sized network than a large-sized one in that, for keeping the provisioning swiftness, the probing latency, which is increased as the network size increases, cannot be too long. Besides, the size of a probing packet, where all the per-wavelength link-state along an alternate path is recorded, should be limited so that the probability of a successful probing without any dropping due to a transmission error can be kept high. As a result, a large-sized network — in which a round-trip-delay and the node processing time may take a few seconds — cannot cooperate with the FAR architecture well. In addition, path-based link metrics[5,6] are important for performing RWA. The FAR can provide a strong path-based characteristic during the path selection process, which encourages us to adopt this framework.

The most commonly used approach in planning alternate paths for each C–S pair is to randomly select a link-disjoint path-pair.[1,2,5] On the other hand, the proposal for preparing all the k-shortest paths for each C–S pair, which can yield a better performance — at the expense of very high maintenance cost and probing traffic — in dealing with a large number of alternate paths for each connection request, can be seen in Ho and Mouftah.[5] The most recent reported scheme for planning alternate paths is presented in Ho and Mouftah; namely, capacity-balanced alternate routing (C-BAR) in which the load-balancing and diversity requirements are imposed to the planning process, and the performance in terms of blocking probability is improved significantly.[6]

As for the wavelength assignment, the reported schemes cooperating with the FAR architecture are briefly surveyed and summarized as follows. The first-fit, random-fit, and min-product[7] select a lightpath by assigning a wavelength to one of the alternate routes according to a predefined rule without considering the dynamic link-state. The FPLC takes the wavelength plane on which the lightpath along this alternate path has the widest bottleneck. The least-used and most-used schemes[8] use custom-defined cost function and standard link-state metrics (i.e., the maximum number of reservable wavelength channels) to select one of the feasible lightpaths for a connection request. With wavelength reservation, a threshold is defined on the residual wavelength channels for a single-hop connection.[9]

Although each of the preceding schemes has its design originality, the performance behaviors are limited due to the ignorance of the influence of network topology and the location of each C–S pair on the traffic distribution. In this chapter, a novel protocol called asynchronous criticality avoidance (ACA) is proposed. With the ACA protocol, every C–S pair in the network keeps track of the number of feasible lightpaths. If the number of feasible lightpaths between the C–S pair is equal to or lower than a predefined threshold, *critical links* are generated between the C–S pair. The critical links along with the associated information (called criticality information) of the C–S pair will be broadcast to all the other coordinators in the network, so that the other C–S pairs can avoid the use of these wavelength channels during an RWA process as much as possible. This is also called a *criticality avoidance routing*.

This chapter is organized as follows. Section 14.2 gives a survey on the C-BAR planning algorithm, which will be adopted in this chapter for the selection of alternate paths. Section 14.3 describes the ACA protocol in detail, which includes the signaling mechanism, criticality information coordination, and the two stages of the RWA processes. Section 14.4 conducts a simulation for verifying the proposed signaling protocol. A comparison is made between the cases with and without adopting the criticality information in the path selection process. Section 14.5 summarizes and concludes this chapter.

14.2 Arrangement of physical routes

14.2.1 Assumptions

We assume that a connection request between node i and node j, $\lambda_{i,j}$ and $\mu_{i,j}$ (connections/time), follows a Poisson arrival and an exponential distribution function of holding time. Without loss of generality, $\mu_{i,j} = 1$ for every (i,j). Therefore, the potential traffic load across (i,j) is also $\lambda_{i,j}$. A link is directional with a fixed number of fibers (i.e., F), each of which accommodates a fixed number of wavelength planes (i.e., K). Wavelength conversion is not allowed in this study.

Because most of the activities in a network may follow a repetitive pattern day after day, the average traffic behavior during a time slot of a day is similar to those at the same time period of the other days. With such premise, $\lambda_{i,j}$ is also varied according to this pattern on a *time-of-a-day* basis. For dealing with this characteristic of traffic pattern, $\lambda_{i,j}$ can be derived by averaging the traffic load in the same period of a day (e.g., the average traffic load from 5:00 p.m. to 7:00 p.m.) over a long duration of time (e.g., the past 2 months). The network planning results based on $\lambda_{i,j}$ are in a semidynamic fashion that needs to be updated for different time slots of a day.

14.2.2 Capacity-balanced alternate routing

Capacity-balanced alternate routing (C-BAR), first introduced by Subramanian and Barry, is a network planning algorithm promised to balance the potential traffic load along each link and economize the management cost of alternate paths.[8] The basic idea of C-BAR is to optimize the distribution of alternate paths for each C–S pair by an integer linear programming formulation. A design methodology was proposed by manipulating the number of alternate paths between each C–S pair so that the blocking performance of C–S pairs with different minimum distance can be as close as possible.

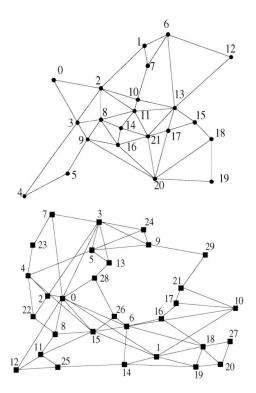

Figure 14.1 Sample networks adopted in this study. (P.H. Ho and H.T Mouftah, in *Optical Networks — Architecture and Survivability,* Kluwer Academic Publishers, Norwell, MA, pp. 88–103, 2003. With permission.)

In this study, the C-BAR algorithm is adopted with the following parameters:

1. F = 5 and W − 16.
2. Two sample networks with 22 and 30 nodes as illustrated in Figures 14.1a and 14.1b are adopted.
3 The potential traffic load is such that the average blocking probability between each C–S pair is ranged from 0.5 to 10% in the sample networks by using a variety of RWA schemes, such as FPLC, with two link-disjoint alternate paths[1] and ADR.[2]
4. The number of alternate paths is taken as 1, 1, 2, 3, and 4 for the C–S pairs with a minimum distance 1, 2, 3, 4, and 5, respectively.

The output of the algorithm is the alternate paths for each C–S pair in the networks.

With the planning algorithm C-BAR, Table 14.1 lists the statistics of alternate paths in the two networks. The symbol "C-BAR" is for the scheme using the C-BAR algorithm for planning alternate paths. Strategy S1 uses all

Table 14.1 Comparison among Four Planning Strategies

		Normalized Deviation (ND)	Percentage of Overlapping (PO)	Normalized Extra Length (NEL)	Average No. of APs for Each S-D
22-node	C-BAR	12.2%	6.39	1.07	2.18
	S1	38.5%	20.3	1.00	2.46
	S2	33.8%	39.2	1.64	11.85
	2D	48.2%	0	1.03	2
30-node	C-BAR	11.6%	6.17	1.06	2.31
	S1	48.5%	16.6	1.00	2.88
	S2	42.2%	41.3	1.61	13.18
	2D	57.7%	0	1.02	2

Source: P.H. Ho and H.T Mouftah, in *Optical Networks — Architecture and Survivability*, Kluwer Academic Publishers, Norwell, MA, pp. 88–103, 2003. With permission.

the shortest paths (in hop count) as alternate paths for each C–S pair. Strategy S2 uses all the shortest and second shortest paths as alternate paths; strategy 2D randomly selects two link-disjoint shortest paths as alternate paths.

In Table 14.1, the three evaluation indexes for the deployment of alternate paths — ND, PO, and NEL — are explained as follows. PO is the *average percentage of overlapping* between alternate paths across each C–S pair. A large value of PO stands for an inefficient control and dissemination in the network. The *normalized deviation (ND)* is defined as:

$$ND = \frac{1}{2 \cdot E} \cdot \sum_{all(a,b)} \frac{|w_{a,b} - \overline{w}|}{\overline{w}}$$

where $w_{a,b}$ is the potential traffic load across link (a,b), $\overline{w}$ is the average potential traffic load of each directional link, and E is the number of edges in the network. The *normalized extra length (NEL)* is defined as:

$$NEL = \frac{1}{no\_SD} \cdot \sum_{all(i,j)} \frac{avg\_length_{i,j}}{H_{i,j}}$$

where no_SD is the number of C–S pairs in the network, $H_{i,j}$ is the minimum distance between node i and j, and $avg\_length_{i,j}$ is the average length of alternate paths between C–S pairs (i,j) in hops.

From the data presented in Table 14.1, it is clear that C-BAR manipulates the distribution of alternate paths for each C–S pair and yields a significant improvement in all the evaluation indexes. Therefore, the use of the C-BAR planning algorithm can achieve the best performance among all the other planning techniques with a wide variety of wavelength assignment schemes.[6] Strategy 2D is simple and also more fault-tolerant in control and management; however, it is outperformed by strategies S1 and S2 by underutilizing the network resources.[5] Compared with strategy S1, strategy S2 provides more working capacity by equipping each C–S pair with more alternate paths. Therefore, strategy S2 gives better performance while at the expense of requiring more control efforts and dissemination traffic.[5,6]

14.3 Asynchronous criticality avoidance protocol (ACA)

This section presents asynchronous criticality avoidance (ACA) protocol, which contains two basic functional elements: inter-arrival criticality coordination and path selection process.

14.3.1 Inter-arrival criticality coordination

The inter-arrival criticality coordination (abbreviated as *criticality coordination* in the following context) is defined as the efforts made between two network events (i.e., a request for setting up or tearing down a lightpath). Successful inter-arrival criticality coordination starts right after the completion of a network event and must be completed before the arrival of the next network event. To implement the ACA protocol, a series of signaling mechanisms is performed.

We define that a wavelength channel (or a set of wavelength channels on the same wavelength plane) is (are) *critical* to a C–S pair if the occupancy of the wavelength channel (or the set of wavelength channels) will block any of the future connection requests arriving at this C–S pair. After the completion of a network event, probing packets are sent by each coordinator to all its subordinators along the corresponding APs defined in the routing table for gathering wavelength availability information. After each coordinator has derived all the per-wavelength link-state along its APs, it examines whether critical channel(s) to any of its subordinators exist. The algorithm for finding out the critical channel(s) between a coordinator (i.e., node i) and one of its subordinators (i.e., node j), according to the gathered link-state, is presented in the following pseudo code, which is invoked whenever an inter-arrival criticality coordination is performed across a C–S pair.

> **Input:** *Critical_Link$^{(j)}$*: a data structure storing the critical channels to node j in the routing table of node i, which is null if there is no critical situation; *Old_Critical_Link$^{(j)}$*: a data structure storing the critical channels of the last criticality coordination, $L^{(i,j)}$: the threshold of the available lightpaths to the node j. A critical situation is raised if the number of available lightpaths to node j is less than or equal to $L^{(i,j)}$.
> Start:
> *num_Cpath* $\leftarrow$ 0;/* the number of *critical* links */
> *Old_Critical_Link$^{(j)}$* $\leftarrow$ *Critical_Link$^{(j)}$*;
> Critical_Link$^{(j)}$ $\leftarrow$ null;
> For $m \leftarrow 1$ to $|M_{i,j}|$ Do

/* for each AP between node i and j */
For $k \leftarrow 1$ to K Do/* for each wavelength plane */
$\{(a,b,k), bn\} \leftarrow$ Derive the bottleneck along the mth alternate path;
/* the bottleneck is on the wavelength plane k of span a-b with the number of available channels bn, where $0 < bn < F$ */
If $((a,b,k) \notin Critical\_Link^{(j)})$ **Then**
Push $\{(a,b,k)\}$ to $Critical\_Link^{(j)}$;
$num\_Cpath \leftarrow num\_Clink + bn$;
End If
End For
End For
If $(num\_Cpath > L^{(i,j)})$ **Then**
/* the number of total available lightpaths is larger than the threshold, so no critical situation between node i and j exists; */
If $(Old\_Critical\_Link^{(j)} \neq null)$ **Then**
/* if there was a critical situation announced between node i and j in the last inter-arrival planning*/
Broadcast $(Old\_Critical\_Link^{(j)}, \textbf{false})$;
/* release the criticality between node i and j */
/* a **"true"** is for raising up the critical links, and a **"false"** is for releasing the critical links */
$Critical\_Link^{(j)} \leftarrow null$;
/* reset the stack since there is no critical situation */
End If
Else If $(Old\_Critical\_Link^{(j)} \neq Critical\_Link^{(j)})$ **Then**
/* Criticality exists, and an update of critical links is needed */
$new\_Clink \leftarrow$ Derive the new critical links that need to be announced now;
$old\_Clink \leftarrow$ Derive the old critical links that need to be released now;
/* For example, in case $Old\_Critical\_Link$ includes $(a-b, k1)$, $(c-d, k2)$, $(e-f, k3)$, and $(g-h, k3)$, while the $Critical\_Link$ includes $(a-b, k1)$, $(g-h, k4)$ and $(m-n, k5)$, then the old_Clink will be composed of $(c-d, k2)$ and $(g-h, k3)$ and the new_Clink will be $(m-n, k5)$ */
If $(new\_Clink \neq null)$ **Then**
Broadcast $(new\_Clink, \textbf{true}, num\_Cpath)$;
End If
/* announce the newly generated criticality links to all the other ingresses */
/* num_Cpath is broadcast for the decision making of path selection by the other coordinators */
If $(old\_Clink \neq null)$ **Then**
Broadcast $(old\_Clink, \textbf{false}, num\_Cpath)$;
End If
/* announce the old critical links that need to be released to all the other ingresses */
End If
End If

With the pseudo code, node i announces *critical channel(s)* to all the other coordinators by broadcasting the specific physical location and wavelength plane(s) of the *critical channel(s)*, and announces a relief of *critical channel(s)* once the criticality is called off after a disconnection of a lightpath. The threshold of the number of feasible lightpaths before a criticality, $L^{(i,j)}$, is announced. It is proportional to $\rho_{i,j}$, which is the parameter that directly determines the probability of the next connection arrival.

If the ACA protocol fails to complete before the arrival of the next network event, the system has no choice but to use the outdated data. The situation is similar to that of the stall link-state in traditional IP networks, which impairs the network dynamics due to the need to reselect lightpaths when resource reservation conflicts occur. A good network design should be such that most of the inter-arrival planning can be successful. The topic of how to adapt the latency taken by the inter-arrival criticality coordination and the traffic dynamics, however, has been out of the scope of this chapter.

14.3.2 Path selection with criticality information

During the criticality coordination, each coordinator node receives the criticality information from some of the other coordinators that need to update their criticality information. An example of critical channels and the criticality information dissemination is illustrated in Figure 14.2. Two alternate paths (i.e., AP1 and AP2), each of which contains five wavelength channels on the *kth* wavelength plane (or there are five fibers in a direction), have been set up between End 1 and End 2. We assume that End 1 is the coordinator, and the threshold of criticality between End1 and End2 is, $L^{(End1,End2)} = 2$. As the criticality coordination starts, the probing packets are sent by End 1 back and forth along the APs, to derive the per-wavelength link-state for

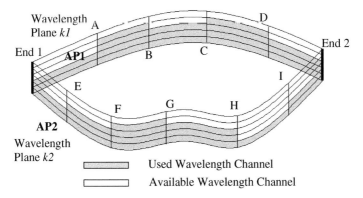

Figure 14.2 An example for defining the critical channels and criticality information dissemination. (P.H. Ho and H.T Mouftah, in *Optical Networks — Architecture and Survivability*, Kluwer Academic Publishers, Norwell, MA, pp. 88–103, 2003. With permission.)

the C–S pair End1-End2 right after a network event is settled. Assume only two available lightpaths are left, and all the others along AP1 and AP2 are not available at this time. The critical links in this case are (A-B, $k1$) and (C-D, $k1$) in AP1, and (F-G, $k2$) in AP2, which will be disseminated by End 1 to all the other coordinators. Note that if another C–S pair routes a lightpath traversing either (A-B, $k1$) or (C-D, $k1$), the lightpath in AP1 is blocked. In this case, the criticality situation along AP1 needs to be called off. In other words, the criticality of (A-B, $k1$) or (C-D, $k1$) has to be called off in the next inter-arrival planning if (C-D, $k1$) or (A-B, $k1$) is occupied.

The cost of taking a wavelength channel that is not critical is determined by the ratio between the number of all wavelength channels and the number of residual wavelength channels in the wavelength plane. The cost function for ranking the feasible lightpaths between the C–S pair (i,j) is as follows:

$$C_{m,k}^{(i,j)} = \sum_{t=1}^{h_m^{(i,j)}} c_{t,m,k}^{(i,j)} \tag{14.1}$$

$$c_{t,m,k}^{(i,j)} = \begin{cases} \infty & \text{if the wavelength channel is announced critical or occupied by a lightpath} \\ \dfrac{F_{t,m,k}^{(i,j)}}{FC_{t,m,k}^{(i,j)}} & \text{otherwise} \end{cases}$$

where $C_{m,k}^{(i,j)}$ is the total cost of the lightpath on the kth wavelength plane along the mth alternate path between node i and j, $c_{t,m,k}^{(i,j)}$ is the cost of taking the tth link on the kth wavelength plane along the mth alternate path between node i and j, $h_m^{(i,j)}$ is the number of hops along the mth alternate path between node i and j, and $F_{t,m,k}^{(i,j)}$ and $FC_{t,m,k}^{(i,j)}$ are the number of fibers and free wavelength channels on the tth link of the kth wavelength plane along the mth alternate path between node i and j. $F_{t,m,k}^{(i,j)}$ (or $F_{a,b}$, as $span(a-b) = span_{t,m,k}^{(i,j)}$) is the number of wavelength channels on the kth wavelength plane along the span a-b. The lightpath to be set up is selected from all the feasible lightpaths along all the alternate paths and all the wavelength planes. For example, the lightpath with a cost:

$$\min_{1 \le k \le K} \min_{1 \le m \le |M_{i,j}|} C_{m,k}^{(i,j)}.$$

If using the cost function in Equation (14.1) blocks the connection request, the algorithm will still select a lightpath in order not to suffer an immediate blocking while yielding a *minimum interference* to the future connection requests. The cost function of a feasible lightpath at this stage is as follows:

$$C_{m,k}^{(i,j)} = \sum_{(r,s) \in CriticalitySet} \frac{L^{(r,s)}}{num\_Cpath^{(r,s)}} \tag{14.2}$$

where $num\_Cpath^{(r,s)}$ is the number of feasible lightpaths between the node r and s, and *CriticalSet* is the set of C–S pairs where the critical lightpath(s) is (are) interfered. The lightpath with the least value of $C_{m,k}^{(i,j)}$ is selected. Eq. (14.2) distinguishes the interference of setting up a lightpath by taking the threshold of criticality $L^{(r,s)}$ into consideration, which is an important reference on how "critical" the occupancy of the wavelength channel is to the C–S pair. The larger $L^{(r,s)}$ is, the more likely the next connection request will be launched between nodes r and s because $\rho_{i,j}$ is proportional to $L^{(r,s)}$. In other words, with the same amount of residual available lightpaths, $num\_Clink^{(r,s)}$, a larger $L^{(r,s)}$ incurs a higher cost proportionally.

14.4 Verification

Simulation is conducted in two different networks (22-node and 30-node) as depicted in Figure 14.1. Every connection request is for a bidirectional lightpath on the same wavelength plane and physical route. To balance the tasks of sending/receiving/coordinating the probing packets, the coordinator of a C–S pair is defined as the end node numbered by v such that $|\upsilon - v| < N/2$ for any v that numbers the other end node, where the network nodes are numbered from 0 to N-1. With this, every network node gets equal chance to behave as a coordinator and a subordinator. Note that all the nodes can be coordinators and subordinators to other different nodes at the same time.

To examine the proposed C-BAR algorithm and the ACA protocol, the simulation adopts the following reported RWA schemes for the purpose of comparison:

1. FPLC with 2 link-disjoint shortest paths as APs, abbreviated as FPLC-2D
2. Adaptive dynamic routing, abbreviated as ADR
3. ADR with a length limitation of one extra hop than the minimum distance on the selection of lightpath, abbreviated as ADR-2S (i.e., only the shortest and the second shortest paths are possibly selected)
4. ADR with a length limitation of two extra hops than the minimum distance on the selection of lightpath, abbreviated as ADR-3S (i.e., only the shortest, second shortest and the third shortest paths are possibly selected) (ADR-xS is also called *length-constraint ADR*)
5. C-BAR with FPLC, abbreviated as C-BAR&FPLC (Alternate paths are arranged for each C–S pair according to the planning design results described in Section 14.2.2)
6. C-BAR with ACA, abbreviated as C-BAR&ACA (Alternate paths are arranged for each C–S pair according to the planning design results described in Section 14.2.2)

For the ADR, the cost function is:

$$C_{a,b}^k = 1 + \frac{F}{RB_{a,b}^k}$$

where $RB_{a,b}^k$ is the number of available wavelength channels on the kth wavelength plane. When a connection request arrives, Dijkstra's algorithm is invoked. The final decision will be made by selecting the best lightpath among all the wavelength planes. ADR-nS is basically the ADR with a length limit. If the length of a derived path exceeds the minimum distance by (n-1) hops, the path will not be taken; instead, a blocking is announced. For FPLC, the lightpath with the widest bottleneck on a wavelength plane is selected. If two or more candidate lightpaths have the same width of bottleneck, the one with a shortest distance has a higher priority to be selected. FPLC is reported to be one of the most efficient wavelength assignment schemes under the FAR architecture in dealing with path selection for multi-fiber WDM wavelength-routed networks.[10] FPLC has an increasingly better performance than the other wavelength assignment schemes (e.g., minimum product, most used, or least used, etc.)[10] when the number of fibers along a link (i.e., F) is getting larger; however, this simulation did not use FPLC to examine the cases with longer APs (i.e., paths with non-minimum-distance) because FPLC does not actually put any weight on the length of a path. As a result, the preparation of long APs may increase the average network resources consumed by each connection request, which impairs performance.

The blocking performance for all the cases is plotted in Figure 14.3. Each fiber contains 16 wavelength channels, and 5 fibers are in each direction. The

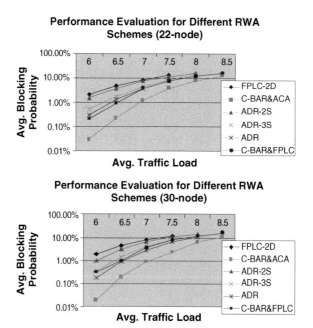

Figure 14.3 Comparison of different RWA schemes with the proposed schemes, using ACA protocol. (P.H. Ho and H.T Mouftah, in *Optical Networks — Architecture and Survivability*, Kluwer Academic Publishers, Norwell, MA, pp. 88–103, 2003. With permission.)

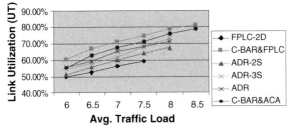

Avg. Link Utilization vs. Avg. Traffic Load between Each C-S pair (22-node)

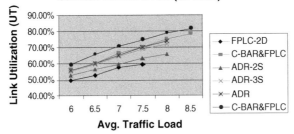

Avg. Link Utilization vs. Avg. Traffic Load between Each C-S Pair (30-node)

Figure 14.4 Link utilization (UT) vs. traffic load in the (a) 22-node network and (b) 30-node network. (P.H. Ho and H.T Mouftah, in *Optical Networks — Architecture and Survivability*, Kluwer Academic Publishers, Norwell, MA, pp. 88–103, 2003. With permission.)

number of connection requests for a trial is 600,000, and averaging the results of four trials derives a final data in each plot.

The simulation results show that C-BAR&ACA outperforms all the other schemes significantly. The reason for the superiority of C-BAR&ACA scheme can be observed in Figures 14.4 and 14.5, in which the average *link utilization (UT)* and the *normalized average extra length (NEL)* in each case are demonstrated. *UT* and *NEL* are defined as follows:

$$UT = \frac{\sum_{f}^{F}\sum_{j}^{J}\sum_{k}^{K}\sum_{e}^{E}\delta_{f,j,k,e} \cdot t_e}{F \cdot J \cdot K \cdot \sum_{e}^{E} t_e}$$

where F is the number of fibers along a span, J is the number of links (directional) in the network, K is the number of wavelengths in a fiber, E is the number of network events during an experimental trial, t_e is the time elapsed between the *eth* and the *(e+1)th* events, and $\delta_{f,j,k,e}$ is a binary param-

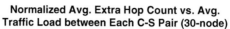

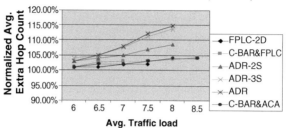

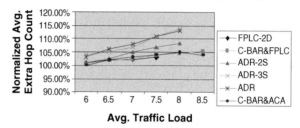

Figure 14.5 Normalized average extra length (NEL) vs. traffic load in the (a) 22-node network and (b) 30-node network. (P.H. Ho and H.T Mouftah, in *Optical Networks — Architecture and Survivability*, Kluwer Academic Publishers, Norwell, MA, pp. 88–103, 2003. With permission.)

eter, which is 1 if the *kth* wavelength channel on the *fth* fiber of *jth* links between the *eth* and the *(e+1)th* events is occupied, and 0 otherwise.

$$
NEL = \frac{\displaystyle\sum_{v}^{V} \sum_{e}^{E} \frac{h_{v}}{h_{v,\min}} \cdot t_{e}}{V \cdot \displaystyle\sum_{e}^{E} t_{e}}
$$

where V is the number of connections launched into the network, h_{v} is the hop count of the *vth* connection, and $h_{v,min}$ is the smallest possible hop count of the *vth* connection.

The best way of improving performance is to increase the *UT* without paying extra cost in *NEL*. As presented in Figures 14.4 and 14.5, C-BAR&ACA gets the highest link utilization among all the schemes while taking relatively lower extra network resources for setting up a connection, and therefore yields a better performance. Figure 14.6 demonstrates the improvement by activating the criticality avoidance mechanism, in which a comparison is made between the cases with and without taking the criticality information into account. The

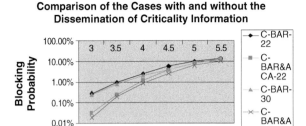

Figure 14.6 Comparison of the cases with and without the dissemination of criticality information. C-BAR-22 and C-BAR-30 are the cases without the criticality information dissemination. (P.H. Ho and H.T Mouftah, in *Optical Networks — Architecture and Survivability,* Kluwer Academic Publishers, Norwell, MA, pp. 88–103, 2003. With permission.)

criticality information dissemination improves the performance especially when the traffic load is low, in which case the chance of blocking can be largely reduced by not taking the critical lightpaths of the other C–S pairs. Because *NEL* should be similar in each case, the adoption of the ACA protocol can effectively increase the link *UT* by reducing the segmentation of network resources. This improvement is at the expense of extensive signaling dissemination and a longer computation time that may be consumed.

14.5 Conclusions

We have demonstrated a novel strategy known as the ACA protocol, which cooperates with the FAR for performing path selection in dynamic WDM wavelength-routed networks. The ACA protocol performs RWA in two stages. It first uses the cost function defined in Equation (14.1) to evaluate the lightpaths without traversing any critical links. If no feasible lightpath is found, the algorithm then uses Equation (14.2) to determine a lightpath with the minimum interference to the future arrivals of connection request. The criticality information coordinated by each coordinator is disseminated to all the other coordinators as an *inter-arrival criticality coordination,* which is defined as the planning efforts conducted between two consecutive network events. This chapter provides a detailed pseudo code with step-by-step comments in the procedure of performing criticality coordination. Simulation is conducted on two networks with 22 and 30 nodes, respectively. The simulation results show that with the criticality avoidance mechanism, the blocking probability can be significantly reduced compared with ADR, length-constraint ADR, and FPLC schemes by having the pattern of alternate paths to be 2D and C-BAR. We conclude that the ACA protocol can improve the network performance especially when the traffic load is light, with which the availability of a certain amount of lightpaths between each C–S pair can be preserved most of the time. With the ACA protocol, both the load-bal-

ancing characteristic of traffic distribution and avoidance of resource segmentation can be achieved at the same stage.

References

7. L. Li and A. K. Somani, A new analytical model for multi-fiber WDM networks, *IEEE J. on Selected Areas in Commun.*, vol. 18, no. 10, Oct. 2000, pp. 2138–2145.
8. J. Spath, Resource allocation for dynamic routing in WDM networks, *Proc. of All-Optical Networking 1999: Architecture, Control, and Management Issues*, Boston, MA, Sept. 1999, pp. 235–246.
9. E. Karasan and E. Ayanoglu, Effects of wavelength routing and selection algorithms on wavelength conversion gain in WDM optical networks, *IEEE/ACM Trans. on Networking*, vol. 6, no. 2, April 1998, pp. 186–196.
10. S. Xu, L. Li, and S. Wang, Dynamic routing and assignment of wavelength algorithm in multi-fiber wavelength division multiplexing networks, *IEEE J. on Selected Areas in Commun.*, vol. 18, no. 10, Oct. 2000, pp. 2130–2137.
11. P.-H. Ho and H. T. Mouftah, Network planning algorithms for the optical Internet based on the generalized MPLS architecture, *Proc. of IEEE Globecom '01*, San Antonio, TX, Nov. 2001, OPC03-2.
12. P.-H. Ho and H. T. Mouftah, Capacity-balanced alternate routing for MPLS traffic engineering, *Proc. of IEEE Int. Symp. on Computer and Commun. 2002 (ISCC 2002)*, Taormina, Italy, July, 2002, pp. 927–932.
13. G. Jeong and E. Ayanoglu, Comparison of wavelength-interchanging and wavelength-selective cross-connects in multi-wavelength all-optical networks, *Proc. of IEEE INFOCOM '96*, San Francisco, CA, March 1996, pp. 156–163.
14. S. Subramaniam and R. A. Barry, Wavelength assignment in fixed routing WDM networks, *Proc. of IEEE International Conference on Commun. (ICC '97)*, Montreal, Canada, June 1997, pp. 406–410.
15. D. Katz, D. Yeung, and K. Kompella, Traffic engineering extensions to OSPF, Version 2, Internet Draft, <draft-katz-yeung-ospf-traffic-09.txt>, work in progress, July 2001.
16. H. Zang, J. P. Jue, and B. Mukherjee, A review of routing and wavelength assignment approaches for wavelength-routed optical WDM networks, *SPIE Optical Networks Mag.*, vol. 1, no. 1, Jan. 2000, pp. 47–60.

chapter fifteen

Distributed lightpath control for wavelength-routed WDM networks

Jun Zheng
University of Ottawa
Hussein T. Mouftah
University of Ottawa

Contents

0-8493-1333-3/03/$0.00+$1.50
© 2003 by CRC Press LLC

15.1 Introduction

Lightpath control is one of the key issues in wavelength-routed wavelength division multiplexing (WDM) networks. To achieve good network performance, the network must employ a lightpath control mechanism to efficiently establish lightpaths for connection requests. In general, lightpath establishment can be performed under either centralized or distributed control. Although centralized control is relatively simple to implement and works well for static traffic, it is not scalable and reliable. For this reason, centralized control is not considered suitable for large networks with dynamic traffic. In contrast, distributed control improves network scalability and reliability, and is highly preferred for large WDM networks; however, distributed control increases the complexity in implementation, which presents a big challenge for network designers.

The emergence of WDM technology has tremendously increased the usable transmission capacity of optical fibers. WDM allows multiple optical signals to be transmitted simultaneously and independently in different optical channels over a single optical fiber, and can thus meet the ever-increasing bandwidth demand of network users. The early deployment of WDM technology was mostly in point-to-point transmission. With the advent of reconfigurable optical devices, such as optical add/drop multiplexers (OADMs) and optical cross-connects (OXCs), WDM is evolving from a point-to-point transmission technology toward a networking technology. Wavelength-routed WDM networks are considered a promising network infrastructure for the next-generation optical Internet.[1]

In a wavelength-routed WDM network, a lightpath must first be established before data is actually transferred. Due to the limitation in the number of wavelengths available on each fiber link as well as the wavelength-continuity constraint in the absence of wavelength converters, a connection request may be blocked because of the unavailability of wavelength resources. For this reason, efficient provisioning of lightpaths has been a crucial issue in wavelength-routed WDM networks.

To achieve good network performance, the network must employ a lightpath control mechanism to efficiently establish lightpaths for connection requests. In general, lightpath control can be either centralized or distributed. Under centralized control, all connection requests are sent to a central controller and are processed sequentially. The central controller maintains global network state information and is responsible for lightpath establishment on behalf of all network nodes. Under distributed control, all connection requests are processed concurrently at different network nodes. Each node makes its decisions based on the network state information it maintains.

Distributed lightpath control has received much attention in recent years and has been widely studied in the literature. This chapter discusses the major issues that are involved in distributed lightpath control with a focus on lightpath establishment, including routing, wavelength assignment, and wavelength reservation. We also review various lightpath establishment mechanisms that have already been proposed in the literature, including various routing approaches, wavelength assignment algorithms, and wavelength reservation protocols.

15.2 Distributed lightpath control

This section introduces the architecture of a wavelength-routed WDM network and discusses the major issues involved in distributed lightpath control in terms of routing, wavelength assignment, and wavelength reservation.

15.2.1 Network architecture

A wavelength-routed WDM network typically consists of routing nodes interconnected by point-to-point WDM fiber links in an arbitrary meshed topology, as depicted in Figure 15.1.

In this architecture, each WDM link consists of a pair of unidirectional fiber links that operate in WDM mode with a certain number of optical channels (or wavelengths) on each fiber. Each routing node consists of an OXC that performs wavelength routing and switching optically, and an electronic controller that controls the OXC. The controller maintains network state information (e.g., network topology and wavelength usage) for routing, which can be either local or global information. An access device which is used as an electronic interface between the optical network and its client networks can be connected to each node. In the context of this chapter, an access node and its associated routing node are, as a whole, referred to as a network node. Logically, the network consists of a data network and a control network. The data network, which is used for transferring data and operates in circuit switching, consists of OXCs and data channels. The control network, which is used for exchanging network

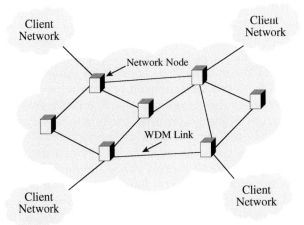

Figure 15.1 A wavelength-routed WDM network.

state information and control information, consists of electronic controllers and control channels, and operates in packet switching. The control channels can be implemented by one or more dedicated wavelengths over each fiber link or through a dedicated IP network.

15.2.2 *Issues in distributed lightpath establishment*

In a wavelength-routed WDM network, data is transferred from one node to another over optical connections called lightpaths. A lightpath is a unidirectional all-optical connection between a pair of source and destination nodes, which may traverse multiple physical fiber links without undergoing any opto-electronic (O/E) and electro-optic (E/O) conversion at each intermediate node. Two lightpaths cannot share the same wavelength on a common fiber link. In the absence of wavelength converters, a lightpath must use the same wavelength on all the links it spans, which is known as the wavelength-continuity constraint. This constraint is unique to WDM networks. It would largely affect the wavelength utilization and thus degrade the network performance in terms of the request block probability. To eliminate this constraint, wavelength converters can be used at network nodes; however, the wavelength conversion technology is not yet mature. The cost of wavelength converters is still considerably high and will remain so in the short term.

To dynamically establish a lightpath for a connection request under distributed control, the network must first decide a route for the connection and select a suitable wavelength on each link of the decided route; it then reserves the selected wavelength and configures the optical switch at each intermediate node along the decided route. In the presence of wavelength conversion, a suitable wavelength can be any available wavelength. Otherwise, all the links of the decided route must use the same wavelength. If no wavelength is available on any of the links or no common wavelength exists on all the links in the absence of wavelength conversion, a connection request is blocked.

Accordingly, distributed lightpath establishment involves three main aspects: routing, wavelength assignment, and wavelength reservation. In the context of routing, the objective is to make a routing decision for each connection request in a manner that can improve the network performance in terms of the request-blocking probability and the connection setup time. Because the number of wavelengths available on each fiber link is limited and the wavelength-continuity constraint exists in the absence of wavelength converters, reducing the request-blocking probability is of great concern. Another concern is the computational complexity of a routing algorithm. A routing algorithm must be performed online; therefore, a higher computational complexity would result in a longer computational time and thus largely degrade the quality of service of the network. For this reason, a routing algorithm must avoid high computational complexity. To achieve good network performance, a variety of routing approaches have been proposed in the literature, such as explicit routing, hop-by-hop routing, source routing, and destination routing. These approaches will be discussed in more detail in the next section.

In the context of wavelength assignment, the objective is to select a suitable wavelength among multiple available wavelengths in a manner that can increase the wavelength utilization. This is particularly important in the absence of wavelength converters. For this reason, a variety of wavelength assignment algorithms have already been proposed in the literature, such as random-order, fixed-order, most-used, least-used, etc.

Under distributed control, routing and wavelength assignment can be based on either local or global network state information. Under local information, it is very likely that a connection request is blocked because of the unavailability of wavelength resources. An effective way to address this problem is to allow each node to maintain global information and use this information to make more intelligent decisions. In this way, the possibility of a wavelength reservation failure can be largely reduced; however, the network state changes constantly under dynamic traffic, therefore, the network must employ an effective signaling protocol to timely update the network state information maintained at each node in order to make correct decisions. In general, this can be implemented either periodically or upon a change,[2] which would cause significant control overheads in the network. Reducing such control overheads is also a concern. It should be noted that due to the propagation delay on each link, a network node might still make an incorrect decision based on updated information.

In the context of wavelength reservation, the objective is to minimize the possibility of a reservation failure and reduce the connection setup time. Because the network state changes constantly under dynamic traffic, a wavelength that is available at a given time may be no longer available at a later time. For this reason, a reservation failure is very likely to occur at an intermediate node along the decided route. On the other hand, it takes a two-way delay to make a reservation between a pair of source and destination nodes; therefore, the reservation time is also a concern. Network designers should take these factors into account in the design of a reservation protocol so as to minimize the possibility of a reservation failure and reduce the reservation time.

It should be pointed out that in many cases the routing, wavelength assignment, and wavelength reservation aspects are closely related to each other and are therefore considered in an integrated manner, especially the routing and wavelength assignment (RWA) aspects. For example, in adaptive routing based on global information, routing and wavelength assignment are often performed together while wavelength reservation is relatively independent.

15.3 Distributed lightpath control mechanisms

This section reviews various routing, wavelength assignment, and wavelength reservation approaches that have already been proposed in the literature.

15.3.1 Routing

Two basic types of routing paradigms are used: explicit routing and hop-by-hop routing.

15.3.1.1 Explicit routing

In explicit routing, the entire route of a connection is decided by a single network node, which can be either the source node or the destination node.

Source routing. The most commonly used explicit routing approach is source routing. In source routing, the entire route of a connection is decided by the source node. If the source node only maintains local wavelength usage information, a routing decision is made without knowledge of the information on other links. As a result, it is very likely that a connection request is blocked at an intermediate node along the decided route because of a wavelength reservation failure. An effective way to address this problem is to allow each node to maintain global network state information and meanwhile employ a signaling protocol to update this information periodically or upon a change. As a result, the source node can make more intelligent routing decisions and the possibility of a wavelength reservation failure can be largely reduced.

Destination routing. Under certain network conditions, the destination node of a connection can also decide the entire route of the connection in order to achieve better network performance than source routing. For example, a destination routing mechanism is proposed in Zheng and Mouftah, which uses global network state information for routing and allows the destination node of a lightpath to make a routing decision.[3] Because the network state changes constantly with dynamic traffic, the idea behind this mechanism is to use the most recent network state information to make a routing decision so that the possibility of a reservation failure can be minimized. The simulation results have shown the improvements in terms of the request-blocking probability and connection setup time as compared with source routing. In the context of explicit routing, three basic types of routing approaches are used: fixed routing, fixed-alternate routing, and adaptive routing.[4]

Fixed routing. In fixed routing, only a single fixed route exists between each pair of source and destination nodes in the network. The fixed route is predetermined, and any connection between a pair of nodes uses the same fixed route. This imposes a strict restriction on routing. A typical example of fixed routing is the fixed shortest-path routing, in which the shortest-path route is used for any connection between a pair of source and destination nodes and is predetermined by using a shortest-path algorithm, such as Dijkstra's algorithm or the Bellman–Ford algorithm. Fixed routing is simple to implement; however, because a strict restriction is placed on routing, it may result in a high request-blocking probability in the network. In addition, it has no capability to handle a network failure.

Fixed-alternate routing. In fixed-alternate routing, a set of alternate routes exists between each pair of source and destination nodes. The set of alternate routes is predetermined and stored in a routing table maintained at each node. In general, they are link-disjoint in the sense that any of the routes does not share any link with any other of the routes. For example, these routes may include the first-shortest-path route, the second-shortest-path route, the third-shortest-path route, etc. The actual route of a connection can only be chosen from the set of alternate routes, which also imposes a restriction on routing to a certain extent. Compared with fixed routing, fixed-alternate routing can significantly reduce the request-blocking probability. In addition, it also provides some degree of survivability capability against network failures.

Adaptive routing. In adaptive routing, routing is not restricted. Any possible route between a pair of source and destination nodes can be chosen as the actual route of a connection. The choice of routes is based on the current network state and wavelength availability. Compared with fixed-alternate routing, adaptive routing can further reduce the request-blocking probability but increases the computational complexity.

A typical example of adaptive routing is the least-cost-path routing.[4] To compute an optimal route, a cost that can measure the link distance and resource usage is assigned to each link and a least-cost routing algorithm is performed upon the arrival of a connection request. Normally, the least-cost-path route from the source node to the destination node is decided as the actual route. If multiple routes have the same cost, one of them is selected randomly. A connection request is blocked only when no route is available between the source node and the destination node.

Another example of adaptive routing is the least-congested-path (LCP) routing.[5] Similar to fixed-alternate routing, a set of predetermined paths exists between each pair of source and destination nodes. The congestion degree of a link depends on the number of wavelengths available on the link. A link with fewer available wavelengths is considered more congested. The congestion degree of a path is determined by the congestion degree of the most congested link along the path. For a connection request, the least-congested path is decided as the actual route. It has been demonstrated by Li and Somani that the LCP routing can improve the network performance as compared with fixed routing and fixed-alternate routing; however, the LCP routing significantly increases the computational complexity and control overheads. To compute the least-congested path, an LCP routing algorithm must examine all links on all the predetermined paths and the network state information maintained at each node must be updated in a timely manner.

To overcome these shortcomings, a neighborhood-information-based LCP routing algorithm is proposed in Li and Somani.[6] Instead of examining all links on all the predetermined paths, this algorithm only examines the first k neighborhood links on each of the predetermined paths. It has been demonstrated that with $k = 2$, this algorithm can achieve similar performance to fixed-alternate routing.

15.3.1.2 Hop-by-hop routing

In hop-by-hop routing, each node independently decides the next hop. A routing decision is made dynamically one hop at a time and is based on either local or global network state information.

An example of hop-by-hop routing based on global information is the distributed-routing approach.[7] In this approach, each node maintains a routing table, which specifies the next hop and the cost associated with the shortest path to each destination node for each wavelength. Hop counts or link distances can measure the cost. A connection request is routed one hop at a time. Each node independently decides the next hop, and the wavelength that results in the lowest cost is selected. The routing table is maintained by using a distributed Bellman–Ford algorithm and must be updated whenever a connection is set up or taken down. It has been demonstrated that the distributed-routing approach yields lower request-blocking probability than the link-state approach in Ramaswami and Segall.[2]

An example of hop-by-hop routing based on local information is the alternate-link routing approach.[8] In this approach, each node maintains a routing table, which specifies one or more alternate outgoing links to each destination node. These links are predetermined and stored in the routing table. A connection request is routed one hop at a time. Each node independently chooses one link from the alternate links based on some criteria, such as the shortest-path first or the least-congested link first, and then forwards the connection request to the next hop. Because a node only maintains local information on its outgoing links, no information update is required and thus the control overheads are greatly reduced. It has been proven that under light load, this routing approach outperforms both fixed routing and fixed-alternate routing. Under heavy load, fixed-alternate routing produces better performance.

15.3.2 Wavelength assignment

Wavelength assignment algorithms can be classified into the following basic categories: random-order, fixed-order, least-used, and most-used.

15.3.2.1 Random-order

The random-order algorithm searches the wavelengths in a random order. It first determines the wavelengths that are available on each link of a decided route and then chooses one randomly among those available wavelengths, usually with uniform probability. This algorithm does not require any global network state information and is thus suitable for distributed control.

15.3.2.2 Fixed-order

The fixed-order algorithm searches the wavelengths in a fixed order. In this algorithm, all the wavelengths are indexed, and are searched in the order of their index numbers. The first available wavelength found is chosen. Similar to the random-order algorithm, this algorithm also does not require any global network state information and is thus suitable for distributed control. Unlike the random-order algorithm, it has a lower computational cost because it does not necessarily search all the wavelengths for each route. As a result, this algorithm is highly preferred in practice for distributed control.

15.3.2.3 Least-used

The least-used algorithm chooses the wavelength that is the least used in the network. The purpose is to balance the load over all the wavelengths. The idea behind this algorithm is that it is more likely that a shorter route is found on the least-used wavelength than a most-used wavelength, which can result in more available links for those connection requests that arrive later. The least-used algorithm requires global network state information to compute the least-used wavelength, which introduces additional signaling overhead. Moreover, it also requires additional storage and computation cost. As a result, this algorithm is more suitable for centralized control and is thus not preferred in practice for distributed control.

15.3.2.4 Most-used

The most-used algorithm is just the opposite of the least-used algorithm. It chooses the wavelength that is the most used in the network. This algorithm also requires global network state information to compute the most-used wavelength. It has similar signaling overhead, and storage and computation cost to those with the least-used algorithm; however, it outperforms the least-used algorithm significantly.[9] As a result, this algorithm is also more suitable for centralized control and is thus not preferred in practice for distributed control.

In addition to the preceding basic algorithms, a variety of other algorithms are proposed. For example, the min-product algorithm,[10] the least-loaded algorithm,[11] the MAX-SUM algorithm,[12] and the relative-capacity-loss algorithm[13] are proposed for multi-fiber networks. The wavelength-reservation algorithm and protecting-threshold algorithm[14] are proposed for achieving a higher degree of fairness. Readers are referred to Zang et al. for a more detailed review of these algorithms.[4]

15.3.3 Wavelength reservation

Wavelength reservation approaches can be classified into two basic paradigms: parallel reservation and sequential reservation.

15.3.3.1 Parallel reservation

In parallel reservation, the wavelengths on each link of a decided route are reserved in parallel. An example of parallel reservation protocols can be found in Ramaswami and Segall.[2] With this protocol, once a route is decided and a wavelength is selected, the source node sends a request packet to each intermediate node along the decided route, respectively. In response, each intermediate node tries to reserve the selected wavelength on the link; then it sends either a positive or a negative acknowledgment packet back to the source node. If the source node receives a positive acknowledgment from all the nodes, the lightpath is established successfully. Otherwise, the connection request is blocked. Parallel reservation can reduce the reservation time but causes more control overheads.

15.3.3.2 Sequential reservation

In sequential reservation, the wavelengths on each link along a decided route are reserved on a hop-by-hop basis. A wavelength may be reserved at each intermediate node either by a forward control packet on its way forward to the destination node or by a backward control packet on its way backward to the source node, which is correspondingly referred to as forward reservation or backward reservation.

Forward reservation. In forward reservation, a wavelength is reserved by a forward control packet on its way from the source node to the destination node along a decided route. A basic forward reservation protocol (FRP) can be found in Yuan et al.[15] With FRP, the source node sends a request (*REQ*) packet to the destination node along a decided route. At each intermediate node, the *REQ* packet tries to reserve a selected wavelength. If the wavelength can be reserved, the *REQ* packet is forwarded to the next hop. Otherwise, a negative acknowledgment (*NAK*) packet is sent back to the source node and the *REQ* packet is dropped. The *NAK* packet releases all the wavelengths already reserved by the *REQ* packet and informs the source node of the reservation failure. If the *REQ* packet can reach the destination node, the destination node sends a positive acknowledgment (*ACK*) packet back to the source node along the reverse route. The *ACK* packet will configure the optical switch at each intermediate node. When the *ACK* packet reaches the source node, the lightpath is established successfully. Figure 15.2 illustrates the forward reservation. The shaded area represents the period during which a wavelength is reserved but not in use. Obviously, a lot of bandwidth on the reserved wavelength is wasted during the reservation period, which would greatly decrease the wavelength utilization. An effective way to address this problem is to use backward reservation.

Backward reservation. In backward reservation, a wavelength is reserved by a backward control packet on its way from the destination node to the source node along the reverse of a decided route. A basic backward

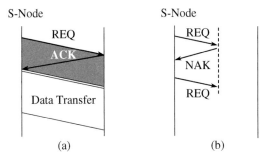

Figure 15.2 Forward reservation: (a) successful; (b) unsuccessful.

reservation protocol (BRP) can also be found in Yuan et al.[15] With BRP, the source node first sends a probe (*PROB*) packet to the destination node along a decided route. The *PROB* packet does not reserve any wavelength at each intermediate node. Instead, it just collects wavelength availability information on each link along the decided route. When the destination node receives the *PROB* packet, it selects an available wavelength and then sends a reservation (*RESV*) packet back to the source node along the reverse route. It is the *RESV* packet that reserves the selected wavelength and simultaneously configures the optical switch at each intermediate node on its way back to the source node. If the *RESV* packet cannot reserve a suitable wavelength at an intermediate node, the node sends a failure (*FAIL*) packet to the destination node and a negative acknowledgment (*NACK*) packet to the source node, respectively. The *FAIL* packet disconfigures the optical switches and releases the wavelengths already reserved by the *RESV* packet while the *NACK* packet simply informs the source node of the reservation failure. If the *RESV* packet can reach the source node, the lightpath is established successfully. Obviously, BRP can reduce the bandwidth waste significantly, as presented in Figure 15.3.

To reduce the possibility of a reservation failure, a variety of reservation policies have been proposed in the literature, such as holding, aggressive, and retrying.

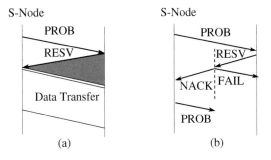

Figure 15.3 Backward reservation: (a) successful; (b) unsuccessful.

Holding. In FRP, each node actually adopts a dropping policy. In the case of a reservation failure at an intermediate node, a *NAK* packet is immediately sent back to the source node and the *REQ* packet is dropped. To increase the possibility of a successful reservation, a holding policy can be adopted.[15] With the holding policy, an intermediate node does not immediately send a *NAK* packet to the source node in the case of a reservation failure. Instead, it will buffer the *REQ* packet for a specified period of time. If the desired wavelength becomes available during this period, the *REQ* packet will continue its trip. Otherwise, a *NAK* will be sent back to the source node to inform of the reservation failure. To implement the holding policy, each node must maintain a queue to buffer the *REQ* packets and a timer to control the holding time. This increases the complexity in implementation. The holding policy can also be adopted in BRP.

Aggressive. In FRP, a conservative policy is actually adopted. The source node only selects one available wavelength at a time and tries to reserve it along the decided route. In the case of a reservation failure, the source node will be informed and then try another wavelength. Obviously, this policy may result in a long reservation time. To reduce the reservation time, an aggressive policy can be adopted.[15] With the aggressive policy, the source node will reserve all available wavelengths on the outgoing link and then send a *REQ* packet that carries the information on the reserved wavelengths. At each intermediate node, the common wavelengths that are available on both incoming link and outgoing link are computed. If no common wavelength is available, a *NAK* packet is sent back to the source node to release the reserved wavelengths and inform of the reservation failure. Otherwise, all common wavelengths are reserved and the *REQ* packet is forwarded to the next hop carrying the information on the reserved wavelengths. If the *REQ* packet reaches the destination node, it will select one wavelength from the reserved wavelengths and send an *ACK* packet back to the source node. The *ACK* packet will release those wavelengths reserved by the *REQ* packet but not selected on its way back. The drawback of this policy is the over-reserved wavelengths during the reservation period, which would greatly reduce the wavelength utilization. The aggressive policy can also be adopted in BRP.

Retrying. In BRP, a *NACK* packet is sent to the source node in the case of a reservation failure at an intermediate node and the request is blocked. To increase the possibility of a successful reservation, a retrying policy is proposed by Saha.[16] With this policy, the destination node will try another available wavelength in the case of a reservation failure. Obviously, retrying increases the connection setup delay. It has been demonstrated, however, that this policy can significantly improve the request-blocking probability without largely increasing the setup delay as long as the number of retries is restricted to two or three.

15.4 Conclusions

Lightpath control is one of the key issues for achieving good network performance in wavelength-routed WDM networks. Distributed control is considered an effective solution to lightpath establishment for large networks with dynamic traffic. This chapter discussed the major issues that are involved in distributed lightpath establishment in terms of routing, wavelength assignment, and wavelength reservation.

We have also reviewed various distributed lightpath control mechanisms that have already been proposed in the literature, including various routing approaches, wavelength assignment algorithms, and wavelength reservation protocols. Under dynamic traffic, the objective of routing and wavelength assignment is to increase the wavelength utilization and minimize the request-blocking probability. Different RWA algorithms may result in different network performance. The performance of an RWA algorithm also depends on the network state information available at each network node. If global information is available, more intelligent decisions can be made, which could significantly improve the network performance. However, this requires timely-updated network state information maintained at each network node, which largely increases the complexity in implementation.

In the context of wavelength reservation, the objective is to minimize the possibility of a reservation failure and reduce the connection setup time. The performance of a reservation protocol depends on whether a reservation is made in parallel or in sequential, and whether forward or backward. In addition, various reservation policies can be adopted in a basic reservation protocol, which could significantly reduce the possibility of a reservation failure.

References

1. R. Ramaswami and K.N. Sivarajan, *Optical Networks — A Practical Perspective*, Morgan Kaufmann Publishers, San Francisco, 1998.
2. R. Ramaswami and A. Segall, Distributed network control for optical networks, *IEEE/ACM Trans. on Networking*, vol. 5, no. 6, Dec. 1997, pp. 936–943.
3. J. Zheng and H.T. Mouftah, Distributed lightpath control based on destination routing for wavelength-routed WDM networks, *SPIE Opt. Networking Mag.*, vol. 3, no. 4, Jul./Aug. 2002, pp. 38–46.
4. H. Zang, J.P. Jue, and B. Mukherjee, A review of routing and wavelength assignment approaches for wavelength-routed optical WDM networks, *SPIE Opt. Networking Mag.*, vol. 1, no. 1, Jan. 2000, pp. 47–60.
5. K.-m. Chan and T.-s. Perter Yum, Analysis of least-congested path routing in WDM lightwave networks, *Proc. IEEE INFOCOM '94*, vol. 2, Toronto, Canada, Apr. 1994, pp. 962–969.
6. L. Li and A.K. Somani, Dynamic wavelength routing using congestion and neighborhood information, *IEEE/ACM Trans. on Networking*, vol. 7, no. 5, Oct. 1999, pp. 779–786.

7. H. Zang et al., Connection management for wavelength-routed WDM networks, *Proc. IEEE CLOBECOM '99*, Rio De Janeiro, Brazil, Dec. 1999, pp. 1428–1432.

8. J.P. Jue and G. Xiao, An adaptive routing algorithm for wavelength-routed optical networks with a distributed control scheme, *Proc. 9th Int. Conf. Comp. Commun. Networking*, Las Vegas, NV, Oct. 2000, pp. 192–197.

9. S. Subramaniam and R.A. Barry, Wavelength assignment in fixed routing WDM networks, *Proc. ICC '97*, Montreal, Canada, vol. 1, June 1997, pp. 406–410.

10. G. Jeong and E. Ayanoglu, Comparison of wavelength-interchanging and wavelength-selective cross-connects in multiwavelength all-optical networks, *Proc. IEEE INFOCOM '96*, San Francisco, CA, vol. 1, Mar. 1996, pp. 156–163.

11. E. Karasan and E. Ayanoglu, Effects of wavelength routing and selection algorithms on wavelength conversion grain in WDM optical networks, *IEEE/ACM Trans. on Networking*, vol. 6, no. 2, Apr. 1998, pp. 186–196.

12. R.A. Barry and S. Subramaniam, The MAX-SUM wavelength assignment algorithm for WDM ring networks, *Proc. OFC '97*, Feb. 1997.

13. X. Zhang and C. Qiao, Wavelength assignment for dynamic traffic in multi-fiber WDM networks, *Proc. ICCCN '98*, Lafayette, LA, Oct. 1998, pp. 479–485.

14. A. Birman and A. Kershenbaum, Routing and wavelength assignment methods in single-hop all-optical networks with blocking, *Proc. IEEE INFOCOM '95*, vol. 2, Boston, MA, Apr. 1995, pp. 431–438.

15. X. Yuan, R. Melhem, and R. Gupta, Distributed path reservation algorithms for multiplexed all-optical interconnection networks, *IEEE Trans. Comp.*, vol. 48, no. 12, Dec. 1999, pp. 1355–1363.

16. D. Saha, A comparative study of distributed protocols for wavelength reservation in WDM optical networks, *SPIE Opt. Networking Mag.*, vol. 3, no. 1, Jan./Feb. 2002, pp. 45–52.

chapter sixteen

Recent advances in dynamic lightpath restoration in WDM mesh networks*

Chava Vijaya Saradhi
Indian Institute of Technology
C. Siva Ram Murthy
Indian Institute of Technology

Contents

* This work was supported by the Department of Science and Technology, New Delhi, India.

16.1 Introduction

Wavelength division multiplexing (WDM) — transmitting many light beams of different wavelengths simultaneously through an optical fiber — and *wave-*

length routing — a network switching or routing node that routes signals based on their wavelengths — are rapidly becoming technologies-of-choice to meet the tremendous bandwidth demand of the new millennium. Several important advantages, such as increased usable bandwidth (nearly 50 THz) on an optical fiber, reduced electronic processing cost, protocol transparency, and low bit-error rates (BER [10^{-12} to 10^{-9}]), have made wavelength-routed WDM optical networks a de facto standard for high-speed backbone transport networks. In the emerging next-generation transport networks, called *intelligent optical networks*, WDM-based optical components such as add-drop multiplexers (ADMs) and optical cross-connects (OXCs) will have full knowledge of the wavelengths in the network, status, and traffic-carrying capacity of each wavelength. With such intelligence, these (intelligent) optical networks could create self-connecting and self-regulating connections *on-the-fly.*

A WDM mesh network consists of wavelength routing nodes interconnected by point-to-point optical fiber links in an arbitrary topology. In these networks, a message can be sent from one node to another using a wavelength continuous path (called a *lightpath*) without requiring any electro-optical conversion and buffering at the intermediate nodes. This process is known as *wavelength routing*. The requirement that the same wavelength must be used on all the links along the selected route is known as the *wavelength continuity constraint.*[1,2] Two lightpaths can use the same wavelength only if they use different fibers (wavelength reuse). A physical route and a wavelength uniquely identify a lightpath. If a lightpath is established between any two nodes, traffic between these nodes can be routed without requiring any intermediate opto-electrical conversion and buffering. Traffic demand can be either static or dynamic.

In *static lightpath establishment* (SLE), traffic demand between node pairs is known *a priori* and the goal is to establish lightpaths so as to optimize certain objective function (maximizing single-hop traffic, minimizing congestion, minimizing average weighted hop count, etc.). The nodes, together with the set of lightpaths at the *optical layer*, form a *virtual topology.*[1,2] An optical layer consists of a set of lightpaths and can be used in wide-area backbone networks between the lower physical layer and higher client (electrical) layers. The *dynamic lightpath establishment* (DLE) problem concerns establishing lightpaths with an objective of increasing the average call acceptance ratio (or equivalently reducing the blocking probability) when connection requests arrive to and depart from the network dynamically. Several heuristic algorithms for the routing and wavelength assignment (RWA) problem are available in the literature.[3–6]

Generally, longer-hop connections are subject to more blocking than shorter-hop connections. A good wavelength routing (WR) algorithm is critically important in order to improve the network performance in terms of average call acceptance ratio. A WR algorithm has two components: route selection and wavelength selection. Different WR algorithms have been proposed in the literature to choose best pair of routes and wavelengths. Based on the restriction (if any) on choosing a route from all possible routes, route selection algorithms can be fixed routing (FR), alternate routing (AR), and exhaust routing (ER).[4,7] Depending upon the order in which wavelengths are searched, the wavelength selection algorithms can be most used (MU), least used (LU), fixed ordering (FX), and random ordering (RN). In Mokhtar and Azizoglu, all these wavelength selection algorithms are compared and the results showed that MU scheme performs best compared with all other wavelength assignment schemes;[4] however, the MU scheme requires the actual or estimated global state information of the network to determine the usage of every wavelength. This scheme is more suitable for centralized implementation and is not easily amenable for distributed implementation.

Similar to any communication network, WDM networks are also prone to hardware (such as routers and/or switches and cable cuts) failures and software (protocol) bugs. As WDM networks carry huge volumes of traffic, maintaining a high level of service availability at an acceptable level of overhead is an important issue. It is essential to incorporate fault-tolerance into quality of service (QoS) requirements. The types of applications being deployed across the public Internet today are increasingly mission-critical, whereby business success can be jeopardized by poor performance of the network. It does not matter how attractive and potentially lucrative our applications are if the network does not function reliably and consistently. Restoration could be provided at the optical layer or at the higher client (electrical) layers, each of which has its own merits. The optical layer has faster restoration and provisioning times and uses the wavelength channels optimally.[8,9] Because of these, many of the functions are moving to the optical layer. The foremost of them are routing, switching, and network restoration. High-speed mesh restoration becomes a necessity, and this is made possible by doing the restoration at the optical layer using optical switches.

This chapter presents recent advances in dynamic restoration of light-paths in WDM optical networks. It explains the operation of these recent schemes and discusses their performance. In Section 16.2, we review some commonly used terms and present a broad classification of restoration schemes designed for dynamic traffic. Section 16.3 discusses centralized restoration methods for dynamic traffic. We present a *segmented protection paths algorithm,* which overcomes the disadvantages of the existing end-to-end detouring and local detouring methods. Because the failure of components is probabilistic, the single-failure model is not realistic, especially for large networks. We present a scheme called *partial backups,* which considers the probabilistic nature of failure of components for establishing reliable connections *(R-Connections).* Section 16.4 provides a brief review of distributed control protocols for lightpath restoration and also describes a *preferred link-based* distributed control protocol for establishing reliability-constrained, least-cost lightpaths, which can be used in conjunction with the existing reactive schemes for restoration. The trend in the current network development is moving toward a unified solution that supports voice, data, and various multimedia services. In this scenario, different applications/end users may need different levels of fault-tolerance, availability, bandwidth, delay, etc., and differ in how much they are willing to pay for the service they get. We discuss differentiated services in fault-tolerant WDM networks in Section 16.5. Finally, a brief summary is provided in Section 6.

16.2 *Terminology and classification of dynamic lightpath restoration methods*

Fault-tolerance refers to the ability of the network to configure and reestablish communication upon failure. A related term known as *restoration* refers to the process of rerouting affected traffic upon a component failure. A network with restoration capability is known as *survivable network* or *restorable network*. The lightpath that carries traffic during the normal operation is known as the *primary lightpath*. When a primary lightpath fails, the traffic is rerouted over a new lightpath known as the *backup lightpath*. The process of assigning the network resources to a given traffic demand is known as *provisioning* a network.[10-14] Given a set of traffic demands, the provisioning problem is to allocate resources to the primary and backup lightpaths for each demand, so as to minimize the spare resources required. The resources in this case are the number of wavelengths for single-fiber networks and the number of fibers for multi-fiber networks. A connection request with a fault-tolerance requirement is called a *dependable connection (D-connection)*.[15,16]

The two primary measures of dependability are reliability and availability. Reliability of a resource (or component) is the probability that it functions correctly (potentially despite faults) over an interval of time, whereas availability of a resource (or component) is the probability that it is being operational at any given instant of time. The restoration methods differ in their assumptions about the mode of network control:

- *Centralized* or *distributed*, the traffic demand
- *Static* or *dynamic*, the functionalities of cross-connects
- *Wavelength selective* or *wavelength convertible*, and the performance metric
- *Restoration guarantee, restoration time,* and *spare resource utilization*

The methods designed for establishing connections with fault-tolerance requirements can be broadly divided into *reactive* and *proactive*. In the reactive method of restoration, when an existing lightpath fails, a search is initiated for finding a new lightpath that does not use the failed components. This has an advantage of low overhead in the absence of failures; however, this does not guarantee successful recovery because an attempt to establish a new lightpath may fail due to resource shortage at the time of failure recovery. In the proactive method of restoration, backup lightpaths are identified and resources are reserved along the backup lightpaths at the time of establishing primary lightpath itself. Proactive or reactive restoration method is either link-based or path-based. The link-based method employs *local detouring*, while path-based method employs *end-to-end detouring*. Local detouring reroutes the traffic around the failed component, while in end-to-end detouring a backup lightpath is selected between the end nodes of the failed primary lightpath. Local detouring is inefficient in terms of resource utilization. Furthermore, handling node failures is very difficult in local detouring. A path-based (end-to-end) restoration method is either *failure dependent* or *failure independent*. In a failure-dependent method, a backup lightpath is associated with every link used by a primary lightpath. When a primary lightpath fails, the backup lightpath that corresponds to the failed link will be used. In a failure-independent method, a backup lightpath that is disjoint with the primary lightpath is chosen.

A proactive restoration method may use a dedicated backup lightpath for a primary lightpath. In a dedicated backup scheme, wavelength channels are not shared between any two backup lightpaths. For better resource utilization, multiplexing techniques can be employed. If two lightpaths do not fail simultaneously, their backup lightpaths can share a wavelength channel. This technique is known as *backup multiplexing (BM)*.[15] A proactive restoration method can employ *primary-backup multiplexing* (PBM)[16] to further improve resource utilization. This technique allows a wavelength channel to be shared by a primary and one or more backup lightpaths. This chapter concentrates on the restoration methods designed for *dynamic traffic* only. A broad classification of restoration methods designed for dynamic traffic is illustrated in Figure 16.1. Though we have not shown the classification of restoration schemes under distributed control, the same classification as under centralized control is applicable in many cases, and several research efforts are being focused in this direction.

16.3 Centralized restoration methods

Several authors have recently proposed some dynamic routing algorithms for fault-tolerant routing in WDM networks.[15–19] In Anand and Qiao, two

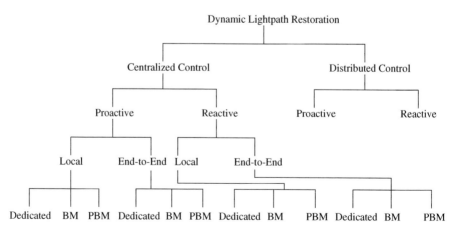

Figure 16.1 Classification of dynamic lightpath restoration schemes.

online routing and wavelength assignment algorithms are presented (i.e., static method and dynamic method).[18] The static method is used to establish primary and protection lightpaths such that once a route and wavelength have been chosen, they are not allowed to change. On the other hand, dynamic method allows for rearrangement of protection lightpaths (i.e., both route and wavelength chosen for a protection path can be shifted to accommodate a new request). Both methods are based on a dedicated path protection scheme, and in both methods the primary paths are not allowed to rearrange. Contrary to intuition, the results show that static strategy performs better than dynamic strategy in terms of the number of connection requests satisfied for a given number of wavelengths. In Bandyopadhyay, a dynamic rerouting scheme in case of fault occurrence for WDM all-optical networks was proposed.[17] Sridharan and Somani support three classes of service: full protection, no protection, and best-effort protection.[19] Krishna et al. and Kodialam and Lakshman tried a tradeoff between local and end-to-end detouring.[20,21] Recently, carrying IP over WDM networks in an efficient manner has generated considerable interest. This is because the rapid pace of developments in WDM technology is now beginning to shift the focus more toward optical networking and network level issues. Survivability provisioning in optical multi-protocol label switching (MPLS) networks is considered in Ghani.[22] In Li and Ramaswami, some methods to detect and isolate faults such as fiber cuts and router failures have been proposed.[23] A comprehensive survey of the restoration schemes is available in the literature.[2,24–26]

16.3.1 Backup multiplexing-based routing

The algorithm presented in Mohan and Murthy uses backup multiplexing technique.[15] It basically uses the alternate routing method, wherein a set of candidate routes for every source–destination pair is pre-computed. The candidate routes of a source–destination pair are chosen to be link–disjoint. In response to a new request, a minimum-cost primary-backup lightpath pair is chosen. The key idea here is to choose the primary-backup lightpath pair that requires minimum free wavelength channels. Note that if a wavelength is assigned to either a primary or backup path, it is no longer available for any other primary or backup lightpaths. If a wavelength channel is already used by one or more backup lightpaths, it can be used by a new backup lightpath (if allowed) with no extra cost. Thus, the algorithm ensures that at the time of routing a new lightpath pair, the network is taken to a new state so as to maximize the total number of free channels in the network. For every wavelength channel on a link, the algorithm maintains a list of links whose failures will lead to the use of the channel by a backup lightpath of a failed primary lightpath. In other words, the list associated with a channel consists of those links used by all the primary lightpaths where backup lightpaths use the channel.

A new backup lightpath can use a wavelength channel only if the corresponding primary lightpath does not use any of the links in the list associated with the channel. Two kinds of wavelength assignment policies were discussed. In the first method, the primary and backup lightpaths use the same channel; hence the name primary dependent backup wavelength assignment (PDBWA). In the second method, no such restriction is made on wavelength usage by the primary and backup lightpaths; hence the name primary independent backup wavelength assignment (PIBWA). Although the first method is computationally simpler than the second one, it results in poor blocking performance. The first method assumes that the nodes are equipped with only *fixed* transceivers, whereas the second method assumes that the nodes are equipped with *tunable* transceivers.

The performance of the algorithm has been verified through simulation experiments on ARPA-2 and mesh–torus networks. The connectivity of ARPA-2 network is denser than the mesh–torus network. The results show that for a given blocking probability, both ARPA-2 and mesh–torus networks are able to carry more traffic load when the proposed algorithm is used when compared with the dedicated backup reservation method. The factor by which the carried load increases is about 3 in the case of mesh–torus and is about 0.8 in the case of ARPA-2. The results show that the usefulness of backup multiplexing increases as the network connectivity increases. This is because in a densely connected network, the candidate routes are usually shorter, and the number of possible link-disjoint candidate routes is greater.

16.3.2 Primary-backup multiplexing-based routing

The algorithm presented in Mohan et al. uses primary-backup multiplexing. It also uses alternate routing method and proactive path-based restoration approach.[16] The objective of this algorithm is to improve the blocking performance while allowing an acceptable reduction in restoration guarantee. Here, a wavelength channel is allowed to be shared by a primary lightpath and one or more backup lightpaths. Because of this, the lightpaths that correspond to the backup lightpaths on this channel lose their restoration capability. A new lightpath pair may cause an increase in the average number of non-restorable lightpaths per link. A lightpath pair is admissible only if its establishment does not take the network to a state where the average number of non-restorable lightpaths per link exceeds a predefined threshold value. For a lower threshold value, the restoration guarantee is higher and the blocking performance is lower. By appropriately choosing a threshold value, a desired tradeoff can be achieved between the restoration guarantee and the network-blocking performance. The key idea of the algorithm is to choose a minimum cost lightpath pair among those that are admissible. The cost of a primary-backup lightpath pair computed by this algorithm is the number of free channels used by the pair plus the number of primary-backup multiplexed channels traversed times a penalty factor. A sufficiently high value is chosen as the penalty factor so that a pair that traverses a minimum number of primary-backup multiplexed channels is chosen, and in case of a tie, the number of free channels used is taken into account.

An important issue here is how to compute the number of non-restorable lightpaths that are created by routing a lightpath pair. When a new backup lightpath traverses a channel currently used by a primary lightpath, only the new primary lightpath becomes non-recoverable; however, it is not trivial to compute the count of non-restorable lightpaths created by routing a new primary lightpath over the channels reserved for some other backup paths. A straightforward solution will be to keep the identity and restorability status of every primary lightpath that corresponds to backup lightpaths on each of the channels. This requires a large storage and also a more complex algorithm to compute the count and update the restorability status. To overcome the preceding shortcomings, a computationally simple heuristic method was proposed to compute the count of non-restorable lightpaths created by routing a new primary lightpath. This heuristic needs to know only the number of backup lightpaths multiplexed on a channel and the number of links used by their primary lightpaths. It also keeps track of the number of backup lightpaths that continue to the next link.

The performance of the algorithm was verified through simulation experiments on ARPA-2 and mesh–torus networks. The performance is measured using two metrics: relative performance gain and reduction in restoration guarantee. The blocking performance of the network when no lightpath is provided with a backup lightpath is taken as the lower limit, and the performance when only backup multiplexing is used is taken as the upper limit. The performance of the proposed algorithm using primary-backup multiplexing is measured with relation to the previous two limits. The results show that the performance gain is attractive enough to allow some reduction in restoration guarantee. In particular, under light load conditions more than 90% performance gain is achieved at the expense of less than 10% guarantee reduction. The results also show that the performance improves as the network connectivity increases.

16.3.3 Segmented protection paths

16.3.3.1 Motivation

In conventional approaches to fault tolerance,[8,10–13,15–19,27] end-to-end protection lightpaths are provided, and they are able to handle any component failure under the *single link failure* model. In the single link failure model only one link in the whole network is assumed to fail at any time. End-to-end detouring has the additional requirement that for a call to be accepted it is essential to find sufficient resources along two totally disjoint paths between the source–destination pair. Even when two disjoint routes in the network run between the source–destination node pair, it is possible for the primary lightpath to be routed (along the shortest hop path or minimum delay path) so that there cannot exist an end-to-end protection lightpath. Second, the end-to-end method of establishing protection lightpaths might be very inefficient for delay-critical applications such as the online video, which require that not only the primary but also the protection paths have delay along them within specified bounds. Hence, it is possible that no protection lightpath found from the source to the destination has its delay within the permissible limit from the shortest path delay between them, despite the network having a considerable amount of free resources (wavelengths). Lastly, local detouring is inefficient in terms of resource utilization. Furthermore, handling node failures is very difficult in local detouring.

This section presents the segmented protection paths[28,29] algorithm instead of an end-to-end protection path from the source to the destination and shows that the algorithm not only solves the problems explained earlier but is also more resource efficient than the end-to-end protection method. The segmented protection paths algorithm[28,29] also has advantages such as faster recovery time and more flexibility for controlling reliabilities of each connection independent of others. We now explain the concept of segmented protection paths.

16.3.3.2 Concept of segmented protection paths

In a segmented protection paths scheme, protection paths are found for only parts of the primary path. The primary path is viewed as smaller contiguous segments which are called *primary segments,* as shown in Figure 16.2. A protection path for each primary segment, which is called a *protection segment,* is found independently. Collectively, all the protection segments are called the *segmented protection path.* Figure 16.2 illustrates these terms, where a primary path with 8 links is shown. Links of the primary path are numbered 1 through 8, while those of the segmented protection path are named *A* through *J.* All the intermediate nodes on the primary path are denoted by *N1* to *N7.* The primary path has three primary segments, each of which has a protection segment covering it.

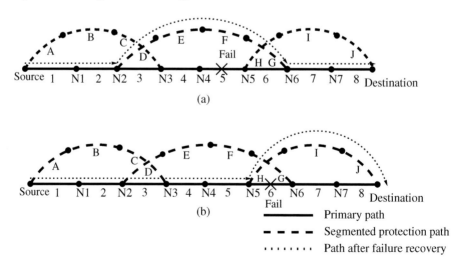

Figure 16.2 Illustration of segmented protection paths.

The first primary segment spans links 1 to 3, and its protection segment consists of links *A* to *C* and covers the first primary segment. The second primary segment spans links 3 to 6, and its protection segment spans links *D* to *G* and covers the second primary segment. The third primary segment spans links 6 to 8, while its protection segment spans links *H* to *J*. Together, these three protection segments constitute the segmented protection path for this primary path.

Note that successive primary segments of a primary *overlap* at least by one link. When a component in a primary segment fails, the data is routed through the protection segment activated rather than through the original path, only for the length of its primary segment as illustrated. If only one primary segment contains the failed component, the protection segment corresponding to that primary segment is activated, as depicted in Figure 16.2a, for the failure of link 5. If two successive primary segments contain the failed component, then any one of the two protection segments corresponding to the primary segments is activated, as shown in Figure 16.2b, for the failure of link 6. It should be noted that the end-to-end protection scheme is a special case of the segmented protection scheme when the number of segments is equal to one. Some of the advantages of the segmented protection paths scheme over end-to-end protection scheme with simple examples are described next.

16.3.3.3 Advantages of segmented protection paths

Consider Figure 16.3. One D-connection is established between *N26 (S1)* and *N5 (D1)*. With the primary path routed as presented in the figure, along one of the shortest paths between them, an end-to-end protection path may not exist but a segmented protection path exists. Another example is illustrated in Figure 16.4 over USANET. For a dependable connection to be established

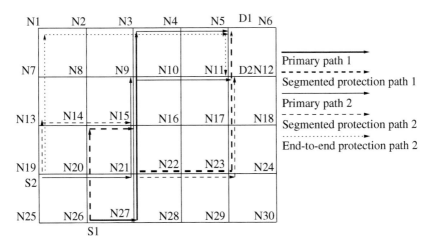

Figure 16.3 An example to show the benefits of segmented protection paths.

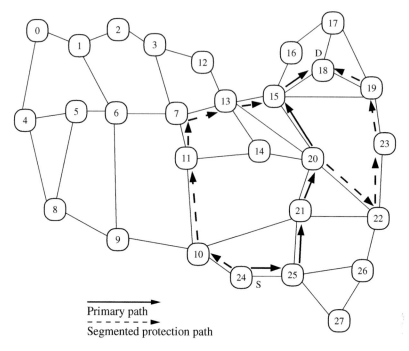

Figure 16.4 No end-to-end protection path but segmented protection path exists.

between nodes 24 (*S*) and 18 (*D*), if the primary path is established along the unique shortest path between them, it is easy to see that there cannot exist an end-to-end protection path but there will be a segmented protection path as presented in the figure.

We illustrate yet another advantage of segmented protection paths in Figure 16.3. A dependable connection will be established between N19 (*S2*) and N11 (*D2*). The primary path, end-to-end protection path, and segmented protection path are routed as shown. We can see that while the end-to-end protection path requires eight hops, all the protection segments together require only seven hops, hence lesser resource reservation. Because the end-to-end protection path is a special case of segmented protection path, we can safely say that the *shortest segmented protection path*, which is defined as the segmented protection path for which the sum of the hop counts of all its protection segments is minimum, results in better spare resource reservation than the end-to-end protection path. Saradhi and Murthy[28,29] presented an algorithm to select this *shortest segmented protection path* and showed that its complexity is the same as that of any shortest-path-finding algorithm.

We now demonstrate how the segmented protection paths offer more flexibility in providing D-connections through Figure 16.5. Assume that each link on the mesh is having only one wavelength. Two D-connections will be established: N19 (*S1*) to N10 (*D1*) and N21 (*S2*) to N12 (*D2*). The primary

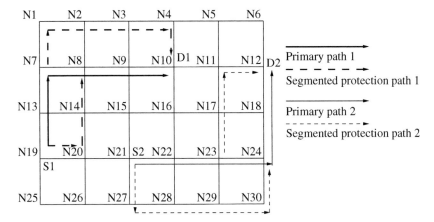

Figure 16.5 Segmented protection paths are more flexible for routing than end-to-end protection paths.

lightpaths (shortest paths) of these connections are shown in the figure. It is not possible to establish end-to-end protection lightpaths for both the connections, as both the protection lightpaths contend for the wavelength along the link from *N15* to *N16*; however, segmented protection lightpaths can be established as depicted in Figure 16.5. We could have also taken an end-to-end protection lightpath for one of the connections and a segmented protection lightpath for the other. We briefly explain yet another advantage of the segmented protection paths scheme later in Section 16.3.3.6.

16.3.3.4 Segmented protection path selection algorithm

Let directed graph $G(V, E)$ represent the given network topology. Every node n in the network is represented by a unique vertex v in the vertex set V and every duplex link l between nodes n_1 (v_1) and n_2 (v_2) in the network is represented in the graph G by two directed edges e_1 and e_2 from v_1 to v_2 and v_2 to v_1, respectively.

Let S and D denote the source and destination nodes in the network between which we need to establish the D-connection. We denote a primary path (working path) in graph G with a sequence of vertices $W = S, i_1, i_2, \ldots, i_n, D$, with S and D denoting source and destination, respectively. In order to find the shortest segmented protection path, construct a weighted directed graph G' by modifying the directed graph G as follows:

1. Every directed edge other than those along the primary path (i.e., edges between any two successive vertices in the sequence W) is assigned a weight given by a cost function determined by the delay or hop count.
2. For edges along the primary path the weights are assigned as follows: edges directed from a vertex in the sequence W to its successor vertex are assigned a weight of infinity. It is equivalent to removing

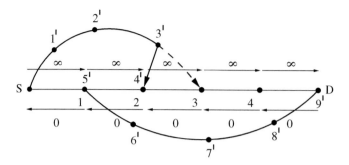

Figure 16.6 Illustration of the construction of modified graph G' and the shortest segmented protection path from the path chosen.

the edges. Edges directed from a vertex in the sequence W to its predecessor vertex are assigned a weight of zero. This is given in Figure 16.6.

3. For every edge $e(v_1, v_2) \in E$, $\ni$ $v_1 \notin W$ and $v_2 \in (W - S)$, replace e with $e'(v_1, v'_2)$, where v'_2 is the predecessor of v_2 in W. That is, replace every edge from any vertex v_1 not in W, directed into any intermediate vertex v_2 in W, with another edge directed from v_1 to v'_2, predecessor of v_2.

4. To find the shortest segmented protection path, on the resulting graph G', run the least cost path algorithm for directed graphs (e.g., Dijkstra's algorithm) from the source to the destination on G'. Let the path (protection path) obtained be denoted by a sequence of vertices $P = S, i'_1, i'_2, ..., i'_m, D$.

5. The segmented protection path consists of protection segments PS_1, PS_2, PS_3, ... As one traverses along the sequence P from S to D, we generate the vertex sequences for these segments PS_1, PS_2, PS_3, ... one after the other (i.e., first PS_1 is generated, then PS_2, and so on). In the following discussion we use the phrase *open a segment* to indicate the beginning of the generation of the protection segment, and *close a segment* to indicate the ending of the generation of the protection segment. So, segmented protection paths algorithm first opens PS_1, generates it, and closes it, then opens PS_2, generates it, closes it, and so on, until all the protection segments are generated. At any stage of the traversal, if an *opened* protection segment is being generated then it is denoted as *current protection segment*. If all the *opened* segments are *closed*, *current protection segment* is NULL. The vertex sequences PS_1, PS_2, PS_3, ... are initialized to be empty when *opened*. The phrase "*add* vertex v to a sequence" means the vertex is appended at the end of the sequence. For constructing protection segments, traverse the sequence P (found in step 4). At every stage of the traversal, let i'_c denote the current vertex. The algorithm performs the appropriate actions as indicated in a through d, for every i'_c. This procedure ends on reaching D:

a. If $i'_c = S$ then open PS_1 and add i'_c to it.
b. If $i'_c \neq i_k$ for any $k \leq n$, (i.e., i'_c does not lie on W) then
   i. If current protection segment $\neq$ NULL then add i'_c to current protection segment.
   ii. If current protection segment = NULL then open next protection segment and add i'_{c-1} and i'_c to it in that order.
c. If $i'_c = i_k$ for any $k \leq n$, (i.e., i'_c lies on W) then
   i.. If current protection segment $\neq$ NULL then add i_{k+1} to current protection segment and close it.
   ii. If current protection segment = NULL do nothing.
d. If $i'_c = D$ then add i'_c to current backup segment and close it.

The resulting vertex sequences define protection segments in G, which form the shortest segmented protection path for the primary path P.

We now explain step 5 of the algorithm through an example in Figure 16.6. In Figure 16.6 we show the primary path between S and D, over nodes numbered 1 through 4. Suppose the path chosen between S and D in G' is over the nodes numbered 1' through 9'. We denote by a dotted line the edge between 3' and 3 in G, which is replaced in step 3 with an edge between 3' and 2(= 4'). Then we generate the protection segments as follows. First, we open PS_1 and add S as given in case a. Then we add 1' through 3' in succession to PS_1, as given in sub-case i. of case b. When we traverse 4'(= 2), we add 3 and close PS_1 as given in sub-case i. of case c.) Then we ignore 5' as given in sub-case ii. of case c. When we come to 6', we open PS_2 and add 5' and 6' to it as given in sub-case ii. of case b. Then we add 7', 8', and 9' as we did previously, before closing PS_2 with D as given in case d.

Regarding the complexity of the algorithm, it is easy to see that the complexity of step 2 is $O(|V|)$, while the complexity of steps 1 and 3 is at most $O(|E|)$. Further, the complexity of step 4 is the same as the complexity of least-weight path algorithm like Dijkstra's algorithm which is $O(|V|^2+|E|)$. The complexity of step 5 is $O(|V|)$ as we just traverse the path chosen and make constant amount of computation at each step. Hence, the overall complexity of the algorithm is $O(|V|^2+|E|)$ which is the complexity of the least-weight path algorithm.

16.3.3.5 Wavelength assignment for primary and segmented protection paths

When a connection request from a source S to a destination D arrives, the wavelength selection algorithm finds all free wavelengths on the predetermined primary path using modified FX algorithm. The idea behind using this algorithm is to achieve the performance closer to that of the MU algorithm but without requiring any global state information. Here wavelengths are not reserved, but the availability of free wavelengths is noted down (remembered). If no free wavelength is available, the connection request is rejected. Note that if a wavelength is assigned to either a primary or segmented protection path, it is no longer available for any other primary or segmented protection lightpaths.

After finding all free wavelengths on the primary path, the algorithm tries to find all free wavelengths on the predetermined segmented protection path, again using modified FX algorithm. If no free wavelength is available, the connection request is rejected. After finding all free wavelengths on the primary and the segmented protection paths, the first free wavelength common to both the primary and segmented protection paths will be chosen and reserved.

Note here that the wavelength selection algorithm uses PDBWA, because all the protection segments should be on the same wavelength as that of the primary wavelength. In other words, the segmented protection path establishment is failure independent, but protection path activation is failure dependent.

A situation may arise where wavelength-continuous routes exist for primary path on one wavelength and for the protection path on some other wavelength, but no wavelength-continuous routes are available on the same wavelength for both the primary and protection paths. In such a case, the request is rejected by this method. This is because the research assumes that the nodes are equipped with only *fixed* transmitters and receivers.[28,29]

16.3.3.6 Failure detection and recovery

When a fault occurs in a component in the network, all the lightpaths passing through it have to be rerouted through their protection lightpaths. This process is called *failure recovery,* and is required only when a component in the primary lightpath fails. Failure recovery is done in three phases: failure detection, failure reporting, and protection lightpath activation or lightpath rerouting. The time taken to reestablish the lightpath is equal to the sum of the times taken by each of the preceding three phases, and is called *failure recovery delay.* This delay is crucial to many mission-critical and real-time applications, and has to be minimized. It is assumed[28,29] that the nodes adjacent to the failed link can detect the failure by monitoring the power levels on the links.[23] After failure detection, the end nodes that have detected the fault will report it to the concerned end nodes. This is called *failure reporting.* Failure reports are sent in both directions: toward the source and destination nodes. After the failure report reaches certain nodes, the protection path is activated by those nodes and is called *protection path activation.* Failure reporting and protection path activation need to use control messages. Control messages carry connection identifier and lightpath information. For carrying these control messages, we assume a *real-time control channel* (RCC),[27] where a dedicated channel is established and maintained for sending control messages.

In end-to-end protection lightpath scheme, the control messages (failure reports) have to reach the source and the destination before they can activate the protection lightpath. In a segmented protection paths scheme, however, it is not necessary. Failures can be handled more locally. The end nodes of

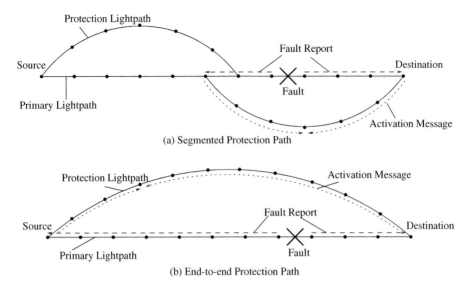

Figure 16.7 Illustration of failure recovery.

the primary segment initiate the recovery process on receiving the failure report. They send the activation message along the protection segment. Activation messages will carry the connection identifier and lightpath information. These messages are used to set the state of the switches such that protection lightpath is switched from an inbound link to an appropriate outbound link. As resources are reserved along the protection lightpath beforehand, the D-connection will be resumed. This failure recovery process is illustrated in Figure 16.7. The time taken for failure reporting and segmented protection path activation is dependent on the lengths of primary and segmented protection path. Hence, if n segments are in the segmented protection path, then this gives about $O(n)$ improvement in the failure reporting and activation times. This could be a very important and substantial improvement, especially for WDM optical networks that carry huge amounts of data and for long distance real-time applications that cannot tolerate long durations of service disruption.

16.3.3.7 Performance study

The performance of the algorithm has been verified through simulation experiments similar to those in Anand and Qiao, and Han and Shin[18,27] on the 8 × 8, 10 × 10, and 12 × 12 mesh networks and three random networks, namely *RandNet1* with 70 nodes and 156 links, *RandNet2* with 80 nodes and 200 links, and *RandNet3* with 90 nodes and 282 links. The performance is measured using three metrics: the number of requests that can be satisfied, average call acceptance ratio (ACAR), and average spare wavelength utilization (ASWU). ACAR denotes the fraction of requested calls that are accepted, averaged over a long duration of time. In experiments, two parameters are introduced:

minimum length (ML) and *maximum delay increment (MDI)*. The parameter ML denotes the length of the shortest path between the source and the destination. A requested D-connection has the shortest path between the source and the destination whose length is greater than ML. The parameter ML is chosen depending on the size and diameter of the network topology. The parameter MDI denotes the restriction on the number of hops along the protection lightpath exceeding that along the primary lightpath. The parameter MDI is essential to have the bit-error rate and delay along both the primary lightpath and segmented protection lightpath to be as low as possible and is set to 3 in all the experiments.

As expected, the segmented protection paths scheme performs well in terms of the number of requests satisfied compared to the end-to-end protection scheme.[18] The percentage of improvement over the end-to-end protection scheme is more (about 0 to 15% for single-fiber) when *incremental* traffic is considered. The scheme also performs well in terms of ACAR. As explained in Figure 16.3, the end-to-end protection scheme reserves more number of wavelengths for D-connections, so the chances of finding a common free wavelength for future D-connections become less; however, the segmented protection paths scheme conserves wavelengths by providing a lesser number of wavelengths for protection lightpaths. By doing so, this scheme enhances the chances of finding a common free wavelength for future D-connections. Because of the preceding two reasons, the ACAR of this scheme is more than that of the end-to-end protection scheme, and the percentage of improvement varies from 3 to 25%. As expected, this scheme requires a lesser amount of spare wavelengths than the end-to-end scheme until around 55% of load. This is because the end-to-end protection scheme reserves a greater number of wavelengths for D-connections. As the load increases, however, the ACAR of the segmented protection paths scheme is greater, so it requires slightly more spare wavelengths. The savings in spare wavelengths reserved increases as we go to large networks. This is because the efficiency of this scheme increases as the number of backup segments increases.

16.3.4 Partial backup lightpaths

16.3.4.1 Motivation

In conventional approaches to fault tolerance,[15,18,27] end-to-end backup lightpaths are provided and they are able to handle any component failure under the *single link failure* model. In the single link failure model, only one link in the whole network is assumed to fail at any instant of time. Because the failure of components is probabilistic, such a model is not realistic, especially for large networks. We note that connections with end-to-end backup lightpaths also have to be reestablished in case more than one link fails simultaneously. In such a probabilistic environment, network service providers cannot give any absolute guarantees but only probabilistic guarantees. End-to-end detouring has the additional requirement that for a call to be accepted it is essential to find sufficient resources along two totally disjoint paths between the source–destination pair. Even when two routes are in the network between the source–destination node pair, it is possible for the primary lightpath to be routed (along the shortest hop path or minimum delay path) so that there cannot exist an end-to-end backup lightpath.

Second, every lightpath does not necessarily need fault tolerance to ensure network survivability. Third, at any instant of time, only a few lightpaths critically require fault tolerance. For such critical lightpaths, full backup lightpaths may be exclusively reserved. Fourth, failures do not occur frequently enough in practice to warrant end-to-end backup lightpath. Fifth, providing protection against fiber network failures could be very expensive due to less number of wavelengths available and high costs associated with fiber transmission equipment.

Lastly, today's applications and services are mostly based on the ubiquitous IP, and the trend is likely to continue. The trend in the current network development is moving toward a unified solution that will support voice, data, and various multimedia services (multi-service providers). This is evidenced by the growing importance to concepts such as QoS and differentiated services that provide various levels of service performance. Recently, there has been considerable interest in providing reliable connections in optical WDM networks. The problem of providing reliable connections in optical ring networks is considered in [30, 31]. Here, in [30, 31], each connection is assigned a maximum failure probability (MFP). The problem of providing the service differentiation is achieved through the primary backup multiplexing.[16] The lower class connections are assigned protection wavelengths used by the higher class connections. The objective is to find the routes and wavelengths used by the lightpaths in order to minimize the ring total wavelength mileage, subject to guaranteeing the MFP requested by the connection (i.e., the problem is considered a provisioning problem). This section explains the concept of partial backup lightpaths,[32,33,34] its advantages, and how the algorithm considers the probabilistic nature of failure of components.

16.3.4.2 Concept of partial backup lightpaths

Applications/end users differ in their willingness to pay for a service that provides fault-tolerance. Considering the requirements of different applications/end users, it is essential to provide services with different levels of reliabilities. In [33, 34], the authors incorporate the reliability of connections as a parameter of QoS and describe a scheme for establishing connections with such QoS requirements. A connection with the reliability requirements is called an *R-Connection* (reliable connection). Reliability of a resource (or component) is the probability that it functions correctly over a period of time. Reliability of an R-connection is the probability that enough resources reserved for this R-connection are functioning properly to communicate from the source to the destination over a period of time. Reliability has a range from 0 (never operational) to 1 (perfectly reliable). It is assumed (with reasonable justification) that reliability comes at cost. Therefore, a more reliable connection comes at a greater cost. The fiber reliability from the point of view of loss variation for various cable-environment parameters (example, temperature, humidity, and radiation) was studied by several authors.[35-38] Even though the majority of fiber failures have been reported due to external factors such as dig-ups, fire, etc., a few failures are also reported due to strength loss of the fiber itself. Despite the low probability of fiber failure, however, the associated economic risk is appreciable because of:

1. The high cost of the fiber repair or replacement
2. Large volumes of data passing through optical networks
3. In recent days, the micro-electro-mechanical (MEM) optical switches are becoming increasingly popular; because these switches work based on the rotation of the mirrors, the reliability of these components is of particular importance

The algorithm presented by [33, 34], establishes an R-connection with primary lightpath and an optional backup lightpath. A backup lightpath is provided when the reliability specified by the application requires that a backup lightpath is provided, and it can be either end-to-end or partial which covers only a part of the primary lightpath. The length of the primary lightpath covered by the partial backup lightpath can be chosen to enhance the reliability of the R-connection to the required level. The length of the primary lightpath for which backup is provided depends on the reliability required by the application/end user but not on the actual length of the primary, network topology, and design constraints. If certain portions of the primary lightpath are considered less reliable (more vulnerable), then the backup lightpaths are provided for only those segments of the primary lightpath. For providing backup lightpaths, resources need to be reserved along the backup lightpaths as well. This is an added cost, and the partial backup scheme preserves resources by using only the required amount of backup lightpaths. By doing so, it reduces the spare resource utilization and thereby increases the ACAR. In this scheme, many R-connections will have only a partial backup lightpath rather than end-to-end backup lightpath. This means that if there is a fault in the part of the primary lightpath which is not covered by the backup lightpath, then the R-connection cannot be restored immediately: the whole path has to reestablished. But, note here that connections with end-to-end backup lightpaths also have to be reestablished in case of more than one link failing simultaneously. Various real-time applications like video-on-demand, video conferencing, scientific visualization, computer-assisted collaborative work, and virtual reality benefit greatly from our scheme, where the disruption of a connection is nuisance which we would like to avoid, but is not mission-threatening.

In [33, 34], the authors assume that none of the nodes have wavelength conversion capabilities (because all-optical wavelength converters are expensive). The wavelength continuity constraint imposed by these networks

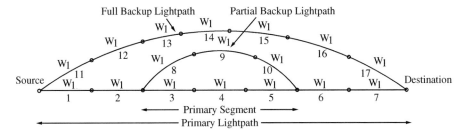

Figure 16.8 Illustration of partial backup and full backup lightpaths.

requires that same wavelength must be allocated on all the links of the chosen route from the source to the destination. A primary segment is a sequence of contiguous links along the primary lightpath. A partial backup lightpath covers only a primary segment (i.e., the backup lightpath can be used when a component along the primary segment encounters a fault). Figure 16.8 illustrates the benefit of a partial backup lightpath. An R-connection has to be established from the source to the destination. The primary lightpath consists of 7 links (i.e., links 1, 2, 3, 4, 5, 6, and 7). Here, links 3, 4, and 5 and their end nodes form a primary segment. The partial backup lightpath, consisting of links 8, 9, 10 and their end nodes covers the preceding primary segment. The end-to-end backup lightpath (which is disjoint from the primary lightpath) consists of 7 links (i.e., 11, 12, 13, 14, 15, 16, 17) and covers the entire primary lightpath.

For simplicity [33, 34] assume that nodes are fully reliable (i.e., only links are prone to faults and all the wavelength channels on a link are assumed to have same reliability). The subsequent discussion can be easily extended to include node failures also. We now explain how to find the reliability of an R-connection from the source to the destination as shown in Figure 16.8 with partial backup lightpath, full backup lightpath, and no-backup lightpath. The reliability of a segment consisting of links with reliabilities r_1, r_2, ... r_n will be $\prod_{i=1}^{n} r_i$. Let r_p denote the reliability of the primary lightpath, r_s denote that of the primary segment which is covered by a backup lightpath, r_b that of the backup lightpath and r_c that of the composite path comprising of primary and backup lightpaths. Here, $r_p = \prod_{i=1}^{7} r_i$, $r_s = r_3 . r_4 . r_5$, and $r_b = r_8 . r_9 . r_{10}$. Now r_c = (reliability of part of primary lightpath not covered by the backup lightpath) × (reliability of primary segment and partial backup lightpath together).

$$r_c = \frac{r_p}{r_s}.(r_s + r_b.(1 - r_s)) \tag{16.1}$$

16.3.4.3 Advantages of partial backup lightpaths

Let r_r denote the reliability requested by an application/end user. We now observe how the partial backup lightpaths are useful. Suppose the reliability of each of the links is 0.9800, and the required reliability r_r is 0.9150. Then, for the R-connection shown in Figure 16.8, using partial backup lightpath, $r_p = 0.8681$, $r_b = r_s = 0.9411$. Then, using Equation (16.1), $r_c = 0.9192$. Thus, having a partial backup for any 3 links is just enough in this case as the required reliability is 0.9150.

Now, consider the same R-connection in Figure 16.8, using end-to-end backup lightpath. Because the entire primary lightpath is covered by backup lightpath, reliability of the primary segment is equal to reliability of the primary lightpath, $r_s = r_p = 0.8681$. The reliability of the full backup lightpath (in this case, which has the same number of links as the primary lightpath), $r_b = 0.8681$. Then, using Equation (16.1), $r_c = 0.9826$, which is much more than the reliability required by the R-connection. Note that the end-to-end scheme is not able to distinguish the R-connections with different reliability requirements. Now, consider the same R-connection in Figure 16.8, using no backup lightpath at all. In this case, the composite reliability $r_c = r_p = 0.8681$, which is less than the reliability required by the R-connection. From the example, it is clear that the partial backup scheme preserves resources by using only the required amount of backup lightpaths. By doing so, it reduces the spare resource utilization and thereby increases the ACAR. It also distinguishes the R-connections with different reliability requirements.

16.3.4.4 *Providing backup lightpaths to reliable connections*

When an application/end user requests an R-connection from a source to a destination, the algorithm tries to provide a lightpath with the requested reliability using no backup lightpath at all, or a partial backup lightpath, or an end-to-end backup lightpath, in that order. Partial backup scheme aims at minimizing the amount of spare resources while establishing the lightpath with the requested reliability. As the resources reserved for backup lightpath are used only when there is a fault in the primary lightpath, providing a backup lightpath means a large amount of extra resource reservation.

Hence, this scheme establishes a backup lightpath only when it is not possible to find a primary lightpath with the required reliability. This scheme assumes that the delay along a path and network resources reserved to be proportional to the length of the path. Therefore, the amount of resources reserved and delay are synonymous with path length. This scheme uses usual algorithms (to minimize resource reservation or delay) for finding the shortest route from the source to the destination, and fixed ordering wavelength assignment policy to select free wavelength. If the reliability of the shortest lightpath so found is below the required reliability, then the algorithm tries to find a route with required reliability, using a modified route-selection algorithm (described later in this section). If this modified algorithm fails to find a suitable route or the wavelength continuity constraint is not satisfied, then the algorithm tries to establish the R-connections by using backup lightpaths (partial or end-to-end) to achieve the required reliability. We now present the algorithm in detail; r_r, r_p, r_b, r_s, and r_c are as described in the previous section:

1. Find a primary route from source to destination, using Dijkstra's shortest path algorithm.
2. Find a common free wavelength along the route found in step 1 using FX (fixed ordering) wavelength assignment policy. If a free common wavelength is available then set flag-1 to *true*, otherwise *return* failure.
3. If $r_p \geq r_r$ and flag-1 is *true, then* accept this R-connection and *return* success. *Else go to* step 4.
4. Use the modified route selection algorithm (described in the next section) to find a primary route from the source to the destination.
5. Find a common free wavelength along the route found in step 4 using FX (fixed ordering) wavelength assignment policy. If a free common wavelength is available then set flag-2 to *true*.
6. If new $r_p \geq r_r$ and flag-2 is *true, then* accept this R-connection and *return* success. *Else go to* step 7.
7. Reconsider the primary route found in step 1.
8. Identify some segments of primary lightpath for which we can give a backup lightpath to enhance their reliabilities. Find backup routes for the identified primary segments using modified route selection algorithm.
9. Find whether the same wavelength on which primary is established is available on the identified segments. If same wavelength is available then set flag-3 to *true*.
10. Select one segment which satisfies the reliability requirement and whose flag-3 set to true (whether a backup satisfies the reliability requirement or not can be decided by evaluating r_c using Equation [16.1]). If such a backup exists, accept that primary and backup and return *success*.
11. *Return* failure.

In step 7, the algorithm reconsiders the shortest path found in step 1 instead of the one found in step 4 in order to decrease the load on links with high reliability, which would be preferentially chosen by the modified path selection algorithm. The main issues involved are given next; they are discussed later in this section:

1. The modified route selection algorithm to find a route with higher reliability in step 4
2. Identification of the segments of the primary lightpath in step 8
3. Selection of a suitable backup lightpath among all the eligible backup lightpaths in step 10
4. Selection of wavelength along the route chosen (primary, partial, or end-to-end backup)

Although the algorithm establishes only one backup lightpath, it can be easily adapted to establish multiple backup lightpaths to further enhance the reliability of an R-connection. For example, in step 11, we can instead have:

Establish one end-to-end backup lightpath and one partial backup lightpath. This primary with two back-up lightpaths might satisfy the reliability requirement. *If* it satisfies, accept this R-connection with two backup lightpaths and *return* success.

In the following discussion, we present some simple solutions to the issues raised in this section.

16.3.4.5 Modified route selection algorithm

Finding a route subject to multiple constraints on routing metrics is NP-complete.[3,39,40] In [33, 34] the authors are interested in minimizing resource utilization and maximizing reliability. There is no provably efficient algorithm for doing this, and so heuristic algorithms are presented by [33, 34]. The algorithm attempts to find routes with higher reliability at the expense of greater path length. To do this, a cost function for each link that is dependent both on its reliability and delay (along it) is defined. Then, it uses Dijkstra's minimum cost algorithm to find a route from the source to the destination. Delay is an additive metric whereas reliability is a multiplicative one (i.e., the delay along a route is the sum of the delays along each link, whereas the reliability of a route is the product of the reliabilities of the links in it). Because Dijkstra's algorithm takes costs to be additive, the logarithm of the reliability is used in the cost function. Thus, a suitable cost function is given as,

$$cost = delay - relWeight * log(reliability) \tag{16.2}$$

where *relWeight* is a parameter. By varying the value of *relWeight*, the network provider can control the tradeoff between reliability and delay along the path chosen.

16.3.4.6 Identification of primary segments

As described earlier in this section, the algorithm identifies some suitable segments of the primary lightpath and finds backup lightpaths for them to enhance the reliability of the R-connections to the desired level. So, it identifies primary segments where the reliability is less than *estRel*, which is calculated next; r_p, r_s, r_r and r_c are as described in Section 16.3.4.2.

$$r_c = \frac{r_p}{r_s}.(r_s + r_b.(1 - r_s)) \geq r_r \Rightarrow r_s \leq \frac{r_p}{r_r}.(r_s + r_b.(1 - r_s))$$

Now, $r_b < 1$. Therefore, $r_s < r_p r_r.(r_s + (1 - r_s))$

$$\Rightarrow r_s < estRel = \frac{r_p}{r_r} \tag{16.3}$$

Among primary segments of a given length, it would be advantageous to provide backup lightpaths for primary segments with low reliability because, as seen from Equation (16.1), r_c increases as r_s decreases assuming $r_b \approx r_s$.

16.3.4.7 Selection of suitable backup

A number of segments, up to a maximum of *segmentTrials* (which is input to the algorithm), are found as described previously and are remembered. It also adds the whole primary lightpath as an alternative in case an end-to-end backup lightpath is very convenient. Then, it tries to find backup lightpaths for those that satisfy the reliability requirement. It uses the modified route selection algorithm to find a backup route between the end nodes of the primary segment, taking care to exclude all the components of the primary other than the end nodes of the primary segment. Among these backups (for different primary segments), the backup reserving the lesser amount of resources is preferable. In the case of backups reserving slightly different amounts of resources, however, it might be better to choose one which gives higher composite reliability. So, it selects a backup based on an *expense* function given next.

$$expense = pathLength - compositeRelFactor * r_c \qquad (16.4)$$

Here, *compositeRelFactor* is a parameter that allows a tradeoff between composite reliability and extra resource reservation. The algorithm chooses the backup with the least *expense* value.

16.3.4.8 Failure detection and recovery

After failure reporting, if the failed component is covered by a backup lightpath, backup activation is done. In that case, the end nodes of the primary segment initiate the recovery process on receiving the failure report. They send the activation message along the backup lightpath. Activation messages will carry the connection identifier and lightpath information. These messages are used to set the state of the switches such that backup lightpath is switched from an inbound link to an appropriate outbound link.

As resources are reserved along the backup lightpath beforehand, the R-connection will be resumed. The delay suffered here is low, as required by most real-time applications. This process is illustrated in Figure 16.9. If the failed component is not covered by a backup lightpath, the source initiates the recovery process upon receiving the failure report. The source again requests a reliable connection to be set up, which may take a much longer time.

16.3.4.9 Performance study

The performance of the algorithm has been verified through the simulation experiments similar to those in Han and Shin on the 8×8 mesh network.[27] The performance is measured using three metrics: ACAR, ASWU, and reliability got by each connection request. In the simulation experiments, reli-

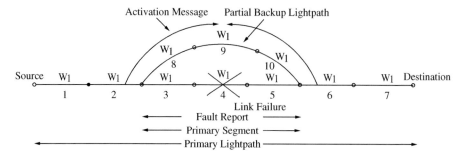

Figure 16.9 Illustration of failure recovery in case of partial backup lightpaths.

ability of the links was set as a uniformly distributed random value between 0.97 and 1.0. Reliability of all the fibers on a link and all the wavelength channels on a fiber are assumed to be equal.

The ACAR is highest for partial backup scheme compared to end-to-end and no backup schemes. The effectiveness (i.e., percentage of improvement over the end-to-end scheme) of this scheme increases as the size of the network increases. The high ACAR observed for partial backup scheme is expected because most of the R-connections have partial backups. The ACAR for end-to-end scheme is less because of longer backups. Generally, longer-hop connections are subjected to more blocking than shorter-hop connections due to wavelength continuity constraint. As the end-to-end backup scheme reserves more wavelengths for R-connections, the chances of finding a common free wavelength for future R-connections become less. Because of this, the ACAR for this scheme is less, but the partial backup scheme conserves wavelengths by providing backup lightpaths to only less reliable segments. By doing this, the scheme enhances the chances of finding a common free wavelength for future R-connections.

The difference in spare wavelengths reserved is quite significant at low and intermediate loads but decreases by a small amount at high loads. For a given number of wavelengths, as the required reliability increases the spare wavelength utilization for a partial backup scheme increases. Whereas, for the end-to-end scheme, it is the same. As the size of the network increases, the partial backup scheme tends to be more effective than end-to-end backup scheme. Partial backup scheme provides R-connections with reliability close to the requested reliability. A good level of service differentiation has been achieved using partial backup scheme. End-to-end backup scheme provides most of the R-connections with higher reliability because end-to-end backup lightpaths are provided for all R-connections.

16.4 *Distributed restoration methods*

Many of the existing algorithms for selecting routes and wavelengths for primary and backup lightpaths assume centralized control. Such algorithms are useful for small networks and are not scalable to large networks. For simplicity and scalability purposes, distributed control protocols are desired. In a distributed implementation, up-to-date network state information is not known to any of the nodes. Also, the possibility exists for reservation conflicts between several simultaneous search attempts. This may result in poor channel utilization. Another important issue is reducing the restoration time, as the search for a backup path is initiated only after failures. This section explains distributed protocols for lightpath restoration.

16.4.1 A distributed control protocol for proactive methods

In Doshi et al., a distributed control protocol for selecting primary and backup lightpaths is presented.[10] It basically uses a proactive path-based restoration approach. It assumes that every routing node is equipped with WIXC (i.e., every node is capable of performing wavelength conversion optically). The same method can be used for point-to-point WDM networks, where every node is capable of performing wavelength conversion by means of electro-optical conversion and buffering; however, this protocol requires suitable changes when the routing nodes are equipped with WSXCs. The procedure for finding a backup path for a given primary path is explained here. It should be noted that this procedure can be extended to find both the primary and backup paths simultaneously. Because the nodes possess conversion capability, a path need not use the same wavelength on its links. Every link has a fixed capacity. Link capacity is measured in terms of wavelength channels. A demand requires a backup path to be established between a source and destination node, given a primary path between the same nodes.

The protocol considers single link and node failures. At every node, link capacity control tables are maintained for each of the outgoing links. A table associated with a link has information about the possible (node or link) failures in the network. The following discussion pertains to one link or table. Associated with every possible failure, there is a list of demands with backup paths that will require the link upon the failure. Each demand in a list requires one wavelength on the link. The maximum over the size of lists in a link capacity control table gives the number of wavelengths used on the link thus far. In other words, this protocol employs backup multiplexing.

From the size of the list associated with a fault, the number of free wavelengths available (residual capacity) on the link upon occurrence of the fault can be determined. A source node initiates the search for a backup path independent of searches by other source nodes. It sends control messages carrying the identity of the primary path on the supervisory channel. A distributed breadth-first search method is used to select a backup path. When a node receives such a probe message, it examines the table that corresponds to the outgoing link. In the table, it looks up the residual capacity of the faults that correspond to the nodes and links traversed by the primary path of the demand. By doing so, it determines if a free wavelength is available for the backup path on the outgoing link.

It is possible that the search process is going on for multiple demands simultaneously. Because of this, resource (wavelength channel) contention may occur on a link, leading to a deadlock situation. This protocol prevents resource contention as follows. Two primary paths that do not share a link or node (except source and destination) will never contend for a wavelength on any link. Therefore a source node, before initiating a search for a backup path, locks the links and nodes on its primary path. Once the backup path search procedure terminates, the locks are released. Because of the distributed nature and lack of global coordination of the search procedure, the resulting routes may need more wavelength channels than a centralized algorithm is able to achieve. In order to improve the channel utilization, optimization procedure is used. This optimization procedure is invoked either periodically or whenever a search fails. Changing the existing backup paths carries out the optimization. The backup paths are used as cold-standby, so the traffic on the network is not disrupted.

The effectiveness of the distributed protocol was evaluated using a simplified simulation model because the complete simulation model is extremely complex. The simplified simulation model captures only the dynamics of demand contention locking. For a sample random network with 28 nodes and 48 links and a set of 50 demands, the number of demands blocked by the distributed protocol is compared with that by an optimal solution. The simulation results show that the performance of the distributed protocol is very close to that of the optimal solution.

16.4.2 A distributed control protocol for reactive methods

A distributed control protocol for reactive methods has been proposed in Ramamurthy and Mukherjee.[9] Upon a link failure, this protocol searches for backup lightpaths for the failed lightpaths. Both the link-based and path-based restorations have been considered. It uses a *two-phase* process to search for a backup path between a pair of nodes on a given wavelength. This pair of nodes includes the end nodes of the failed link in the case of link-based restoration, and includes the end nodes of the failed path in the case of path-based restoration. The two-phase process is described next.

The source node sends broadcast messages on all outgoing links. A message reserves (locks) the wavelength on a link, while traversing it. This message carries the hop count of the path it has traversed so far. When a broadcast message reaches the destination node, it sends a confirm message along the path to the source node. Upon receiving a confirm message, intermediate nodes configure their cross-connects. When the source node receives a confirm message, the backup path establishment is done. All the wavelengths that were reserved during the broadcast-phase are released by cancel messages that are sent by a node upon a timeout or upon receiving a message with hop-count exceeding a predefined value. The link-based restoration requires that the backup path uses the same wavelength used by the corresponding primary path. Since the failed paths are on different wavelengths, the distributed path selection for the failed paths are carried out in parallel. In case of the path-based restoration, the free wavelengths on the failed link are partitioned into a set of wavelengths, one set for each of the failed path. This results in parallel path search on different wavelengths, avoiding the conflicts between them.

The performance of both methods was evaluated on an interconnected-ring topology. The performance metrics considered were restoration efficiency and restoration time. Restoration efficiency is defined as the ratio between the number of lightpaths restored and the total number of lightpaths failed. Suitable values for propagation delay and message processing delay were assumed. The results show that the link-based method has a better restoration time than the path-based method. Two reasons can be attributed to this. First, no control messages need be sent to the end nodes of the failed path in a link-based method. Second, backup paths around the failed link are likely to be shorter. The results show that the path-based method has better restoration efficiency. This is because a path-based method can use any wavelength for the backup path. It has been further observed that restoration time for both the methods are in the order of milliseconds only. Also, restoration efficiency decreases with the increasing load. This is because at a high load, fewer channels are free and also the reservation conflicts are more as many paths fail.

16.4.3 Distributed control for reliability-constrained least-cost routing

In this section, we explain a distributed control scheme presented in [41, 42] for establishing reliability-constrained least-cost lightpaths in WDM optical networks. One of the problems studied in this class of constrained optimization problems is the least-cost delay-constrained routing and is presented in [43, 44].

16.4.3.1 Network model

In [41], the network is modeled as an undirected graph $G = (V,L)$, where V is a set of nodes and L is a set of interconnecting links. Let R^+ be a set of real numbers. Each physical link $l \varepsilon L$, is characterized by the following four functions.

Reliability function $R : L \rightarrow R^+$

Cost function $C : L \rightarrow R^+$

Total wavelength function $Tset : L \rightarrow \{\lambda_1, \lambda_2, ... \lambda_n\}$

Available wavelength function $Aset : L \rightarrow \{\lambda_1, \lambda_2, ... \lambda_n\}$, $Aset \subseteq Tset$ and
$$| Aset | \leq | Tset |$$

A path $P = (v_0, v_1, v_2, ..., v_n)$ in the network has two associated characteristics:

$$CostC(P) = \sum_{i=0}^{n-1} C(v_i, v_{i+1})$$

$$ReliabilityR(P) = \prod_{i=0}^{n-1} R(v_i, v_{i+1})$$

16.4.3.2 Problem formulation

A lightpath establishment request (also referred to as a connection or a call) in the network described previously is represented as a 5-tuple: $Req = (conid, s, d, \Delta, n_w)$, where id is the connection request identification number; $s \varepsilon V$ is the source node for the connection; $d \varepsilon V$ is the destination node for the connection; Δ is the reliability constraint to be satisfied; n_w is the number of wavelengths required for the connection (i.e., number of lightpaths to be established between the nodes s and d).

Let P_{sd} denote the set of all paths of the form $P = (s = v_0, v_1, v_2, ..., v_n = d)$ between the source s and the destination d that satisfy the following two conditions:

C1 : $Aset(v_0, v_1) \cap Aset(v_1, v_2) \cap ... \cap Aset(v_{n-1}, v_n) \geq n_w$

C2 : $R(P) \geq \Delta$

Then, the reliability-constrained least-cost lightpath establishment problem can now be formulated as:

$$Find\ P' \varepsilon P_{sd}\ such\ that\ C(P') = min\{C(P): P \varepsilon P_{sd}\}$$

16.4.3.3 The preferred link routing approach

To establish a lightpath between a source node s and a destination node d, a route between them and also the free wavelengths present on the route are to be found. The scheme in [41] uses the backward reservation method along with the preferred link-based routing algorithm to establish reliability-constrained least-cost lightpath. The preferred link-based routing algorithm finds a route between s and d, and also the free wavelengths on it, simultaneously. The preferred link routing framework [43, 44] is fundamentally a backtracking-based route selection method. This framework describes a set of actions to be performed by each node whenever it receives a connection setup or connection reject packet.

When a node v receives a connection setup packet, it forwards it along the first preferred link (preferred links are ordered depending on the heuristic values computed and are discussed in the next section). The connection setup packet includes the connection identifier (*conid*), the path taken by the packet up to this point (*P.path*), the product of the reliabilities of the links in this *P.path* (*P.reliability*), the set of available wavelengths on this *P.path* (*Aset*), the reliability required by the connection (Δ), and the number of wavelengths required by the connection (n_w). Before forwarding the connection setup packet on the first preferred link, three tests are performed (as discussed later in detail), and the available wavelength set, *Aset*, is updated by taking the intersection of it and the set of free wavelengths on that preferred link. If a reject packet is received from the node at the other end of this link, then the node v attempts to forward the packet along the next preferred link and so on until a specified number of links has been tried out. If all such attempts result in failure, then v sends back a reject packet to the node from which it received the connection setup packet. If the connection setup packet reaches the destination, then a path is found between the source and the destination satisfying the given reliability and wavelength constraints.

If the source gets a reject packet from all the nodes attached to its preferred links, then it queues the packet in its local buffer and retransmits it after GAP time. If the number of retransmissions reaches MAX_TRIES, the connection request will be rejected. When the connection setup packet reaches the destination, the set S_f is formed by taking a subset of the collected free wavelengths. A *LOCK* message is sent from the destination to the source to lock the set of wavelengths, S_f, along the path. The size of the set S_f is greater than or equal to the number of wavelengths (n_w) required by that connection request. For preparing the wavelength set, S_f, we generate a random number between 0 and the maximum number of wavelengths available, W; starting from this random number we choose $\delta * n_w$ (where $\delta \geq 1$) of free wavelengths in a cyclic manner. When the *LOCK* message reaches the source, a *RES* message is sent from the source to the destination with the required number of wavelengths, n_w (these are selected randomly from the set, S_f). The *RES* message moves toward the destination, updating the status of wavelengths at the intermediate nodes and releasing all the wavelengths locked except for the wavelengths in the set n_w. When the data transmission on the allocated lightpath is complete, the source node prepares a message called *REL* message to release the connection. The *REL* message traverses toward the destination releasing the wavelengths (n_w) used by the connection. When the *REL* message reaches the destination, the release operation is complete.

To implement the proposed heuristics in conjunction with the preferred link routing framework, each node in the network is equipped with two data structures namely, a *connection status buffer* and a *preferred link table*.

The *connection status buffer (CSB)* at each node v contains one entry for every connection for which v has received a connection setup packet. Each entry contains a pair of elements (*packet, tried*) where *packet* is the connection setup packet received by the node and *tried* is the number of preferred links on which v has tried to forward the packet. Therefore, the CSB at a node v contains the complete status information for every connection that was handled by v. The entry corresponding to a connection is removed when the connection is either accepted or rejected.

The structure of the *preferred link table (PLT)* to be maintained at each node depends upon the nature of the heuristic function that is employed to construct the table. For describing the structure of the PLT, we classify all heuristic functions into two major categories: *destination-specific heuristics* and *connection-specific heuristics*.

Destination-specific heuristics are those where the computation is specific to each destination. Therefore if the destination nodes of two different connection requests arriving at a given node are the same, then the two connections will share an identical list of preferred links. Each node v in the network is equipped with a PLT that contains one row for every destination. Each row contains the preferred links for that particular destination in terms of decreasing preference. The maximum number of entries per row is denoted by k, the maximum number of preferred links. Obviously k is upper bounded by the maximum degree of any node in the network. The preference for the link will be decided based on the value of the heuristic function that is computed for each (link, destination) pair.

Connection-specific heuristics are those where the computation depends on the particular parameters (such as reliability and n_w) carried by a connection setup packet arriving at the node. In such cases, the list of preferred links is individually computed for each connection request. As a result, the ordering of the links will be connection-specific instead of destination-specific. For such heuristic functions, the number of rows in the PLT will vary dynamically depending on the number of connections currently being handled by the node. The table entries corresponding to a node are removed when the connection is accepted or rejected.

16.4.3.4 Tests before forwarding control packet

Before forwarding any packet along a link, each node conducts three tests on the link parameters. The link is used for forwarding the packet only if all the tests are successful. These tests are described later in this section. Let $Req = (id, s, d, \Delta, n_w)$ be a connection request and P be a connection request packet arriving at a node v. Let $P.path$ denote the path taken by the packet up to this point and $P.reliability$ denote the product of the reliabilities of the links in this $P.path$. Before forwarding the packet along link $l = (v, v')$, node v conducts the following three tests:

(T1) Reliability Test: Verify that $P.reliability \times R(l) \geq \Delta$
(T2) Wavelength Availability Test: Verify that $|Aset(P.path) \cap Aset(l)| \geq n_w$
(T3) Loop Test: Verify that v' is not a node in $P.path$

16.4.3.5 Heuristic functions to compute preferred links

The *cost-reliability product (CRP)* heuristic is a destination-specific heuristic. Then the CRP value of a link $l = (i,x)$, corresponding to the destination d, is given by

$$CRP = \frac{C(l)}{R(l) \times MRELIABLE(x,d)}$$

where $C(l)$ and $R(l)$ denote the cost and reliability of the link l, respectively; and $MRELIABLE(x, d)$ the maximum reliable path from node x to node d in the network. To load the PLT entries corresponding to node d the following steps are performed. The links adjacent to node i are arranged in increasing order of their *CRP* values and first k links are chosen and used to populate the PLT entries for destination d.

The *residual reliability maximizing (RRM)* heuristic is a connection-specific heuristic. Let a connection setup packet belonging to a connection request $Req = (conid, s, d, \Delta, n_w)$ arrive at node i. For each link $l = (i,x)$ at i, let $RRM(l, Req)$ denote the value of the heuristic for a link l corresponding to a connection Req. Then $RRM(l, Req)$ is given by

$$RRM(l, Req) = P.reliability \times R(l) \times MRELIABLE(x,d) - \Delta$$

where $R(l)$ denotes the reliability of link l; Δ is the reliability required by the connection; $MRELIABLE(x, d)$ the maximum reliable path from node x to node d in the network; and $(P.reliability)$ is the product of the reliabilities of the links in the $P.path$. If in the calculation of the heuristic function, a particular link produces a negative value, then that link is not included in the preferred link list. The links are arranged in the preferred list in decreasing order of their RRM values, so that the links with higher *RRM* values are given greater preference. The intuitive idea underlying this function is to maximize the residual reliability (i.e., the reliability available for setting up the rest of the path).

The *cost-residual reliability tradeoff (CRRT)* heuristic is a connection-specific heuristic. Let a connection setup packet belonging to a connection request $Req = (conid, s, d, \Delta, n_w)$ arrive at node i. For each link $l = (i,x)$ at i, let $CRRT(l, R)$ denote the value of the heuristic for link l corresponding to a connection Req. Then $CRRT(l, Req)$ is given by

$$CRRT(l, Req) = \alpha \times C(l) + \frac{(1-\alpha)}{(P.reliability \times R(l) \times MRELIABLE(x,d) - \Delta)}$$

where α is a parameter. By varying the value of α, the network provider can control the tradeoff between the reliability and cost along the path chosen. If, in the calculation of the heuristic function, a particular link produces a negative value, then that link is not included in the preferred list. The links are arranged in the preferred list in increasing order of their *CRRT* values, so that the links with higher *CRRT* values are given greater preference. The intuitive idea underlying this function is to maximize the residual reliability (i.e., the reliability available for setting up the rest of the path) at the same time minimizing the cost of the link chosen.

The *partition-based (PB)* heuristic is destination-independent and connection-independent. Let $avg(i)$ denote the average cost of all the links adjacent to node i. The links adjacent to node i are partitioned into two sets *below* and *above*, where

$$below(i) = l : C(l) \leq avg(i)$$

$$above(i) = l : C(l) \geq avg(i)$$

The links in the two sets are then separately sorted in the decreasing order of their reliability values. Now, a new list is created containing the sorted *below* set, followed by the sorted *above* set. The first k links from the new list are chosen and used to populate the table.

16.4.3.6 Performance study

The performance of the different heuristics has been verified through the simulation experiments similar to those in [45], on the randomly generated networks. The performance metrics considered are: ACAR, average cost (AC), average routing distance (ARD), and average call setup time (ACST). The effect of parameters such as the reliability requirement of connections (Δ), the wavelength requirement of connections (w), the number of preferred links (k), and the connection arrival rate (λ) on the performance metrics is studied. For each of the randomly generated networks, physical links with single-fiber having different number of wavelengths is considered. All the links are assumed to have the equal number of wavelengths. Lightpaths are assumed to be bidirectional.

Effect of reliability constraint.   The ACAR is high for all heuristics compared to alternate link routing [45]. The ACAR decreases as the reliability value increases in all the cases. The AC decreases as the reliability required by the connection requests increases for all the heuristics. This is because as the reliability required by the connections increases the ACAR decreases and hence the drop in AC. Generally, the reliability of longer paths will be less and hence they will be rejected due to reliability constraint. The ARD of heuristic CRP is smallest of all. The ARD decreases when the reliability required by the connection requests increases because of decrease in ACAR. The ACST of different heuristics decreases as the reliability required by the connections increases. At low reliability requirements RRM performs well with respect to ACST. The ACST of the CRP heuristic is always less than that of the alternate link routing.

Effect of wavelength requirement.   Even though sub-rate multiplexing (traffic grooming) allows us to use bandwidth more efficiently, some services (like virtual private network) may require dedicated wavelengths. Service providers can offer *optical leased lines (leased lambdas)* by providing dedicated wavelengths to customers. This new and revolutionary type of service delivers enhanced flexibility to customers because of the bit rate independence of the wavelength service. In [41], the authors assume that traffic grooming is done at the higher client layers because:

1. As all-optical wavelength conversion and all-optical grooming devices are not presently commercially available, we resort to electronic methods of implementation to incorporate these features into the network.
2. It is very likely that networks of the near future will employ a hybrid, layered architecture using both electronic-switching and wavelength-routing technologies.

The ACAR of all heuristics is high compared to that of alternate link routing. The ACAR for the heuristic RRM is highest among all heuristics. As the wavelength requirement increases the ACAR for all the heuristics decreases, as it is increasingly difficult to find links with a greater number of free wavelengths. For a given number of fibers and wavelengths, reservation conflicts also increase when wavelength requirement increases. As the wavelength requirement increases, the average path cost decreases. Moreover, when the number of wavelengths required is more the reservation failures also will be more. The probability of reservation failure occurrence is more in longer paths than in shorter paths, since longer paths have more links. So when the number of wavelengths required increases, shorter hop connections have more chances to get accepted compared to longer hop connections. Due to this, the average cost of paths decreases. The ARD of CRP is smallest of all the heuristics, and the ARD of RRM is highest of all the heuristics. For the same reasons explained earlier, the ACST decreases with increase in wavelength requirement.

Effect of connection arrival rate.   As the connection arrival rate increases, there is not much drop in ACAR, AC, ARD, and ACST. This is attributed mainly to two reasons:

1. The network is at equilibrium, where the arrival and departure of the connections from the network is almost equal.
2. The network is admitting a greater number of smaller hop connections compared to longer hop connections. The RRM heuristic performs better with respect to ACAR and ACST; the CRP heuristic performs better with respect to AC and ARD.

Effect of number of preferred links.   The ACAR increases as k increases in the case of all the heuristics. The ACAR of RRM heuristic in all cases lies above 0.9. As we observed during the simulation studies, the reason for this is if a connection is not admitted with the initial entries in PLT, the connection may not be admitted with the other entries in the PLT as these entries may not satisfy the reliability constraint of the connection. As k increases, there is scope for a larger number of links to be attempted at each node. AC and

ARD also increase with increase of k because of the reasons explained previously. The effect of the number of preferred links on the performance metrics for alternate link routing and CRRT is almost the same.

16.5 Differentiated services in survivable WDM networks

Recently, there has been growing interest about service differentiation based on lightpath protection. Sridharan and Somani support three classes of service: full protection, no protection, and best-effort protection.[19] Two approaches in the best-effort are considered:

1. All connections are accepted and the network tries to protect as many connections as possible.
2. A mix of unprotected and protected connections are accepted and the goal is to maximize the revenue.

Sridharan and Somani presented integer linear programming (ILP) equations, which capture lightpath protection and also service disruption aspect into the problem formulation.[19] To reduce the computational complexity of the equations, the authors also presented heuristics. In [46, 47], a service classification based on fault-tolerance, availability, bandwidth, and delay is presented. In this section, we first motivate the readers toward the need for service differentiation in survivable WDM networks and then present different classes of lightpaths [46, 47].

16.5.1 Need for differentiated services

Due to the extremely high data rates (10 to 40Gbs or beyond) at which WDM networks are expected to operate, a network failure or malfunction has the potential to severely impact the mission critical and real-time applications. It is essential to incorporate fault-tolerance into QoS requirements. The types of applications being deployed across the public Internet today are increasingly mission-critical, whereby business success can be jeopardized by poor performance of the network. It does not matter how attractive and potentially lucrative our applications are if the network does not function reliably and consistently. In such scenarios (explained earlier) optical transport networks will not be a viable alternative unless they can guarantee a predictable bandwidth (wavelength), fault-tolerance, availability, and delay (response time), to users. Widely scattered users of the network do not usually care about the network topology and implementation details. What they do care about is something fundamental, such as:

- Do I get services with guaranteed timeliness and fault-tolerance with an acceptable restoration time at an acceptable level of overhead?
- Do I have my connection available when I want to access mission-critical applications from a remote location?

- Do I get acceptable response times when I access my mission-critical applications from a remote location or server?
- Do I have certain reliability and security to my data passing through the network?

CoS/QoS (class of service/quality of service) aims to ensure that mission-critical traffic has acceptable performance. In the real world, where bandwidth (resources such as wavelengths) is finite, distributed real-time multimedia applications such as video conferencing, scientific visualization, virtual reality, and distributed real-time control must all vie for scarce resources. CoS/QoS becomes a vital tool to ensure that all applications can coexist and function at acceptable levels of performance. CoS mechanisms reduce flow complexity by mapping multiple flows into a few service levels. Network resources are then allocated based on these service levels, and flows can be aggregated and forwarded according to the service class of the packet. Instead of the fine grain control of QoS, not easily achievable with current technology, CoS applies bandwidth and delay to different classes of network services. CoS easily scales with network expansion.

16.5.2 Different classes of lightpaths

The classification in [46, 47] separates different connections in the optical domain by providing lightpaths with different fault-tolerant capabilities (availability and resource utilization) and with different response times to meet the different connection requests and their traffic QoS/CoS requirements. Diverse *classes* of service are possible depending on the QoS parameters discussed previously, of these [46, 47] considered some service *classes*. We now describe different heuristic algorithms presented in [46, 47] to establish different connection requests, which differ from each other in the number and type of primary and backup lightpaths. The number of backup lightpaths determines the level of fault-tolerance and availability required by the connection. The type of lightpath determines the response times and bandwidth required by the connection.

16.5.2.1 Single dedicated primary and single dedicated backup lightpath (SDPSDB) — class I

In SDPSDB, both primary and backup lightpaths are all-optical, and the traffic is carried entirely in the optical domain. Primary and backup lightpaths are called dedicated because primary lightpath will carry exactly one connection request traffic and backup lightpath is not shared between any two (or more) backup lightpaths and acts as cold standby for the corresponding primary lightpath. Because the traffic is entirely carried in optical domain and is not shared by any other connection request traffic, the response times are very low. Because every lightpath has its own dedicated backup lightpath, it can tolerate a single failure. By providing a single backup lightpath, we can increase the availability of the connection.

I. *Primary Route Selection:* When a connection request from a source S to a destination D arrives, the algorithm finds the shortest path (minimum hop path) using Dijkstra's shortest-path-finding algorithm for use as the primary path.

II. *Backup Route Selection:* For finding backup path all the links along the shortest path are removed, and again the shortest path from source S to the destination D is found for use as the route for backup path using Dijkstra's shortest-path-finding algorithm.

III. *Wavelength Assignment for Primary and Backup Paths:* After finding all the free wavelengths on the primary and the backup paths the following two cases are considered:

Case I: The first free wavelength common to both the primary and backup paths will be chosen and reserved. Note here that primary dependent backup wavelength assignment is used. Because of this, all the backup paths should be on the same wavelength as that of the primary wavelength. This is because we assume that the nodes are equipped with only *fixed* transmitters and receivers.

Case II: In this case, no restriction is made on the use of wavelength for primary and backup paths. Therefore, the best possible primary-backup lightpath pair can be chosen to satisfy the request. Note here that we are using primary independent backup wavelength assignment, because backup path may be on different wavelength than that of primary wavelength. This method has the advantage of better network performance, but it assumes that the nodes are equipped with *tunable* transmitters and receivers.

16.5.2.2 Single dedicated primary and multiple dedicated backup lightpaths (SDPMDB) — class II

In SDPMDB, both primary and all k backup lightpaths are all-optical and the traffic is carried entirely in the optical domain. Because k-disjoint paths are used for backup paths, this guarantees on k failures. Having multiple backup lightpaths increases the availability of the connection compared to single backup lightpath. This is the most expensive class of service and can be allotted to connections that require the highest QoS in terms of response time, availability, and fault-tolerance for multiple failures and bandwidth.

I. *Primary Route Selection:* Same as SDPSDB primary route selection.

II. *Backup Route Selection:* For finding j^{th} backup route for a connection request from source s to destination d all the links and intermediate nodes along the primary and all j-1 backup paths are removed, then the shortest path is found. The resulting path is the jth shortest disjoint path for the connection request. Here, k is a parameter and depends on the connection requirement.

Case I: The first free wavelength common to the primary and all the k backup paths will be chosen and reserved. If no free wavelength is available on any of the k backup paths, the connection request is rejected.

Case II: Here, there is no restriction on availability of the same wavelength along the primary and k backup paths (i.e., any free wavelength along the route can be used).

16.5.2.3 *Single dedicated primary and single shared backup lightpath (SDPSSB) — class III*

In SDPSSB, both primary and backup lightpaths are also all-optical and the traffic is carried entirely in the optical domain. However, the primary path is dedicated and will carry exactly one connection request traffic, whereas backup lightpath is shared between any two (or more, depending on connection bandwidth and response time requirement) backup lightpaths and acts as cold standby for the corresponding primary lightpath. Because backup lightpaths are shared among multiple connections during the failure recovery period, this class of connections will experience additional delay (queueing delay) at the access nodes, and will get less bandwidth. As far as availability and fault-tolerance are concerned, it is the same as SDPSDB. Note here that sharing of backup paths is different from backup multiplexing. Backup multiplexing insists on the link disjointness of the primaries whose backup paths can share a wavelength channel.

  I.   *Primary Route Selection:* Same as SDPSDB primary route selection.
 II.   *Backup Route Selection:* Same as SDPSDB backup route selection.
      Case I: For all free wavelengths along the primary path, check whether the *same* wavelength along the backup path can be shared with any other backup path; if yes, reserve that wavelength along the primary and mark it for sharing along the backup path. If no such wavelength (which satisfies the condition) is available, then reserve first common free wavelength along both primary and backup paths. Otherwise, reject the connection request.
      Case II: Check whether *any* wavelength along the backup path can be shared with any other backup path. If yes, reserve the first free wavelength along the primary path, and mark for sharing the wavelength that can be shared along the backup path. If no such wavelength (which satisfies the condition) is available, then reserve first free wavelength along the primary, and mark for sharing the first free wavelength along the backup path. Otherwise, reject the connection request.

16.5.2.4 *Single shared primary and single shared backup lightpath (SSPSSB) — class IV*

In SSPSSB, both primary and backup lightpaths are also all-optical, and the traffic is carried entirely in the optical domain. In this class, primary and backup lightpaths are shared between any two lightpaths (or more, depending on the connections bandwidth and response time requirement). Because primary and backup lightpaths are shared among multiple connections, this class of connections will experience additional delay (queueing delay) at the access nodes and will get less bandwidth. As far as fault-tolerance and availability are concerned, it is the same as SDPSDB.

   I.   *Primary Route Selection:* Same as SDPSDB primary route selection.
   II. *Backup Route Selection:* Same as SDPSDB backup route selection.
      Case I: All the wavelengths are indexed and searched in the order of their index numbers. The first common wavelength along both the primary and backup paths that can be shared among other connections (either primary or backup) is chosen and marked for sharing. If no such wavelength is available, then in increasing order of index, the wavelength that can share with other connections along the backup path is taken and is checked whether the same wavelength is free on the primary; if free, reserve it along the primary and mark it sharable backup paths. If such a wavelength is not available, then the first free wavelength common to both the primary and backup path will be chosen and reserved. If no common free wavelength is available, then reject the connection request.
      Case II: In this case, the best possible primary-backup lightpath pair that can be shared among other connections (either primary or backup) is chosen to satisfy the request. That is, we use primary independent backup wavelength assignment because backup path may be on different wavelength than that of primary wavelength.

16.5.2.5 Single shared primary and multiple shared backup lightpaths (SSPMSB) — class V

In SSPMSB, both the primary and backup lightpaths are also all-optical, and the traffic is carried entirely in the optical domain. In this class, primary and all *k* backup lightpaths are shared between any two lightpaths (or more depending on the connections bandwidth and response time requirement). Because primary and backup lightpaths are shared among multiple connections, this class of connections will experience additional delay (queueing delay) at the access nodes and will get less bandwidth. As far as fault-tolerance and availability are concerned, this is same as SDPMDB.

   I.   *Primary Route Selection:* Same as SDPSDB primary route selection.
   II. *Backup Route Selection:* Same as SDPMDB backup route selection.
      Case I: The first common wavelength along the primary and *k* backup paths that can share among other connections (either primary or backup) is chosen. If no such wavelength is available, then take in increasing order of index, the common wavelength among *k* backups that can share with other connections and check whether

the same wavelength is free on the primary; if free, mark it as sharable along the primary and k backup paths. If such a wavelength is not available, then the first free wavelength common to the primary and k backup paths will be chosen. Otherwise, reject the connection request.

Case II: In this case, the best possible primary and k backup lightpaths which share among other connections (either primary or backup) is chosen to satisfy the request. In this case, no restriction is made on the use of wavelength for primary and k backup paths (i.e., k backup paths may be on different wavelengths than that of primary wavelength). If not, reserve the first free wavelength along the primary and k backup paths. If no free wavelength is available, reject the connection request.

16.5.2.6 Single dedicated multi-hop primary and single dedicated multi-hop backup lightpaths (SDMHPSDMHB) — class VI

In SDMHPSDMHB, both primary and backup lightpaths are multi-hop lightpaths (depending on the availability of resources). That is, a message may reach its destination via one hop or it may be routed via a multi-hop path with opto-electronic conversions and processing at intermediate nodes. This incurs additional delay due to electronic-optical, optical-electronic conversions at junction of two lightpaths. This class of connections assumes that nodes are equipped with *tunable transceivers*.

I.   *Primary Route Selection:* Same as SDPSDB primary route selection.
II. *Backup Route Selection:* Same as SDPSDB backup route selection.
III. *Wavelength Assignment for Primary and Backup Paths:* Check whether any wavelength is free along the primary and backup paths. If on some link no free wavelength is available reject the connection. Otherwise, the first free wavelength along the primary and backup paths (may not be the same) is chosen and reserved. If free wavelengths are not available along the complete paths, then take in increasing order of free wavelengths on the first link of the paths and reserve the wavelength that results in maximum wavelength continuous path. Do it recursively for complete path.

16.5.2.7 Single shared multi-hop primary and single shared multi-hop backup lightpaths (SSMHPSSMHB) — class VII

In SSMHPSSMHB, both primary and backup lightpaths are multi-hop lightpaths (depending on the availability of resources). That is, a message may reach its destination via one hop or it may be routed via a multi-hop path with opto-electronic conversions and processing at intermediate nodes. In this class of connections, delay is due to electronic-optical, optical-electronic conversions at the junction of two lightpaths and also due to queueing delay. This class of connections also expects that nodes be equipped with *tunable transceivers*.

I. *Primary Route Selection:* Same as SDPSDB primary route selection.

II. *Backup Route Selection:* Same as SDPSDB backup route selection.

III. *Wavelength Assignment for Primary and Backup Paths:* When a connection request from a source S to a destination D arrives, the algorithm finds the first sharable wavelength (i.e., the wavelength for which sharing limit is not crossed) on the predetermined primary and backup paths using modified FX algorithm. If no sharable wavelength is available, on a particular link of either primary or backup path, the algorithm searches for the first free wavelength on that particular link. If no free wavelength is available on a particular link, then the connection is rejected.

16.5.2.8 Performance study

The performance of the different heuristics has been verified through simulation experiments on the 24-node regional network. For this network, single-fiber and multi-fiber physical links are considered. Lightpaths are assumed to be bidirectional, and all the links are assumed to have the same number of fibers. All the fibers are assumed to have the same number of wavelengths. The delay of each link was set to one unit. The connections are requested between node pairs chosen randomly, with a condition that any node is chosen with the same probability. In simulation experiments we consider *incremental* traffic as in Anand and Qiao.[18] In incremental traffic once a connection is admitted, the primary and backup lightpaths stay until the end of simulation. At any time, if a connection request cannot be satisfied using the heuristics, the connection request is rejected. The performance metrics used are blocking probability and the spare wavelength utilization. The effects of design parameters such as, F, W, k, and S on the performance metrics for different networks are studied.

The blocking probability is high for *Case I* wavelength assignment compared with *Case II* wavelength assignment for all the classes. The blocking probability of *Class II* and *Class V* is high compared with all other classes. For a given number of wavelengths (or fibers) as the number of connections increases, the blocking probability increases. For a given number of wavelengths as the number of fibers increases, the blocking probability decreases. For a given number of connections as the number of wavelengths increases, the blocking probability decreases. For a given number of connections as the number of fibers increases, the blocking probability decreases.

As the hot-pair percentage increases, the blocking probability of *Class III*, *Class IV*, and *Class V* decreases. As the sharing limit increases, the blocking probability of *Class III*, *Class IV*, *Class V*, and *Class VI* decreases. The effect of sharing increases as the hot-pair percent increases. As the number of backup lightpaths (k) required increases, the blocking probability increases in the case of *Class II* and *Class V*.

The spare wavelength utilization is high for *Case II* wavelength assignment compared to *Case I* wavelength assignment for all the classes. The spare wavelength utilization of *Class II* and *Class V* is high compared with all other classes. As the number of backup lightpaths required (k) increases, the spare wavelength utilization increases. As the sharing limit increases, the spare wavelength utilization of *Class III, Class IV, Class V*, and *Class VI* decreases. As the hot-pair percentage increases, the spare wavelength utilization of *Class III, Class IV*, and *Class V* decreases. For a given number of wavelengths (or fibers) as the number of connections increases, the spare wavelength utilization increases. For a given number of wavelengths as the number of fibers increases, the spare wavelength utilization increases. For a given number of connections as the number of wavelengths increases, the spare wavelength utilization increases. For a given number of connections as the number of fibers increases, the spare wavelength utilization increases.

16.6 *Summary*

This chapter presented a survey on recent advances in dynamic lightpath restoration schemes for WDM optical networks. We explained the algorithms used by these schemes and also discussed their performance. The performance results can be summarized as follows. The restoration time for the reactive methods is longer, and the restoration is not guaranteed when compared to the proactive methods. In the absence of failures, however, the resource utilization is more efficient in reactive methods. Although the link-based methods result in shorter restoration time, they do not utilize the resources efficiently when compared to the path-based methods. Employing a backup multiplexing technique results in significant performance improvement when compared with a dedicated backup reservation.

In a dynamic traffic environment, proactive methods employing primary-backup multiplexing technique yield significant improvement over backup multiplexing at the expense of reduction in restoration guarantee. The segmented protection paths scheme not only improves the number of requests that can be satisfied but also helps in providing better QoS guarantees on bounded failure recovery time and propagation delay. It is highly flexible to control the level of fault tolerance of each connection, independently of other connections, to reflect its criticality. It also does not insist on the availability of end-to-end disjoint paths. The partial backup scheme is neither a proactive nor a reactive scheme. It acts as a proactive scheme when a component in a path, which is covered by a backup path, fails. Otherwise, it acts as a reactive scheme. It is very resource-efficient and distinguishes the lightpaths with different reliability requirements. Two distributed control protocols for lightpath restoration are presented. These protocols are scalable compared to centralized protocols. A preferred link-based distributed control for establishing reliability-constrained, least-cost lightpaths, and different heuristics are presented. The fact that the route is not pre-computed and is essentially found by probing as in the case with distance vector-based algorithms, the preferred link-based distributed control is more responsive to the network changes. These heuristic functions do not use any global state information. This is a reactive scheme, but at the same time heuristic functions will try to reduce the failure probability to a level required by the end user. This protocol can be used in conjunction with any reactive distributive control scheme available in the literature.

Finally, we explained the need for service differentiation and presented different classes of lightpaths based on fault-tolerance, availability, bandwidth, and delay.

References

1. I. Chlamtac, A. Ganz, and G. Karmi, Lightpath communications: an approach to high bandwidth optical WANs, *IEEE Trans. on Commun.*, vol. 40, no. 7, pp. 1171–1182, July 1992.
2. C. Siva Ram Murthy and M. Gurusamy, *WDM Optical Networks: Concepts, Design, and Algorithms*, Prentice Hall, N.J., Dec. 2001.
3. R. Ramaswami and K.N. Sivarajan, Routing and wavelength assignment in all-optical networks, *IEEE/ACM Trans. on Networking*, vol. 3, no. 5, pp. 489–500, Oct. 1995.
4. A. Mokhtar and M. Azizoglu, Adaptive wavelength routing in all-optical networks, *IEEE/ACM Trans. on Networking*, vol. 6, pp. 197–206, April 1998.
5. K. Bala, T. Stern, and K. Simchi, Routing in Linear Lightwave Networks, *IEEE/ACM Trans. on Networking*, vol. 3, pp. 459–469, August 1995.
6. D. Banerjee and B. Mukherjee, A practical approach for routing and wavelength assignment in large wavelength-routed optical networks, *IEEE J. on Selected Areas in Commun.*, vol. 14, no. 5, pp. 903–908, June 1996.
7. H. Harai, M. Murata, and H. Miyahara, Performance of alternate routing methods in all-optical switching networks, *Proc. IEEE INFOCOM 1997.*
8. S. Ramamurthy and B. Mukherjee, Survivable WDM mesh networks, part I — protection, *Proc. IEEE INFOCOM 1999*, pp. 744–51.
9. S. Ramamurthy and B. Mukherjee, Survivable WDM mesh networks, part II — restoration, *Proc. ICC 1999.*
10. B.T. Doshi, S. Dravid, P. Harshavardhana, O. Hauser, and Y. Wang, Optical network design and restoration, *Bell Labs Tech. J.*, vol. 4, no. 1, pp. 58–84, Jan./ March 1999.
11. N. Nagatsu, S. Okamoto, and K. Sato, Optical path cross-connect system scale evaluation using path accommodation design for restricted wavelength multiplexing, *IEEE J. on Selected Areas in Commun.*, vol. 14, no. 5, pp. 893–902, June 1996.
12. M. Alanyali and E. Ayanoglu, Provisioning algorithms for WDM optical networks, *IEEE/ACM Trans. on Networking*, vol. 7, no. 5, pp. 767–78, Oct. 1999.
13. S. Baroni, P. Bayvel, R.J. Gibbens, and S.K. Korotky, Analysis and design of resilient multi-fiber wavelength-routed optical transport networks, *IEEE/OSA J. of Lightwave Technol.*, vol. 17, no. 5, pp. 743–58, May 1999.
14. E. Modiano and A. Narula-Tam, Survivable routing of logical topologies in WDM networks, *Proc. IEEE INFOCOMM '01*, pp. 348–357, April 2001.
15. G. Mohan and C. Siva Ram Murthy, Routing and wavelength assignment for establishing dependable connections in WDM networks, *Proc. IEEE Int. Symp. Fault-Tolerant Computing*, pp. 94–101, June 1999.
16. G. Mohan, C. Siva Ram Murthy, and A.K. Somani, Efficient algorithms for routing dependable connections in WDM optical networks, *IEEE/ACM Trans. on Networking*, vol. 9, no. 5, pp. 533–566, Oct. 2001.

17. S. Bandyopadhyay, A. Sengupta, and A. Jaekel, Fault-tolerant routing scheme for all-optical networks, *Proc. SPIE Conf. on All-Optical Commun. Syst.*, 1998.

18. V. Anand and C. Qiao, Dynamic establishment of protection paths in WDM networks, part I, *Proc. IEEE IC3N*, Oct. 2000.

19. M. Sridharan and A.K. Somani, Revenue maximization in survivable WDM networks, *Proc. SPIE Optical Networking and Commun.*, vol. 4233, pp. 291–302, 2000.

20. G.P. Krishna, M.J. Pradeep, and C. Siva Ram Murthy, A Segmented Backup Scheme for Dependable Real-Time Communication in Multihop Networks, *Proc. 8th IEEE Int. Workshop on Parallel and Distributed Real-Time Syst.*, pp. 678–684, May 2000 (LNCS Series, Springer-Verlag, vol. 1800).

21. M. Kodialam and T.V. Lakshman, Dynamic routing of locally restorable guaranteed tunnels using aggregated link usage information, *Proc. IEEE INFO-COM 2000.*

22. N. Ghani, Survivability provisioning in optical MPLS networks, *Proc. 5th European Conf. on Networks and Optical Commun. (NOC)*, June 2000.

23. C.S. Li and R. Ramaswami, Automatic fault detection, isolation, and recovery in transparent all-optical networks, *IEEE/OSA J. of Lightwave Technol.*, vol. 15, no. 10, pp. 1784–1793, Oct. 1997.

24. G. Mohan and C. Siva Ram Murthy, Lightpath restoration in WDM networks, *IEEE Network Mag.*, vol. 14, pp. 24–32, Nov./Dec. 2000.

25. T. Wu, Emerging technologies for fiber network survivability, *IEEE Commun. Mag.*, vol. 33, pp. 58–74, Feb. 1995.

26. Y. Ye, S. Dixit, and M. Ali, On joint protection/restoration in IP-centric DWDM-based optical transport networks, *IEEE Commun. Mag.*, vol. 38, pp. 174–183, June 2000.

27. S. Han and K.G. Shin, Efficient spare resource allocation for fast restoration of real-time channels from network component failures, *Proc. IEEE Real-Time Syst. Symp., RTSS*, 1997.

28. C. Vijaya Saradhi and C. Riva Ram Murthy, Dynamic establishment of segmented protection paths in single and multi-fiber WDM mesh networks, *Proc. SPIE Optical Networking and Communications*, vol. 4874, pp. 211–222, Aug., 2002.

29. C. Vijaya Saradhi and C. Siva Ram Murthy, Dynamic establishment of segmented protection paths in single and multi-fiber WDM mesh networks, *Communicated to Optical Networks Magazine*, 2002

30. A. Fumagalli and M. Tacca, Optimal design of differentiated reliability (DiR) optical ring networks, *Proc. Int. Workshop on QoS in Multiservice IP Networks (QoS-IP) 2001*, Jan. 2001.

31. A. Fumagalli and M. Tacca, Differentiated reliability (DiR) in WDM ring without wavelength converters, *Proc. ICC '00*, 2000.

32. M.J. Pradeep and C. Siva Ram Murthy, Providing differentiated reliable connections for real-time communication in multihop networks, *Proc. High Performance Computing*, pp. 459–468, Dec. 2000 (LNCS Series, Springer-Verlag, vol. 1970).

33. C. Vijaya Saradhi and C. Siva Ram Murthy, Routing differentiated reliable connections in single and multi-fiber WDM optical networks, *Proc. SPIE Optical Networking and Commun.*, vol. 4599, pp. 24–35, August 2001.

34. C. Vijaya Saradhi and C. Siva Ram Murthy, Routing differentiated reliable connections in WDM optical networks, *Optical Networks Mag.*, vol. 3, no. 3, pp. 50–67, 2002 (Special issue: Protection/Restoration Meets the Reliability Challenge for the Optical Internet).

35. G.S. Glaesemann and D.J. Walter, Method for obtaining long-length strength distributions for reliability predictions, *SPIE J. on Optical Engineering*, vol. 30, no. 6, pp. 746–748, June 1991.

36. G.S. Glaesemann and S.T. Gulati, Design methodology for the mechanical reliability of optical fiber, *SPIE J. on Optical Eng.*, vol. 30, no. 6, pp. 709–715, June 1991.

37. G.S. Glaesemann, Advancements in mechanical strength and reliability of optical fibers, *Proc. SPIE Reliability of Optical Fibers and Optical Fiber Syst.*, vol. CR73, Sept. 1999.

38. M.J. Matthewson, Optical fiber reliability models, *Proc. SPIE Fiber Optics Reliability and Testing, Critical Reviews of Optical Science and Technol.*, vol. CR50, Sept. 1993.

39. Z. Wang, On the complexity of quality of service routing, *Information Processing Lett.*, vol. 69, pp. 111–114, 1999.

40. Z. Whang and J. Crowcroft, Quality of service routing for supporting multimedia applications, *IEEE J. on Selected Areas in Commun.*, vol. 14, no. 7, pp. 1228–1234, Sept. 1996.

41. C. Vijaya Saradhi and C. Siva Ram Murthy, Distributed network control for establishing reliability-constrained least-cost lightpaths in WDM mesh networks, *Technical Report,* Department of Computer Science and Engineering, Indian Institute of Technology, Madras, India, March 2002.

42. Chava Vijaya Saradhi and C. Siva Ram Murthy, A distributed algorithm for reliability-constrained least-cost routing in multihop networks, *Technical Report,* Department of Computer Science and Engineering, Indian Institute of Technology, Madras, India, March 2002.

43. N. Huang, C. Wu, and Y. Wu, Some routing problems in broadband ISDN, *Computer Networks and ISDN Syst.*, vol. 27, no. 1, pp. 101–116, 1994.

44. R. Sriram, G. Manimaran, and C. Siva Ram Murthy, Preferred link-based delay-constrained least cost routing in wide area networks, *Computer Commun.*, vol. 21, no. 8, pp. 1655–1669, 1998.

45. J.P. Jue and G. Xiao, An adaptive routing algorithm for wavelength-routed optical networks with a distributed control scheme, *Proc. IEEE IC3N*, Oct. 2000.

46. C. Vijaya Saradhi and C. Siva Ram Murthy, A framework for routing differentiated survivable optical virtual private networks (OVPNs), *Photonic Network Commun.*, vol. 4, no. 3/4, pp. 457–487, 2002 (Special issue: Routing, Protection, and Restoration Strategies and Algorithms for WDM Optical Networks).

47. C. Vijaya Saradhi and C. Siva Ram Murthy, Dynamic establishment of differentiated survivable lightpaths in WDM mesh networks, *Proc. ICCCAS, 2002*.

chapter seventeen

Restoration in optical WDM mesh networks

Shirshanka Das
University of California
Mario Gerla
University of California

Contents

0-8493-1333-3/03/$0.00+$1.50
© 2003 by CRC Press LLC

17.1 Introduction

Optical networks have been slated to meet the challenge posed by the ever-increasing demand for bandwidth. By using wavelength division multiplexing (WDM), we divide the vast transmission bandwidth available on a fiber into several nonoverlapping wavelength channels and enable data transmission over these channels simultaneously. Along with the huge bandwidth comes the realization that even if traffic is disrupted for a few seconds, a lot of data is going to be lost. Besides, the kind of traffic that is being carried over optical networks may be mission critical, so such huge losses may be catastrophic.

Along with WDM come the accessories that have made optical networking evolve from being just a point-to-point freeway to becoming an overlay network of fast huge freeways. The main components for optical networking are:

- Wavelength add-drop multiplexers (WADMs), which allow a node to add or drop a particular wavelength from an optical signal
- Wavelength cross-connects (WXCs), which allow a node to switch a wavelength from one incoming fiber to another outgoing fiber on the same wavelength
- Wavelength interchanging cross-connects (WIXCs) or wavelength converters, which allow a node to change the wavelength of the stream as it is switched from one incoming fiber to another outgoing fiber.

New physical devices such as Erbium doped fiber amplifiers (EDFAs) and faster, higher-quality lasers and receivers have had a major impact on the maturing of WDM technology. The introduction of networking within the optical layer creates the need for routing wavelengths (demands) over optical networks. Consequently, this also creates a need for link or path restoration in the optical domain.

Restoration typically involves rerouting the demand around the failure on an alternate path. Restoration at the optical layer as opposed to that offered by higher layers gives us shorter restoration times, efficient resource utilization and greater protocol transparency. Until recently, Synchronous Optical NETwork (SONET) self-healing rings have provided very low restoration times on the order of 50 milliseconds; however, the price that must be paid for this is large capacity penalties compared to regular mesh-based networking. Another bothersome fact is that current makeshift solutions for carrying IP traffic over WDM networks have ended up encapsulating IP packets by AAL5 and then encapsulating into SONET frames, which are then sent point to point over WDM. The kind of header overhead that this builds up is quite significant, and ISPs would just love to do away with the intermediate layers and do IP/WDM directly.

Debates are raging about whether restoration should be provided at the optical layer or should be entrusted to the service layer, and about the race conditions that may arise when more than one layer tries to recover from the same fault. Many people, however, believe that restoration at the optical layer will definitely benefit a lot of services. Another debate rages on about what kind of topologies make more sense. Ring topologies are more tractable in terms of routing and wavelength provisioning, while it is extremely difficult to come up with optimal solutions for mesh-based architectures. On the other hand, a lot can be said for the capacity efficiency of mesh topologies, the inherent redundancies and thus improved robustness of such architectures. Restoration paths for ring topologies are also typically longer than those in mesh networks. Simple heuristic algorithms seem to be more attractive for mesh topologies, while centralized optimal algorithms are the choice for ring topologies. An abundance of research has been conducted, and doing justice to all kinds of schemes would require a much longer chapter than this one. Therefore, this chapter just considers optical domain restoration and mesh networks. It looks at the different types of restoration algorithms for mesh networks that have been proposed and analyzes them with regard to their attributes of capacity efficiency, restoration time, scalability, and restorability.

17.2 Types of restoration

17.2.1 The problem

Before we discuss restoration schemes, we need to examine the problems they set out to solve. Some basic assumptions can be made about the system that needs to be protected; this makes a big difference in the selection of the most appropriate scheme. The most common assumptions required by restoration schemes include:

1. The functionality of the cross-connects
   a. Strict WXC implies wavelength continuity constraint (i.e., the same wavelength must be used for a path on each link)
   b. Assuming the presence of WIXCs means that the wavelength continuity constraint can be relaxed

2. The traffic demands
   a. Static, where the traffic matrix is known in advance
   b. Dynamic, in which connections are set up as and when they arrive without any previous knowledge of the arrival times and durations
3. Type of network control
   a. Centralized, which makes sense for smaller networks with a controlled traffic matrix
   b. Distributed, in huge networks where connections arrive in a distributed manner and each node does its own computation for routing and provisioning
4. The primary performance metric (i.e., the metric to be optimized): restoration speed, blocking probability, link utilization, scalability, or restorability.

17.2.2 The solutions

The ubiquitous figure that illustrates the different kinds of methods used to solve the various components of the restoration problem is presented next (see Figure 17.1).

Let us take a broad view of what each kind of restoration method really does. Proactive schemes pre-calculate the alternate routes. Thus after the fault occurs, the connection is simply rerouted to the previously calculated route. Reactive schemes calculate alternate routes after the actual fault occurs. Thus, typical reactive schemes flood packets into the network after the fault to look for free capacity and to set up the new path. Proactive

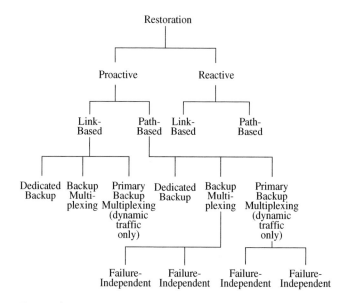

Figure 17.1 Types of restoration.

schemes typically give 100% restoration guarantee and have faster restoration times than reactive schemes. Moreover, reactive schemes cannot guarantee successful recovery.

Link-based schemes reroute the demand around the failed link, while path-based schemes typically use a whole new path for the whole demand. Schemes with dedicated backup reserve the backup route for the particular demand. Thus, primary or other backup paths will not be able to share any link of the backup path. Schemes with simple backup multiplexing allow other backup paths to share the same wavelength on one or more links.

Central to this technique is the single fault assumption, which says that at one time only one link or node will fail. So if two backup paths are fault disjoint, they can share the same backup wavelength under the single fault assumption, since only one of the sharing backup paths will be activated and thus no conflict will occur. An extension of this idea uses primary and backup multiplexing; that is, a primary path may be set up on a wavelength that is backup for one or more demands. So during the life of the primary demand, the backup becomes unusable and comes back into service once the primary demand ends.

Failure-dependent schemes are a further classification of path-based schemes where the backup path is dependent on the particular failure (link/node). Thus a backup path is available for each possible failure on the path. Failure-independent schemes find a link and node disjoint path and make it the backup path. Thus regardless of which link or node has failed, this path can be used. This is attractive for optical networks, since failure isolation can be a time-consuming process.

The rest of this chapter examines a few representative schemes that have appeared in the published literature. These schemes will serve to illustrate the previously mentioned solutions. They will offer us the opportunity to compare different approaches and analyze the performance tradeoffs.

17.3 Case studies

In the next two case studies, we shall look at a mesh network with numerous wavelengths on each fiber and wavelength interchange capability at the cross-connects. Doshi et al. look at two restoration problems in this scenario and propose solutions for them.[1] Let us look at Problem 1. The solution proposed is a *path-based, distributed, dynamic, proactive, heuristic restoration strategy*. A look at Figure 17.1 will tell the reader exactly which path we traverse in the taxonomy of the solutions.

We look at a path restoration strategy in which there exists a pre-computed restoration path for each primary path. This pre-computed path protects the primary path against all failures and therefore has to be node-link disjoint with the primary path. Thus the network is protected from all single component failures.

17.3.1 Optical network design and restoration: problem 1

B. Doshi, S. Dravida, R. Harshavardhana, O. Hauser, and Y. Wang[1]

17.3.1.1 Problem definition

Given a network described by nodes and links, a set of point-to-point demands and their service routes in an optical network, and spare capacity on each link, find restoration routes for as many demands as possible under the link capacity constraints, or else find restoration routes for all demands so that the capacity requirement is minimized.

The problem can be easily shown to be NP-complete by demonstrating that it reduces to the disjoint path problem in a special case. The disjoint path problem is to find mutually node disjoint paths between multiple source destination pairs in a network. Consider the source destination pairs as demands, and add an additional node to the network, which is a neighbor for every node in the original network. Now, select the two hop routes through the additional node as the primary routes for the demands. The problem of finding node disjoint protection paths for the demands is the same as finding multiple node disjoint paths between the multiple source destination pairs since all the demands share the same fault (the additional node), so they can never share the same protection path. Therefore, assigning a spare capacity of 1 on each of the links makes this exactly equivalent to the disjoint path problem. Thus, the problem can be shown to be NP-complete, and we can only have a heuristic solution for this problem.

The algorithm proposed has four basic procedures: *route search, link capacity control, concurrent resource locking,* and *capacity optimization.* These components are used in the following fashion: The source node of each demand executes the concurrent resource locking procedure to lock out contenders. If unsuccessful, the source node waits for a random amount of time and retries; otherwise, it starts the route search procedure to search for the restoration route. Within the route search procedure, a distributed breadth first search procedure is invoked to find a route from the source to the destination, and the link capacity control is queried at every step. On finding a route, it is stored at the destination as the restoration route for the demand; otherwise the optimization procedure is activated to release some critical link capacities. Let us look at the components in more detail.

17.3.1.2 Route search

This is a simple distributed breadth first search. Once contention is locked out, the source sends out packets which visit the nodes in breadth first fashion, looking at whether the link has sufficient capacity or not and going to the next hop depending on that. Since we are looking for node disjoint paths, the packets do not visit nodes that lie in the primary path.

17.3.1.3 Link capacity control

This procedure resides on each link, takes as input a primary route, and determines if this link has sufficient capacity to serve as part of a restoration path for this route. A table is maintained with information about faults, demands affected by them, and the resulting capacity requirement. Obviously, demands that share faults cannot share the capacity on the restoration wavelength, so if there are two demands affected by fault F and both want to get restored through this link, they will need two spare wavelengths. Therefore when a new demand comes in, the primary route is broken up into all possible faults (all nodes and links) and the table entries are updated (for faults that already exist in the table) and added (for faults that do not exist in the table). Accordingly, if the required capacity in any scenario goes higher than the spare capacity on the link, the route search packet is stopped.

17.3.1.4 Resource locking

When a demand is doing a route search, if another demand which shares a fault also does a route search then we can run into race conditions, since the two will be contending for resources on the network. To lock out all contenders, using a token-like approach the demand sends out a token packet that will turn up a flag at each node and link and will return back if it encounters an upturned flag on the way, which clears up the path. If it reaches the destination, then we can be sure that no other demand sharing the primary path is doing a route search, so we can go ahead with our route search. If the lock procedure fails, we have to wait for a random time and try again.

17.3.1.5 Optimization procedure

Because of the ad hoc way in which restoration routes are selected, we may end up with a situation in which moving some of the restoration routes may make other demands that were originally denied restoration get their restoration paths. Thus when a demand fails to find a restoration path, it activates the optimization procedure. The procedure is described in detail in the appendix of Reference 1 and for brevity we leave it out here.

17.3.1.6 Results

The following conservative estimates are made about various parameters:

- Message travel between two nodes is 8 ms per 1000 miles
- Node message processing is performed at a speed of 20 ms per message
- Cross-connect instructions can be executed in 10ms
- Traffic switching and selection can be done in 10ms, and failure detection requires 5ms

The authors use the parallel architecture (described later) as well as three nationwide networks. Restoration times are 130ms for the 28-node network, 505 ms for the 70-node network, and 572 ms for the 301-node network. The distributed solution turns out to be quite close to the optimal solution in all cases.

17.3.1.7 Discussion

The distributed algorithm proposed is quite simple, and effective. The concept of maintaining tables indexed by fault i.d. is intelligent, and the locking scheme to lock out concurrent searches is also nice; however, detractors can point fingers at a few places.

First, all demands are given a unique demand i.d., identified by simply the source and the destination. More than one demand could exist between the same source and destination, and so we need an extra dimension to distinguish between them. The locking procedure is quite conservative — we could have paranoid situations where demands could be contending in a cyclic fashion and backing off together, and no one would be getting through. It would probably be more sensible to just assume a link-state protocol and global topology information at the nodes that would greatly simplify things.

17.3.2 Optical network design and restoration: problem 2

B. Doshi, S. Dravida, R. Harshavardhana, O. Hauser, and Y. Wang[1]

17.3.2.1 Problem definition

Given a network in terms of nodes and links as in problem 1 and a set of point-to-point demands, find both the primary routes as well as the restoration route for each demand, so that the total required network capacity is minimized.

The solution proposed is a *path-based, centralized, static, proactive, integer linear programming (ILP)-based, restoration strategy.*

Notice that the problem is now static in the sense that we assume knowledge of the demands to be set up beforehand. Although this may not be true in general, in certain cases this problem becomes relevant. For example, while migrating from a legacy network to an optical one, we have to transfer existing demands to an optical network. Also, we may have periodic cleaning, when a heuristic distributed algorithm like the one in the previous section may have chosen nonoptimal assignments to primary and restoration routes. Of course, we may also have long-term planning using projected demands so that we can estimate capacity needs at a time point in the future.

17.3.2.2 Problem formulation

Viewing the network as a directed graph, let N be the set of nodes, $L \subset N \times N$ be the set of links, $D \subset N \times N$ be the set of demands. For each L_{ij}, let C_{ij} be the capacity weight for L_{ij}, which can be regarded as a measure of capacity consumption per wavelength on the link. If W_{ij} is the resulting capacity requirement on L_{ij} in terms of number of wavelengths, then the objective function can be given by minimizing $\Sigma C_{ij} W_{ij}$. Basically, this is a two-parameter characterization of the network, where C is the per-wavelength cost and W is the per-fiber cost.

Let us now formulate the ILP for the problem. Let F be the set of all possible single faults. Obviously F = N U L. For each f F, let L_f be the set of links affected by that fault. If f is a link failure then L_f contains only the failed link; if it is a node failure, then L_f contains the set of links that are incident to the node. For each d D, let s and t be the source node and the destination node of the demand d, respectively. We need to define the following variables:

X_{ij}^d : 1 if demand d's *primary path traverses link (ij)*, 0 otherwise
Y_{ij}^d : 1 *if demand d's alternate path traverses link (ij)*, 0 otherwise
Z_{ijf}^d : 1 *if demand d is rerouted through link (ij) under fault f, 0 otherwise*
W_{ij}: total *number* of wavelengths required on link (ij)

The ILP looks like:

$$\min \sum C_{ij} W_{ij} \tag{17.1}$$

$$\text{s.t. } \sum_{j:(ij)\in L} X_{ij}^d - \sum_{j:(ji)\in L} X_{ji}^d = \begin{cases} 1, i = s_d \\ -1, i = t_d \\ 0, otherwise \end{cases} \quad d \in D \tag{17.2}$$

$$\text{(IP) } \sum_{j:(ij)\in L} Y_{ij}^d - \sum_{j:(ji)\in L} Y_{ji}^d = \begin{cases} 1, i = s_d \\ -1, i = t_d \\ 0, otherwise \end{cases} \quad d \in D \tag{17.3}$$

$$\sum_{j:(ij)\in L,(ij)\notin L_f} Z_{ijf}^d - \sum_{j:(ji)\in L,(ij)\notin L_f} Z_{jif}^d = \begin{cases} 1, i = s_d \\ -1, i = t_d \\ 0, otherwise \end{cases} \quad d \in D, f \in F \tag{17.4}$$

$$Z_{ijf}^d \leq Y_{ij}^d, \quad d \in D, (ij) \in L, f \in F, (ij) \notin L_f \tag{17.5}$$

$$\sum_{d\in D} X_{ij}^d + \sum_{d\in D} Z_{ijf}^d \leq W_{ij}, \quad (ij) \in L, f \in F, (ij) \notin L_f \tag{17.6}$$

The objective function is the total weighted capacity requirement. Constraints (17.2) and (17.3) are the flow conservation constraints for demand d's service route and restoration route respectively. Constraint (17.4) enforces the fact that the restoration route consumes link capacity if and only if the service route is affected by the fault in question. Constraint (17.5) ensures that the restoration route of a demand is independent of the failure. Constraint (17.6) determines the link capacity requirement. Constraint (17.4) and the use of the Z variables ensure that the service and restoration route are disjoint.

Note that Constraint (17.6) really ties all the demands together. All the constraints are separate for each demand. So, without Constraint (17.6), the problem can be decomposed to a number of independent sub-problems, one for each demand. The authors use the Lagrangian relaxation method to relax Constraint (17.6) and reduce the problem IP to several (IP_d)s. Many intricacies should be considered, such as adding a penalty term to the objective function, which we shall not explore. The inclined reader is referred to the article for the details.

The overview of the approach is the following: split the IP into (IP_d)s, one for each demand, which are actually NP-complete. Obtaining a lower bound for (IP_d) is easier, however, and because the objective function of (IP) is the sum of the objective functions of (IP_d), we can obtain a lower bound for IP. An algorithm R is proposed for getting the lower bound to (IP_d). Having obtained that, the overall algorithm G is proposed, which uses this algorithm R to find the lower bound to (IP). The outline of algorithm G is as follows:

1. Initially, set all λ values equally and scale them properly so that

$$\sum_{f \in F, (ij) \in L_f} \lambda_{ijf} = C_{ij} .$$

   For each demand d D, call algorithm R to solve sub-problem (IP_d). All the sub-problem solutions collectively become an initial feasible solution S_O to the original problem (IP). Set the best feasible solution thus far $S^* = S_O$.

2. At the kth iteration, update λ values as $\left|\lambda_{ijf}^{k+1}\right| = \left|\lambda_{ijf}^k\right| + \alpha \nabla_p v(\left|\lambda_{ijf}^k\right|$, where α is the step size. α is chosen so that λ^{k+1} is still nonnegative and adaptive to the improvement of the last iteration. Solve all sub-problems (IP_d), and obtain a feasible solution S_{k+1}. If $C(S^*) > C(S_{k+1})$, set $S^* = S_{k+1}$.

3. Keep repeating step 2 until either $v(\lambda)$ cannot be improved or the value of $C(S^*)$ does not decrease for a predefined number of iterations.

4. S^*, the best feasible solution obtained thus far, can be further improved by a local rerouting procedure, for which the reader is again referred to the literature.

17.3.2.3 Restoration architecture

Let us now look at the kind of architecture required to make the restoration possible. This discussion is common to both problems 1 and 2 because the solution to those problems simply associate each primary route with a restoration route. The activation of that route is another problem altogether. We assume unidirectional links and unidirectional connections or demands. The restoration route is not reserved as such; therefore, the actual cross-connect setup at each of the intermediate nodes of the alternate route will have to be done after the fault. The restoration routes for each demand are stored at the destination node. What happens is this: the destination detects failure of the service route, initiates cross-connect operations at every node on the route, signals the source to bridge traffic to this path, and starts receiving the signal for the demand from the new route. Two kinds of architectures are proposed to implement this procedure.

The sequential architecture consists of the following steps: the destination node on detecting failure first determines an incoming port number for the restoration route and switches the signal selector to receive a signal on this port. Then it determines the corresponding outgoing port number at the preceding neighbor node on the restoration route and sends a packet with the outgoing port number and the restoration route with the demand i.d. On getting this message, each intermediate node determines the incoming port number and thus the preceding neighbor's outgoing port number, and sends a message to the preceding neighbor with similar information. This step is repeated one by one until it reaches the source, which executes the cross-connect and bridges the signal to the new route.

You can also have a parallel architecture in which you can have two types of messages: a type 1 message, which is sent from the destination to each of the nodes on the route, and a type 2 message, which is sent to the upstream neighbor by each node after receiving the type 1 message. The type 1 message serves to fix the incoming port of the node and generates the type 2 message, which tells the preceding node which to use as the outgoing port. Once every node gets a type 1 and a type 2 message, both the incoming and outgoing ports are known, so the cross-connect can be executed in parallel over the route.

17.3.2.4 Results

The centralized algorithm G is evaluated for two network topologies, with different demand sets ranging from 100 to 1000 demands. Parameters evaluated are the primary capacity used, and the restoration capacity needed in terms of wavelengths on links, using backup path sharing and without using sharing. The solutions of algorithm G are 20% better than those of the shortest pair and are within 10% deviation of the lower bounds. The distributed form of algorithm G was also evaluated and performed only 7% lower than G, and had much less computational complexity.

17.3.2.5 Discussion

The ILP proposed is detailed, and the algorithm follows after much analysis. The distributed form of the algorithm is especially useful. The problem with such approaches remains that they assume batch arrivals, which holds true only in certain cases. Also, optimization algorithms like these are quite computationally intensive, so these algorithms are better off running in the background, optimizing heuristic algorithms running in the foreground.

17.3.3 Partial path protection for WDM networks: end-to-end recovery using local failure information

H. Wang, E. Modiano, and M. Medard[2]

This case study also looks at an optical mesh network with unidirectional links and demands and comes up with a *segment-based, failure dependent, centralized, dynamic, proactive, ILP-based protection strategy.*

A protection scheme is proposed loosely based on the segment protection concept, which is rechristened as partial path protection (PPP). In this scheme, the network identifies a specific protection path for each link along a considered primary path. Thus, just like path protection, end-to-end protection paths are assigned to primary paths. Each of these paths, however, protects only for one specific link failure instead of the whole primary path. Simple link protection suffers from the problem of significantly longer backup paths, while simple disjoint path protection cannot guarantee a backup path always (e.g., when there is no node-link disjoint alternate path available). PPP is a good hybrid of both. Also, instead of a system where the calls are known in advance, the authors consider a call-by-call system with random arrivals, which is more realistic and useful to study. A central dynamic programming solution which figures out the optimal assignment to minimize the call-blocking probability is impracticable, since the state space of the dynamic program becomes huge for any reasonable time horizon and network size. Thus, a heuristic approach would be more attractive in this case.

Let us first look at a sample case to illustrate path protection vs. partial path protection. Path protection for link failures allocates a link disjoint path from the primary path as the alternate path for any link failure on the path. In PPP, the system reserves the protection resources while setting up a primary path. The major difference with the path protection scheme is that the system now specifies a specific protection path for *each link* along the primary path. Thus each protection path, instead of being associated with a single path (as for path protection) or a single link (as for link protection), is associated with a link/ primary path pair. In the event of a link failure, the call is rerouted along the protection path corresponding to the failed link (see Figure 17.2 and Table 17.1).

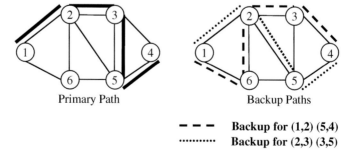

Figure 17.2 Backup path assignment using PPP.

Table 17.1 Illustration of Protection Paths for Primary Path in Figure 17.2

Link on Primary Path (1–2–3–5–4)	Corresponding Protection Path
(1,2)	1–6–2–3–4
(2,3)	1–2–5–4
(3,5)	1–2–5–4
(5,4)	1–2–3–4

Table 17.1 lists the protection paths chosen for each link failure scenario. When the PPP for a particular failure scenario for connection C1 is considered, the protection wavelength can be shared as a protection wavelength for some other connection C2, which is not primary path disjoint with C1, as long as the set of overlaps does not contain the link for which the protection path was originally found. In the case of path protection, this cannot be done because the protection wavelength is protecting against all link failures on the primary path of C1, so sharing is only possible between primary path disjoint connections in path protection.

17.3.3.1 Path assignment

Two approaches to path assignment are considered. The first approach is a greedy approach in which the system jointly allocates primary and protection paths simultaneously so as to minimize the total wavelengths being used. Solving an ILP, which we will encounter shortly, does this. The other approach is to assign the shortest path as the primary path for the incoming call and then assign protection paths to this primary path so as to minimize the cost of the resulting system, where sharing a wavelength is zero cost while allocating a new wavelength for the protection path is unit cost. Obviously a protection path can use a wavelength on a correctly functioning link in that primary path at no extra cost.

17.3.3.2 ILP for greedy approach

Let us look at the ILP that is generated for the greedy approach. Such ILPs regularly come up in such optimization for network routing scenarios. We have seen what an ILP for path protection looks like; let us look at an ILP for partial path protection.

Let L denote the set of all possible links, S denote the source node and D denote the destination node,

$$c_{ij} = \begin{cases} 1, & \text{if at least one wavelength is available on link}(i,j) \ L \\ \infty, & \text{otherwise} \end{cases}$$

$$d_{ij}^{lk} = \begin{cases} 0 & \text{if at least one wavelength on link } (l,k) \text{ other than } (i,j) \\ & \text{is already reserved to protect links other than } (i,j) \\ 1 & \text{if at least one wavelength is available on link } (l,k) \\ \infty & \text{otherwise} \end{cases}$$

$$x_{ij} = \begin{cases} 1 & \text{if the primary path rests on an available wavelength in link } (i,j) \\ 0 & \text{otherwise} \end{cases}$$

$$y_{ij} = \begin{cases} 1 & \text{if the system reserves a wavelength in link } (i,j) \text{ for protection} \\ 0 & \text{otherwise} \end{cases}$$

$$v_{ij}^{lk} = \begin{cases} 1 & \text{if a wavelength on } (l,k) \text{ is reserved to protect} \\ & \text{its associated protection path on } (i,j) \\ 0 & \text{otherwise} \end{cases}$$

The variable **d** indicates which links have wavelengths available to protect some specific link on which the primary path may reside; **v** indicates the assignment of wavelengths to protection. For the ILP for the random call arrival, we need to minimize
if at least one wavelength on link (l, k) other than (i, j)
is already reserved to protect links other than (i, j)

$$c_{ij} x_{ij} + y_{ij} \tag{17.7}$$

Equation (17.7) represents the objective function and the constraints look like this:

$$\sum_{(S,j)\in L} x_{Sj} - \sum_{(jS)\in L} x_{jS} = 1 \tag{17.8}$$

$$\sum_{(D,j)\in L} x_{Dj} - \sum_{(j,D)\in L} x_{jD} = -1 \tag{17.9}$$

$$\sum_{(i,j)\in L} x_{ij} - \sum_{(j,i)\in L} x_{ji} = 0, \forall i \neq S, D \tag{17.10}$$

$$\sum_{(S,l)\in L} v_{ij}^{Sl} - \sum_{(l,S)\in L} v_{ij}^{lS} \geq x_{ij}, \forall (i,j) \in L \tag{17.11}$$

$$\sum_{(l,D)\in L} v_{ij}^{lD} - \sum_{(l,D)\in L} v_{ij}^{Dl} \geq x_{ij}, \forall (i,j) \in L \qquad (17.12)$$

$$\sum_{(l,k)\in L} v_{ij}^{lk} - \sum_{(k,l)\in L} v_{ij}^{kl} = 0, \forall (i,j) \in L, \forall k \neq S, \forall k \neq D \qquad (17.13)$$

Equations (17.8) to (17.10) are flow conservation equations for the primary path, while Equations (17.11) to (17.13) give the flow conservation for the protection path. Equations (17.11) and (17.12) come into play only when $x_{ij} = 1$ (i.e., the primary path passes through the link [i,j]).

$$v_{ij}^{ij} + v_{ji}^{ij} = 0, \forall (i,j) \in L \qquad (17.14)$$

Equation (17.14) enforces the path disjoint property.

$$y_{lk} \geq d_{ij}^{lk} (v_{ij}^{lk} - x_{lk}), \forall (i,j),(l,k) \in L \qquad (17.15)$$

Equation (17.15) indicates whether a unoccupied wavelength on link(l,k) will be reserved for protection.

$$x_{ij} \geq v_{ij}^{lk}, \forall (i,j),(l,k) \in L \qquad (17.16)$$

Equation (17.16) prevents the possibility of assigning a protection path for a link that is not used by the primary path.

$$x_{ij}, y_{ij}, v_{ij}^{lk} \in \{0,1\}, \forall (i,j),(l,k) \in L \qquad (17.17)$$

The difference of the partial path formulation with the path protection formulation lies in Equation (17.15), which considers the situation where the protection path overlaps part of its links with the links on its associated primary path; hence the $(-x_{lk})$ term. Also, we do not need an extra equation to enforce the condition that a link should protect the entire path in PPP.

17.3.3.3 Shortest-path approach
The other approach proposed is a *shortest-path approach*. We can see clearly that the greedy approach will be nonoptimal because it only takes into account present resource usage. It may also choose protection paths, which is not amenable to protection sharing. Another point to note is that a requests primary path cannot be shared with other requests, so the logical thing to do is to assign the fewest possible resources to a calls primary path. Therefore, we

assign the shortest path available as the primary path for the request. After this, protection paths are searched for. Link costs are updated according to the current resource usage. Wavelengths that are in use by other protection paths have a cost of 0. In the case of path protection, links used by the primary path are not available, while in case of PPP, the correctly functioning links are available.

It is quite intuitive to see that for a given request, the SP method may require greater resources, but in the long run, the greedy method may lose out because the protection paths being chosen may not be very amenable to sharing, while resources are being wasted anyway for the non-shortest primary paths. This is because the greedy method will be endeavoring to serve each request using the minimum number of previously unused wavelengths; however, in doing so, the greedy approach happens to choose paths with no protection sharing, harming network resource utilization.

17.3.3.4 Results

The results are plotted in Figures 17.3 and 17.4. The two schemes are simulated over various topologies. The conclusion is that the PPP scheme implemented using the SP approach performs best. This is easy to understand because SP with PPP is intrinsically more flexible than SP–PP in both the protection scheme and the implementation approach. Other interesting observations include the fact that the performance is highly related to the network topology. For a highly connected network, blocking events are very rare. On sparser graphs, the SP–PPP algorithm shows better performance

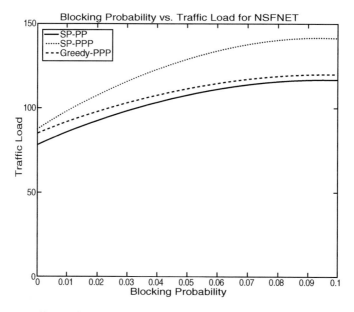

Figure 17.3 Traffic load vs. blocking probability: NSFNET.

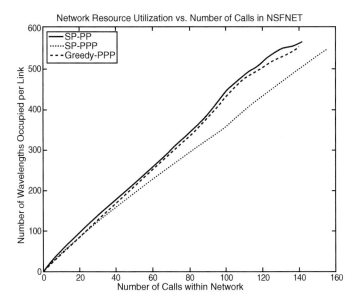

Figure 17.4 Network resource utilization: NSFNET.

because longer primary paths, which enhance the usefulness of capacity sharing, occur more.

17.3.3.5 Discussion

The partial path protection approach proposed can be called a path-based link protection scheme. We are basically trying to find protection paths for every link on the path. The results are quite impressive and show the better performance of PPP over simple path protection because it is more flexible. It is definitely a step in the right direction: that of segment-based protection, in which one can protect sections of the path at once; however, issues such as the architecture to support this type of schemes, the scalability of fault-dependent schemes, and the potential difficulty in isolation of faults in optical networks have to be resolved.

17.3.4 ZRESTORE: a link restoration algorithm with high aggregation and no reservation

S. Das, P. Verma, B.N. Jain, and M. Gerla[3]

This last case study looks again at a mesh-based optical network with unidirectional demands and links, and comes up with a *link-based, distributed, dynamic, proactive, primary backup multiplexing restoration strategy*.

The primary assumptions are unidirectional links with multiple wavelengths or channels and a control channel that is set aside for control message exchange and wavelength converters at each node. Each link is directed and therefore is associated with a "sender" node and a "chooser" node. Each link

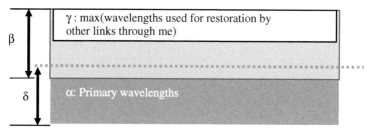

Figure 17.5 Structure of a link.

tries independently to be able to preselect bypass routes for a target number of wavelengths over and above the primary wavelengths on it. Each link has a maximum capacity expressed in number of wavelengths ($_{max}$) (see Figure 17.5). Each link keeps track of:

α — the number of primary channels on the link
β — the number of free channels = $(\lambda_{max} - \alpha)$
γ — the maximum number of channels that this link has promised other links for restoration
δ — the number of channels for which bypasses have been preselected and confirmed

As long as the number of restorable channels on the link (δ) is higher than the primary channels (α) by a threshold (**T**), the link stays in *stable state*. When the primary channels get too close to the restorable channels either due to new paths being set up (primary channels increase) or primary paths on the bypasses kicking out the restorable channels (restorable channels decrease), the link enters a *critical state* and starts searching for bypasses again. Bypass search is a simple procedure of sending out of θ route-capacity-checker packets over the θ bypasses calculated by the sender node of the link using the topology information. These packets have the number of required wavelengths stamped on them, and each link looks at the number of free channels (β) that it has and stamps it on the packet if this number is lower than the requested capacity. The chooser node of the link receives θ route-capacity-checker packets from the source, which report a combined bandwidth of, for instance, **B**, which it intelligently splits among the bypasses so that the total is (the requested bandwidth) and each bypass has the same relative load. If the reported bandwidth **B** is lower than the required bandwidth, then all the reported bandwidth is selected. If the link still remains in a critical condition, then it will initiate another bypass search with the bottleneck links removed. The bypass-confirm messages are sent back from the chooser to the sender node of the link. They pass through each of the intermediate links on the bypasses and set up the entries in the table mentioning which link has been promised how many wavelengths in the form <L,N>. Doing a max over all the wavelength terms in the rows of the table gives us the value of γ for that link (see Figures 17.6a and 17.6b).

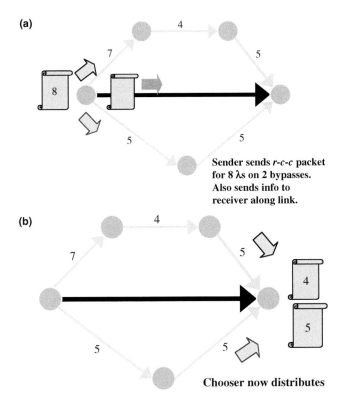

Figure 17.6 Bypass search.

Choosing the target number of wavelengths to search for, given the values for α β δ γ can be done in a number of ways. We divide the total channel number (λ_{max}) into levels in a binary way. So the first level is $\lambda_{max}/2$, the second level is $3*\lambda_{max}/4$ and so on. These levels are static, and the closest level higher than both α and δ is selected as the target number of wavelengths .

When a link fails, the chooser node detects it using the signal power decrease and activates the restoration procedure. The chooser has a table with the bypasses and the number of channels on each available. It checks the number of primary channels to restore, allocates them to the bypasses, and sends out restore packets with the cross-connect port information. The restore packets are processed at each intermediate node, the cross-connect is executed, the new cross-connect information is stamped on the packet and forwarded to the upstream node. Thus, the bypass is activated in the reverse direction. When the sender node gets the restore packet, it just bridges the traffic to the new output ports as specified by the restore packet. The cross-connects are typically executed in parallel; that is, at a particular node, a number of wavelengths are cross-connected simultaneously, since the restore packet contains aggregated information for a number of wavelengths.

17.3.4.1 Results

The protocol is simulated, and representative experiments for 40 nodes and 1200 to 1800 connections are shown. The total restorability, restoration time, and control overhead are calculated. The total restorability stays close to 100% most of the time and drops to around 50% in extreme load conditions, staying at an average of 80%. A useful parameter is *perceived capacity,* which serves as a measure of how much extra capacity this scheme is adding to the network by virtue of the sharing. This comes out to around 40% above the primary capacity. Restoration times are very low, around 4 to 8ms, but the benchmarks that they choose are quite low compared with the Doshi article.[1] The control overhead in the most loaded cases is around 1mbps per node, which is a little large but not much compared with the huge amounts of data traffic that will be carried.

17.3.4.2 Discussion

This algorithm has quite a few novel points. First, it assumes only topology information at each node (and not traffic engineering extensions). This means that the nodes will not be out of sync with the network state frequently. Assuming knowledge of wavelength usage at each link in the network is dangerous and can lead to very wrong estimates by links due to staleness of the information, especially when there are short-lived connections and the traffic matrix is not steady.

Second, this algorithm is very scalable with respect to the number of wavelengths on each link. The number of wavelengths per fiber is expected to reach around 200 very soon and it is very important that algorithms scale well to such numbers. ZRESTORE searches for bypasses using numbers and in aggregate, thus reducing both the control message size as well as number. Also, these searches look for more capacity than is strictly necessary so when new primary channels get set up, their bypasses are already selected and no extra work has to be done for them. Thus, the amortized cost of search is quite low.

Third, the backup channels are not reserved, just remembered, so calls are never blocked and they kick out the backed up channels when they are set up. The control overhead is a little more due to the kickout messages that propagate back to the source of the affected link. Another drawback is the assumption of wavelength conversion at each node. Extending this concept to networks without wavelength conversion is nontrivial, since in that case the search for the bypasses would have to be wavelength specific, so a lot of the benefits that this scheme derives from flexibility would be lost.

17.4 Conclusions and future directions

We considered a particular problem regarding restoration in optical mesh networks and studied a gamut of effective schemes that try to solve the problem in different ways. We reviewed centralized ILP-based formulations, distributed heuristic versions of these algorithms, and even simple distributed algorithms with novel ideas such as aggregation and over-protection. Several other solutions are available but we did not cover them due to space limitations; however, the cases reported here are representative of the types of solutions that one typically encounters while solving the restoration problem. These techniques apply to mesh networks, and their potential will be fully exploited only in large-scale optical mesh networks deployment. The latter will become a reality only under the following conditions:

1. Decommissioning of SONET boxes
2. The maturing of optical technology

The main motivation to decommission SONET is to do away with the overhead introduced by it and thus streamline the system. On the other hand, we will have to rely on IP for multiplexing. Time division multiplexed (TDM) traffic and guaranteed service may have to be handled by multi-protocol label switching (MPLS). The huge TDM user base may be unwilling to go in for such a drastic change. As for condition 1, we expect the state of the art in optical technology to evolve and become commercially viable. Today, optical wavelength converters and fibers with hundreds of wavelengths exist only in research laboratories. Without the accessories that enable optical mesh networking in its truest sense, most of the benefits of the restoration schemes cannot be realized.

Putting all that aside, however, both link and path restoration and protection strategies have shown immense promise. Path restoration strategies have been more efficient in terms of resource usage but lack the flexibility that link restoration has. The way to go, therefore, is a hybrid solution in which sections of the path are independently protected. This solution would provide the right mix of flexibility and resource efficiency and is aptly called segment protection.

Segment protection has already been proposed in MPLS, and a simple mapping of the labels to lambdas would do the trick for optical networks too. The MPLS layer and its inter-working with the optical layer is a very active research area. Very soon, clear trends about what should be done and what should be avoided are going to emerge. Until then, we must rely on 50ms SONET protection.

References

1. B. Doshi, S. Dravida, P. Harshavardhana, O. Hauser, and Y. Wang, Optical network design and restoration, *Bell Labs Technical J.*, Sept. 1999, pp. 58–84.
2. H. Wang, E. Modiano, and M. Medard, Partial path protection for WDM networks: end-to-end recovery using Local failure information, *IEEE ISCC*, July 2002.
3. S. Das, P. Verma, B.N. Jain, and M. Gerla, Zrestore: a link restoration scheme with high aggregation and no reservation, *IFIP ONDM 2002*, Feb. 2002.

4. S. Ramamurthy and B. Mukherjee, Survivable WDM mesh networks, part I
    — protection, *IEEE Infocom '97*, 1997.
5. S. Ramamurthy and B. Mukherjee, SurvivableWDM mesh networks, part II
    — restoration, *IEEE Infocom '97*, 1997.

chapter eighteen

Shared alternate-path protection with multiple criteria in all-optical wavelength-routed WDM networks

Bin Zhou
Queen's University
Hussein T. Mouftah
University of Ottawa

Contents

0-8493-1333-3/03/$0.00+$1.50
© 2003 by CRC Press LLC

18.1 Introduction

This chapter investigates a path protection design problem-based alternate routing for all-optical wavelength-routed wavelength division multiplexing (WDM) networks. We develop a generic formulation for the so-called shared alternate-path protection (SAPP) problem, which uses alternate-path routing for both primary and protection lightpaths, deals with single failure as well as multiple simultaneous failures, and considers the shared risk link group (SRLG) constraint and multiple practical optimization objectives. The formulated problem is an integer programming problem, which is NP-hard. Therefore it is not practical to have exact solutions, especially under multiple contradictory optimization criteria. For this reason, we use heuristics to obtain approximate optimal solutions and propose a survivable alternate routing (SAR) algorithm based on a kth shortest-path routing algorithm and a genetic algorithm. We assume wavelength converters available at each network node and approximately divide the optimization process into two relatively independent stages: candidate primary and protection route computation, and survivable lightpath routing. Through simulation experiments, we show the effectiveness of SAR in solving the SAPP problem and compare the solutions obtained from SAR with those from using a path protection scheme based on shortest-path routing.

18.1.1 Background

WDM, which allows a single fiber to carry multiple signals simultaneously, is becoming the core technology to cope with the rapidly increasing demand for bandwidth in the next-generation Internet. An adverse consequence of this advancement of boosting network capacity is the increased network vulnerability in the sense that a single network failure can significantly reduce the capability to deliver services in large-scale information systems. Therefore it is essential to provide network survivability in the optical network design.

Survivable network architectures are based on either dedicated resources or shared resources. With dedicated protection, network service is restored utilizing the dedicated network resources; for example, 1+1 and 1:1 protection. In the case of a failure, the traffic is switched to the preconfigured protection paths or links. Dedicated protections have very low recovery latency and guaranteed full restoration against single network failure; however, this approach results in inefficient use of network resources. The major research efforts for shared protection scheme have focused on reducing the resource overhead by means of techniques such as sharing of resources across connections or between disjoint paths for the same connection. Unlike the dedicated scheme, a connection can be established over the shared protection resources only after a network failure is detected.

According to the rerouting schemes used after a network failure occurs, shared protection approaches can be classified into link protection and path protection. In link protection, all the connections that traverse the failed link are rerouted around that link. The source and destination nodes of the connections traversing the failed link are oblivious to the link failure. In path protection, when a link fails, the source node and the destination node of each connection that traverses the failed link are informed of the failure. The source and the destination nodes of each connection independently reroute the traffic to backup routes on an end-to-end basis. Link protections are expected to have shorter recovery latency than path protections. Link protection schemes depend on fault localization, however, while no fault localization is necessary for path protection.

It has been reported that shared path protection schemes are more economical than link protections in the sense that they need much less redundant network resources. Furthermore, path protections are more feasible than link protections with available technologies. Path rerouting performed at the edge of the network may allow some or all of the recovery functions to be moved into the end-system. This simplifies network design and allows applications to make use of application-specific information such as tolerance for latency in making rerouting decisions. Accordingly, shared path protection is currently favored in Internet Engineering Task Force (IETF) deliberations for multi-protocol label switching (MPLS)-layer protection and MPLS-controlled optical path protection.

To address the shared protection problem, the SRLG concept is considered one of the most important constraints concerning the constraint-based route computation of optical channel routes. An SRLG is defined as a group of links that share a component, the failure of which causes failure of all links of the group. We define sharable protection link group (SPLG) as a collection of protection routes in which corresponding primary routes do not belong to any other SRLG. SRLG constraints can be defined as follows: protection paths can share links if and only if they belong to a common SPLG. By applying the SRLG constraint to the constraint-based path computation, one can select a route taking into account resource and logical structure disjointness that implies a lower probability of simultaneous light-path failure.

18.1.2 Related work

Many studies have been carried out to provide protection schemes for WDM networks.[1-13] Alanyali and Ayanoglu evaluated two iterative methods, independent and coordinate designs, through simulation for three restoration schemes: full reconfiguration, path-based, and link-based.[6] The numerical results showed that coordinate design with full reconfiguration achieves the most economical solutions in terms of fiber requirements. They also demonstrated that coordinated planning of several failure scenarios provides more efficient designs than those obtained by considering the failures independently. The iterative methods need much shorter computational time than centralized optimization method using CPLEX,[6] but the solutions obtained from the iterative methods are not approximately optimal. This is because increasing the number of iterations cannot result in improved solutions in any case.

A p-cycle protection scheme is proposed in Grover and Stamatelakis, and Iraschko and Grover to speed up the restoration of failures within mesh-based SONET or WDM transport networks.[11,12] The p-cycle scheme chooses an optimal set of p-cycles using integer linear programming optimization to provide spare resources. The recovery path in the p-cycle scheme is between two adjacent nodes of the failure. Therefore, the p-cycle scheme is essentially a failure-dependent protection scheme, and operates in the link protection mode of mesh-based scheme.

Typically, the path protection problem is to optimally determine two disjoint alternate paths — a primary path and a protection path — for each source-destination pair so as to minimize the required network capacity and cost; however, these algorithms unnecessarily restrict that all traffic for each node pair follows the same primary path and protection path after failures. In contrast, an optimal design scheme for survivable WDM transport networks is proposed in Miyao and Saito, which focused on optimally distributed traffic among several candidate primary path and protection path pairs.[3] However, the algorithm requires that a single restoration path is predetermined for each primary path on a one-to-one basis.

Most of the previous work on spare capacity allocation of mesh WDM networks modeled the static protection design as an integer linear programming (ILP) problem. A general adopted design objective is to minimize the total spare capacity required for the restoration in specific failure scenarios. Unfortunately, the resulted ILP formulation is NP-hard.[13] To obtain the optimal solution for even a small-sized network, such as a few tens of nodes, is very time-consuming using available mathematical tools. Relaxation methods are proposed to approximate the IP solution. Lagrangian relaxation, which decomposes the original complex problem into several easier sub-problems, was used by Doshi et al.[9] Simulated annealing and Tabu searching-based methods were proposed in Caenegem et al., Crochat and Boudec, and Liu et al.[1,2,13]

Although extensive research work has been done in protection design of optical network, certain issues still need to be addressed. Several implementations of path protection scheme have been proposed. If the primary paths are identified before protection design (usually using shortest path routing algorithm), the path protection is to find an optimal disjoint path for each primary path; however, simultaneously optimizing primary path routing and backup path routing can provide more cost-effective solutions. There are several algorithms proposed to provide shortest disjoint path pairs between a node pair in a network, but the problem of how to distribute primary traffic and backup traffic over the disjoint paths to obtain an overall optimal network performance needs further investigation.

Most protection design considers routing protection lightpaths to maintain survivability of optical connection against single network failure based on the assumption that multiple failures occur much less frequently than single failure. When multiple-failure protection is required, however, most proposed protection design methods cannot adapt to solve the new problem.

Although SRLG is not a new concept in protection design, to the best of our knowledge there is little work on explicitly modeling the SRLG constraint in protection design. Without considering SRLG constraints, the reserved restoration capacity is either over-reserved or under-reserved for protection. The typical example of the former case is 1:1 protection, which results in costly design because of capacity redundancy. In Davis et al., protection paths for a single demand can share reserved bandwidth under the condition that their primary paths are mutually disjoint,[5] but the method can be improved by applying more general SRLG constraints. In the latter case, protection paths can be multiplexed by other primary paths or protection paths to reduce network resource redundancy without SRLG constraints; however, these methods result in partial restoration after failures occur. Another effect of neglecting SRLG constraints at the protection design stage is that wavelength allocation for backup paths has to be done dynamically after network failures occur.[9]

In previous work, a single optimization objective is considered in protection design, such as minimizing total network cost including both primary resources and backup resources, minimizing network redundancy rate, or minimizing total capacity used for backup. However, simply using a single optimization objective may result in an undesired "optimal" solution. For example, when minimizing network redundancy rate is the only objective considered, increasing the resource utilization on primary path or reducing the reserved resources on the backup paths can reduce the network redundancy rate. The former should be avoided, however, because it will increase the propagation delay of primary lightpaths. Therefore, in practice, multiple optimization objectives need to be considered simultaneously, especially when the design consists of routing working traffic and deploying backup resources.

18.1.3 Our work

This chapter investigates a SAPP problem to provide cost-efficient and survivable optical network design. The SAPP problem can be addressed as follows. Given the traffic demand between each node pair, the physical topology of the optical network, as well as a set of optimization criteria and constraints, decide a set of routes to connect each node pair and the bandwidth (number of wavelengths) to be reserved on the routes. At the same time, compute protection routes and reserve bandwidth correspondingly for each primary lightpath to protect the traffic from any failures defined in a set of failure scenarios. A failure scenario can be a single network component failure, such as a link failure, or a node failure. It can be also a combination of multiple simultaneously occurred failures.

The protection lightpaths can share backup bandwidth if their corresponding primary lightpaths do not belong to the same SRLG. The backup lightpath should be link-disjoint or node-disjoint with its primary lightpath. Therefore, once the failure occurs, traffic on the primary lightpath can be rerouted to backup path. We formulate the SAPP as an integer programming problem with multiple objectives, including minimizing average propagation delay of primary lightpaths, minimizing the network redundancy rate, and minimizing the traffic load on the most congested link. Because the addressed SAPP is an NP-hard problem, a heuristic, survivable alternate routing is developed using fixed alternate routing and a multi-objective genetic algorithm (GA). We test our algorithm on a simplified SAPP problem by relaxing the wavelength continuity constraint (e.g., assuming that wavelength converters are available at each switching node). Simulations are carried out over two network topologies. The results show the effectiveness of the proposed algorithm and influence the number of alternate routes.

The remainder of this chapter is organized as follows: In Section 18.2, the SAPP is formulated as an integer programming problem with multiple objectives. In Section 18.3, the fixed-alternate routing and a genetic algorithm is used to solve a simplified SAPP problem. Section 18.4 presents simulations and numerical results of the proposed algorithms. Section 18.5 summarizes and concludes the chapter.

18.2 Formulation of SAPP

We consider an all-optical WDM network, where the traffic from user networks is aggregated at access nodes. To accommodate the aggregate traffic, multiple lightpaths are established between a pair of source and destination nodes, and the aggregate traffic is allocated among these lightpaths. In the following, we present a generic formulation of the SAPP problem.

In the following formulation, we use s and d to denote the source and destination nodes of a lightpath, and i and j to denote the end nodes of a physical link. The other notations used in the formulation are as follows.

18.2.1 Given parameters

- N: nodes in the network, number 1 through N
- W: the number of wavelengths each fiber supports
- L: the set of network links in the network
- F: the set of failures
- K_w: the number of alternate routes between a pair of source and destination nodes

- K_p: the number of alternate protection routes between a pair of source and destination nodes
- d_{ij}: the propagation delay on a physical fiber link from node i to node j, $(i,j) \in L$
- p_{ij}: the number of fibers on link (i,j), $(i,j) \in L$
- $\Lambda(s, d)$: an $N \times N$ traffic matrix that gives the aggregate traffic between each pair of source and destination nodes in terms of the number of wavelengths required, which is defined as:

$$\Lambda(s,d) = [\lambda_{sd}; \ s,d = 1,2,\cdots,N]_{N \times N}$$

where λ_{sd} is the aggregate traffic from node s to node d in terms of the number of wavelengths required. It should be noted that the aggregate traffic from node s to node d does not necessarily have to be equal to that from node d to node s.

18.2.2 Parameter variables

- $U_{sd}(k)$: the wavelength usage on the kth route from node s and node d, which is a $1 \times W$ vector defined as

$$\mathbf{U}_{sd}(k) = \{u_{sd}(k,w); \ w = 1,2,\cdots,W\}_{1 \times W} \quad k = 1,2,\cdots K$$

where

$$u_{sd}(k,w) = \begin{cases} 1, & \textit{if the kth route uses } w \\ 0, & \textit{otherwise} \end{cases}$$

- $L_{ij}^{sd}(k)$: an indicator that indicates if the kth route from node s and node d is routed through the link from node i to node j, which is defined as

$$L_{ij}^{sd}(k) = \begin{cases} 1, & \textit{if the kth route is routed through} \\ & \textit{the link from node } i \textit{ to node } j \\ 0, & \textit{otherwise} \end{cases}$$

- $L_{ij}^{sd}(k,w)$: an indicator that indicates if the kth route from node s to node d is routed through the link from node i to node j and uses wavelength w, which can be expressed as:
  - $v_{sd}(k)$: the number of lightpaths on the kth route from node s to node d

- $d_{sk}(k)$: the propagation delay on the *k*th route from node *s* to node *d*, which can be expressed as:
    - $v_{ij}^{sd}(k)$: the number of lightpaths on the *k*th route from node *s* to node *d* being routed through the link from node *i* to node *j*

Similarly, we define the following variables about protection routes:

- $\tilde{U}_{sd}(k)$: the wavelength usage on the *k*th protection route from node *s* and node *d*, which is a $1 \times W$ vector defined as:

$$\tilde{U}_{sd}(k) = \{\tilde{u}_{sd}(k, w); \ w = 1, 2, \cdots, W\}_{1 \times W} \quad k = 1, 2, \cdots K$$

where

$$\tilde{u}_{sd}(k, w) = \begin{cases} 1, & \text{if the kth protection route uses } w \\ 0, & \text{otherwise} \end{cases}$$

- $\tilde{L}_{ij}^{sd}(k)$: an indicator that indicates if the *k*th protection route from node *s* and node *d* is routed through the link from node *i* to node *j*, which is defined as:

$$\tilde{L}_{ij}^{sd}(k) = \begin{cases} 1, & \text{if the kth protection route is routed through} \\ & \text{the link from node } i \text{ to node } j \\ 0, & \text{otherwise} \end{cases}$$

- $\tilde{L}_{ij}^{sd}(k, w)$: an indicator that indicates if the *k*th protection route from node *s* to node *d* is routed through the link from node *i* to node *j* and uses wavelength *w*
- $\tilde{v}_{sd}(k)$: the number of lightpaths on the *k*th protection route from node *s* to node *d*
- $\tilde{v}_{ij}^{sd}(k)$: the number of lightpaths on the *k*th protection route from node *s* to node *d* being routed through the link from node *i* to node *j*
- $H_{sd}(k, f)$: the number of survivable lightpaths on the *k*th route from node *s* to node *d* *after failure f*; explicit description of $H_{sd}(k, f)$ depends on the definition of *f*. For example, if *f* is defined as a failure on link (i,j), $H_{sd}(k, f) = V_{sd}(k) - V_{ij}^{sd}(k)$. Obviously, $H_{sd}(k, f) = 0$, if the kth route use the failed network resources defined by *f*; $H_{sd}(k, f) = V_{sd}(k)$, otherwise. Take another example: If *f* is defined as the simultaneous failures of link (i,j) and link (m,n), $H_{sd}(k, f) = [V_{sd}(k) - V_{ij}^{sd}(k)][V_{sd}(k) - V_{mn}^{sd}(k)] / V_{sd}(k)$.

- $\tilde{H}_{sd}(k,f)$: the number of available lightpaths reserved on kth protection routes from node s to node d after failure f
- $G_{sd}(k, f) = 1$: indicates that the *kth* route from node s to node d is influenced by the failure f; $G_{sd}(k, f) = 0$, otherwise $T_{ij}^{sd}[f,k,z,(m,n)]$: the number of wavelengths reserved on the link (i,j) on the zth protection route to protect the kth primary route from the failure f, which traverses link (m,n) lightpaths
- $\varphi(k, z, w, f) = 1$: indicates that the kth primary lightpath shares the same wavelength w with its zth protection lightpath, which can be used as a backup lightpath when the failure f occurs, $\varphi(k, z, w, f) = 0$, otherwise

In our definition if two primary routes share a same link, for instance (m,n), they belong to an SRLP. Obviously, a finer definition of the shared risk can be adopted if the locations of the physical network components are defined previously.

18.2.3 Constraints

- Definition

$$d_{sd}(k) = \sum_{i,j} d_{ij} L_{ij}^{sd}(k)$$

$$v_{ij}^{sd}(k) = v_{sd}(k) L_{ij}^{sd}(k)$$

$$\tilde{v}_{ij}^{sd}(k) = \tilde{v}_{sd}(k)\tilde{L}_{ij}^{sd}(k)$$

$$L_{ij}^{sd}(k,w) = u_{sd}(k,w)L_{ij}^{sd}(k)$$

$$\tilde{L}_{ij}^{sd}(k,w) = \tilde{u}_{sd}(k,w)\tilde{L}_{ij}^{sd}(k)$$

$$G_{sd}(k,f) = [V_{sd}(k) - H_{sd}(k,f)] / V_{sd}(k)$$

$$T_{ij}^{sd}[f,k,z,(m,n)] = G_{sd}(k,f)L_{mn}^{sd}(k)\tilde{H}_{sd}(z,f)\tilde{L}_{ij}^{sd}(z)$$

$$\varphi(k,z,w,f) = u_{sd}(k,w)G_{sd}(k,f)\tilde{u}_{sd}(z,w)\tilde{H}_{sd}(z,f) / \tilde{V}_{sd}(z)$$

- Traffic

$$\sum_j \sum_k v_{ij}^{sd}(k) - \sum_j \sum_k v_{ji}^{sd}(k) = \begin{cases} \lambda_{sd}, & \text{if } i = s \\ -\lambda_{sd}, & \text{if } i = d \\ 0, & \text{otherwise} \end{cases} \quad (18.1)$$

$$\sum_j \sum_k \tilde{v}_{ij}^{sd}(k) - \sum_j \sum_k \tilde{v}_{ji}^{sd}(k) \begin{cases} \geq \lambda_{sd}, & \text{if } i = s \\ \leq -\lambda_{sd}, & \text{if } i = d \\ = 0, & \text{otherwise} \end{cases} \quad (18.2)$$

$$\sum_k v_{sd}(k) = \lambda_{sd} \quad (18.3)$$

$$\sum_k \tilde{v}_{sd}(k) \geq \lambda_{sd} \quad (18.4)$$

$$V_{sd}(k) - H_{sd}(k,f) - \sum_m \tilde{V}_{sd}(m,f) \leq 0 \quad \forall (s,d), k, f \quad (18.5)$$

$$T_{ij}^{sd}[f,k,z,(i,j)] = 0 \quad (18.6)$$

$$\sum_{s,d} \sum_k V_{ij}^{sd}(k) + \underset{mn}{Max} \sum_{sd} \sum_k \sum_z T_{ij}^{sd}[f,k,z,(m,n)] \leq W \times p_{ij} \, \forall f, (i,j) \quad (18.7)$$

Flow-conservation constraints are represented in Constraints (18.1) and (18.2), respectively. Constraints (18.3) and (18.4) specify traffic demand of primary capacity and backup capacity for each node pair s–d, respectively. Constraint (18.5) ensures enough backup resource reserved for each primary path once a failure occurs. Constraint (18.6) ensure that no share between any primary paths and backup paths for each node pair. Constraint (18.7) guarantees that link capacities are not exceeded.

- Wavelengths

$$\sum_{s,d} \sum_k L_{ij}^{sd}(k,w) \leq p_{ij} \quad \forall (i,j), w \quad (18.8)$$

$$\sum_w \sum_k \sum_z L_{ij}^{sd}(k,w) \tilde{L}_{ij}^{mn}(z,w) \leq p_{ij} \quad \forall (s,d), (m,n), (i,j) \quad (18.9)$$

$$\sum_{s,d}\sum_{k}\sum_{z} \tilde{L}_{ij}^{sd}(z,w)\varphi(k,z,w,f)G_{sd}(k,f) \le p_{ij} \quad \forall (i,j) \qquad (18.10)$$

$$\sum_{w}\sum_{z} \varphi(k,z,w,f) \ne 0 \quad \forall (s,d),k \qquad (18.11)$$

Constraint (18.8) indicates the wavelength continuity constraint for primary lightpaths. Constraint (18.9) ensures that primary lighpaths do not share a wavelength with any backup lightpaths on the same fiber link. Constraint (18.10) indicates the wavelength continuity constraint for backup lightpaths where the corresponding primary lightpaths belong to the same SRLG. Constraint (18.11) ensures the existence of the backup lightpath on the same wavelength that the primary lightpath is using; therefore, the wavelength continuity constraint can still be guaranteed after a failure occurs.

18.2.4 *Optimization objectives*

- Propagation delay O_a: minimize the average propagation delay over a primary lightpath

$$Min[\frac{1}{\sum_{s,d}\lambda_{sd}}\sum_{s,d}\sum_{k} v_{sd}(k)d_{sd}(k)] \qquad (18.12)$$

- Link Load O_b: minimize the maximum link load

$$Min\{\sum_{s,d}\sum_{k} V_{ij}^{sd}(k) + \underset{mn}{Max}\sum_{sd}\sum_{k}\sum_{z} T_{ij}^{sd}[f,k,z,(m,n)]\} \qquad (18.13)$$

- Redundancy rate O_c: minimize the ratio of the total reserved capacity for failure protection over the total allocated primary capacity

$$Min\left\{\sum_{ij}\underset{mn}{Max}\sum_{sd}\sum_{k}\sum_{z} T_{ij}^{sd}[f,k,z,(m,n)] \Bigg/ \sum_{s,d}\sum_{k} V_{sd}(k)\right\} \qquad (18.14)$$

The formulated SAPP problem is an InP problem with multiple objectives. If we only consider Objective (18.12), the problem is an ILP problem which can be solved using a popular LP-solver such as CPLEX. The formulation addresses a generic and realistic protection design problem with consideration of multiple alternate routes for each node pair, the SRLG constraint, the wavelength continuity constraint, multiple simultaneous failures, failure-dependent protection, as well as multiple optimization objectives; however, the formulated SAPP problem is very complex involving hundreds of variables and constraints even for a small-sized network. Furthermore, the multiple objectives are mutually and inherently conflicted. For example, according to Objective O_a, short paths should be selected as primary paths, while under the objective O_c, primary paths prefer to long paths so that reducing the redundancy rate. Therefore, it is practically impossible to obtain an exactly optimized solution for all objectives at the same time. Tradeoffs among the multiple objectives have to be made to search an approximated solution.

18.3 Survivable alternate routing algorithm

Obviously, the formulated SAPP problem is an NP-hard problem, which would involve a large amount of computational time and high computational intensity in order to obtain optimal solutions. In this case, heuristics become an effective way to obtain good approximate solutions. For this reason we propose an SAR algorithm. Because wavelength assignment can be relatively independent of routing and routing usually has a larger impact on the network performance than wavelength assignment, we focus on routing and do not consider wavelength assignment in SAR. For this purpose we assume wavelength converters at each network node and thus do not need to consider the wavelength-continuity constraint. Even though the wavelength-continuity constraint is considered, a variety of wavelength assignment algorithms can be found in the literature. Under this assumption, wavelength Constraints (18.8) through (18.11) are relaxed.

To make the problem more tractable, we approximately divide the optimization process into two relatively independent stages: route computation and survivable lightpath routing.

18.3.1 Candidate route computation

At the first stage, a set of alternate routes is computed for each pair of source and destination nodes in the physical topology. The candidate path computation is only based on the physical topology. We define a primary route set (PRS) between a source node and a destination node as a set of pre-computed candidate routes through which wavelength requests between the node pair are routed. A backup route set (BRS) for a primary route is a set of candidate disjoint routes of the primary route, through which traffic on a primary route is rerouted according to the pre-reserved bandwidth on each backup route if the primary routes are broken. Primary routes and backup routes can be either inactive or active. A route is active when bandwidth (wavelengths) is reserved along the route for a traffic request.

Figure 18.1 illustrates two working routes, A-B-C-H and A-D-E-H from node A to node H. For each working route, two backup routes are assigned

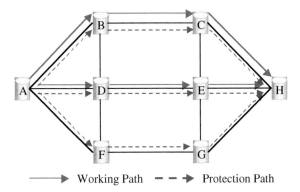

Figure 18.1 WRS and PRS. (B. Zhou and H.T. Mouftah, *IEEE Proceedings of International Symposium on Computers and Communications* (ISCC 2002), Taormina, Italy, July 2002, pp. 732–738. With permission.)

(e.g., A-D-E-H and A-F-G-H are backup routes for A-B-C-H; A-B-C-H and A-F-G-H backup A-D-E-H).

To reduce the average propagation delay, we use k_w shortest paths as candidate primary routes for each node pair, which is computed by using the route-computing algorithm proposed in Lee and Wu.[16] For each failure scenario, k_p protection paths for each primary path are computed over the residual physical network topology by deleting the failed nodes or links. We develop a breadth-first searching algorithm to search a set of protection routes for each primary path. We assume that after the failure of the primary path it is not necessary to provide the same quality of service in terms of propagation delay on the protection path. One of the main objectives of designing shared path protection is to reduce the bandwidth redundancy by improving the sharing of protection bandwidth among different SRLGs. Accordingly, we select protection paths based on their diversity rather than their cost (delay).

The diversity of protection routes can efficiently balance the traffic load over the optical links in the network and therefore reduces the influence of a failure, especially a link failure. For this reason, instead of using the k-shortest path algorithm to generate PRS, a breadth-first diverse routing algorithm is developed. To search protection paths for a working path, a tree is generated from the original network topology graph by deleting the intermediate nodes and links in the working path. Let the source node be the root node of the tree. A breadth-first searching method is used to search the protection paths from root node to destination node. Let $E_{ij}^{sd}(k)$ denote the traversed times of backup link (i,j) by backup paths in the BPS for kth primary path of node pair s-d. Thus, once a new backup path which traverses link (i,j) is found, $E_{ij}^{sd}(k)$ increases its counts. To make the obtained backup paths as diverse as possible, we confine $E_{ij}^{sd}(k)$ between one and a certain pre-defined value K'_p, which is less than K_p. If $E_{ij}^{sd}(k) > K'_p$, then the newly found protection path will not be included in the backup route set, and the searching procedure continues until K_p protection paths are found or the last leaf node is reached.

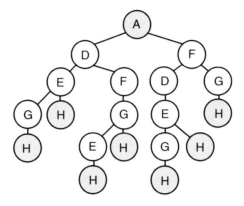

Figure 18.2 Breadth-first backup path searching tree. (B. Zhou and H.T. Mouftah, *IEEE Proceedings of International Symposium on Computers and Communications* (ISCC 2002), Taormina, Italy, July 2002, pp. 732–738. With permission.)

Figure 18.2 depicts the tree generated for working path A-B-C-H. Protection paths, A-D-E-H, A-F-G-H, A-D-E-G-H, A-D-F-G-H, A-F-D-E-H, A-D-F-G-E-H, and A-F-D-E-G-H, are obtained sequentially by using broad first searching algorithm. If $K_l = 2$ and $K_p = 4$, only the A-D-E-H, A-F-G-H, and A-D-E-G-H are eligible solutions.

Obviously, the more candidate primary and protection paths are computed, the better solutions could be obtained. However, with the increase of the number of candidate primary and protection paths the searching time of optimal solutions also increases. In practice, the size of the route sets K_w and K_p is below ten. Consequently, the SAPP problem is simplified further by enforcing the primary and backup routes using routes selected from a pre-computed candidate route set, which identify all $L_{ij}^{sd}(k)$, and $\tilde{L}_{ij}^{sd}(k)$. Constraints (18.1) and (18.2) are accordingly satisfied.

18.3.2 Survivable lightpath routing

At the second stage, "optimal" routes are decided for each of the primary lightpaths between a pair of source and destination nodes, which can only be chosen from the set of candidate alternative routes between the source and destination nodes. For each primary lightpath, a protection lightpath has to be established on one of the candidate protection routes to protect the primary path from any failures, which are defined in the set of failure scenarios. The decisions are subject to Constraints (18.3) through (18.6) as well as Optimization Objectives (18.11) through (18.14). Obviously, it is still an InP problem, although much simplified from the original formulation. For a small-sized network, the optimal solution can be obtained using permutation search. However, the permutation search is not applicable for a large-sized network with tens of nodes and links, since it is inherently computational intensive. Therefore, we develop a genetic algorithm for the survivable lightpath routing. In the following, we focus on the design and implementation aspect of survivable lightpath routing using a multi-objective genetic algorithm (GA).

18.3.2.1 Multi-objective optimization

It is rarely the case that there is a single optimal solution that simultaneously optimizes all the objective functions. Therefore, we normally look for "tradeoffs" instead of single solutions when dealing with multi-objective optimization problems. The notation of "optimum" is therefore different. The most commonly adopted notation of optimality is Pareto optimum.[18] We can generalize our multi-objective problem as follows:

> Find the vector $x^* = [x_1^*, x_2^*, ..., x_n^*]^T$, which will satisfy the m inequality constraints: $\varphi_i(x) \geq 0$, $i = 1, 2, ..., m$; the n equality constraints: $\varphi_i(x) \geq 0$, $i = 1, 2, ..., n$; and optimizes the vector function $f(x) = [f_1(x), f_2(x), ..., f_k(x)]^T$, where $x = [x_1, x_2, ..., x_n]^T$ is the vector of decision variables. We say that a vector of decision variables x^* is Pareto optimal if there does not exist another x so that $f_j(x) \leq f_i(x^*)$ for all $i = 1, ..., k$ and $f_j(x) < f_j(x^*)$ for at least one j, where "$\leq$" means "no better than," and "$<$" means "worse than."

Unfortunately, this concept almost always does not give a single solution, but a set of solutions called the Pareto-optimal set. The vectors x corresponding to the solutions included in the Pareto-optimal set are called non-dominated. The plot of the objective functions where the non-dominated vectors are in the Pareto-optimal set is called the *Pareto front*.

18.3.2.2 GA for multi-objective optimization

It is clear from the previous discussion that a multi-objective optimization problem usually has a set of Pareto-optimal solutions instead of one single optimal solution as in the case of single-objective optimization. In multi-objective optimization the goal is to find as many different Pareto-optimal solutions as possible.

GAs appear to be particularly suitable for solving multi-objective optimization problems because they deal with a set of possible solutions (the so-called population). This allows us to find a number of Pareto-optimal solutions in a single run of the algorithm instead of performing a series of separate runs, as in the case of the traditional mathematical methods.

A GA uses a stochastic searching method that mimics the metaphor of natural biological evolution. A GA operates on a population of potential solutions applying the principle of survival of the fittest to produce better and better approximations to a solution. At each generation, a new set of approximations is created by the process of selecting individuals based on their level of fitness in the problem domain and breeding them together using operators (i.e., crossover and mutation) borrowed from natural genetics. This process leads to the evolution of populations of individuals that are better suited to their environment than the individuals from which they were created, just as in natural adaptation.

A number of multi-objective GAs have been proposed to solve real-world engineering optimization problems, which can be classified into non-Pareto-optimization GAs and Pareto-optimization GAs. Readers interested in a more comprehensive introduction of multi-objective GA are referred to Coello.[18]

As a widely used non-Pareto-optimization GA, the weighted sum genetic algorithm (WSGA) approach assigns a weight for each objective according to the relative importance of it, and thus converts a multi-objective optimization problem into a single objective optimization problem. Although the WS approaches are simple and computationally efficient, their main drawback is the difficulty to determine the appropriate weights that can appropriately scale the objectives without comprehensive information about the problem.

In contrast to the WSGA approach, most proposed Pareto-optimization-based GAs treat the multiple objectives equally. The goal is to find as many different Pareto-optimal solutions as possible. Pareto-based fitness assignment strategies have been developed which usually exhibit fast convergence toward the vicinity of the Pareto-front, but further enhancements to maintain diversity in difficult problems are necessary.

In the addressed SAPP problem, the preferences on the different objectives may be different, therefore a network designer may only be interested in a certain region of the Pareto-front. In the implementation of SAR, a weighted Pareto-based GA is developed to efficiently guide the population toward the interesting area.

18.3.2.3 Weighted pareto-based GA (WPGA)

This section describes the implementation detail of a weighted Pareto-based GA. The programming diagram for survivable alternate routing using GA is shown in Figure 18.3. The GA begins with generating a random initial population of candidate solutions in which primary lightpaths and backup lightpaths are assigned based on randomly chosen routes so that the Constraints (18.3) through (18.5) for each s-d pair can be satisfied. Genetic operators are applied to the initial population in order to produce successive populations of good quality. After every generated population, each individual of the population must be evaluated to be able to distinguish between good and bad individuals. This is done by mapping the objective function to a "fitness function" — a non-negative figure of merit. In each iteration the WPGA evaluates the current generation according to fitness value of each individual solution, then proceeds to create the next generation of candidate solutions. The next generation consists of clones, mutations, crossbreeds from the current generation, and new random candidates. There are three basic generic operators: selection,

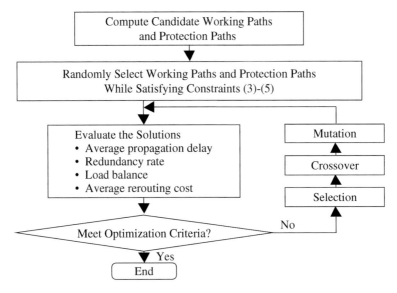

Figure 18.3 Flowchart of SAR.

crossover, and mutation. The optimization criteria in the genetic algorithm can be an acceptable fitness value and/or the prespecified maximum number of generations.

Encoding.   Each routing solution is encoded using a string of $N(N-1)$ $K_w(K_p + 1)$ integer units. The encoding format is presented in Figure 18.4. Each unit can assume integer value from 1 to W, which indicates the bandwidth allocated on the candidate primary route or protection route in the number of wavelength; α_k^{st} indicates the number of wavelengths allocated on the kth candidate primary path of node pair s-t. For each primary path, β_z^{st} indicates the number of wavelengths reserved on the zth candidate protection path. In practice, only the node pairs that have traffic requests are considered. Thus, a compact chromosome coding can be used.

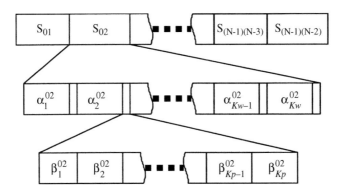

Figure 18.4 The encoding formate of SAPP solutions.

Fitness assignment.   A fitness function is defined to evaluate the quality of each candidate solution. In the implementation of the GA in SAR, a fitness function consists of three parts:

1.   An objective function that combines the multiple objectives
2.   A penalty function that revises the fitness of illegal solutions, which do not meet the constraints defined in 18.2.3
3.   A Pareto-optimal rank

The multiple optimization objectives must be considered simultaneously in the fitness function. Some compromises have to be made according to preference of the objectives. In general, minimizing average propagation delay of primary lightpath and minimizing redundancy rate are more preferable than other objectives because the network works in normal condition in most of the time. Because reducing recovery latency is the main concern of protection design, however, as a substantial part of recovery latency, reconfiguration delay should be minimized. We assign each objective a weight and thus combine the multiple objectives in one objective. Let O_a^{max*}, O_b^{max*}, and O_c^{max*} denote the estimated maximum value of O_a, O_b, and O_c respectively. The objective function (F_o) can be formulated as follows:

$$F_o = Min\,(w_a\,\frac{O_a}{O_a^{max*}} + w_b\,\frac{O_b}{O_b^{max*}} + w_c\,\frac{O_c}{O_c^{max*}})$$

where w_a, w_b, and w_c are the weights for optimality criterion O_a, O_b, *and* O_c, respectively. The weights show the preference of the optimization objectives, and thus guide the search toward interesting region of the solution space.

Besides the multiple objectives, constraints that are not considered in the generation of initial solutions have to be considered in the design of fitness function. In the genetic algorithm, the initial population is chosen from a random global space, which satisfies Constraints (18.3) through (18.5), because they can easily be satisfied by properly assign traffic among the candidate primary paths and backup paths for each node pair; however Constraint (18.6) involves the traffic distribution of the whole network and thus is not considered in solution representation. A penalty function (F_p) is defined to revise the fitness of illogical solution in which the maximum link load is over the link capacity.

$$F_p = \begin{cases} W \times p_{ij} - Max\{\sum_{s,d}\sum_{k} V_{ij}^{sd}(k) + Max\sum_{mn}\sum_{sd}\sum_{k}\sum_{z} T_{ij}^{sd}[f,k,z,(m,n)]\}, \\ \qquad\qquad \textit{if } \max \textit{link load is over the link capacity} \\ 0, \qquad\qquad \textit{otherwise} \end{cases}$$

Ranking non-dominated solutions and sharing techniques used in Srinivas and Deb is adopted in the GA to guide the search toward the global Pareto-optimization region and maintain population diversity.[17] By temporarily removing the non-dominated solutions from the original set and applying a dominance-identification procedure iteratively, each individual x can be assigned a rank $r(x)$.

Diversity along the Pareto-optimal front is supported by fitness sharing, which is introduced in Srinivas and Deb[17] and has been applied successfully to a number of real-world problems. As in the standard sharing procedure, for each individual x the niche count $m(x)$ is calculated, a quantity proportional to the number of other individuals in the neighborhood of the individual x. We use a triangular sharing function. Then, if $r(x)$ is the rank assigned to individual x, the rank is modified according to r'(x).

$$r'(x) = \begin{cases} r(x) + \dfrac{m(x) - \min\limits_{k} m(k)}{\max\limits_{k} m(k) - \min\limits_{k} m(k)} , & if \max\limits_{k} m(k) > \min\limits_{k} m(k) \\ r(x), & otherwise \end{cases}$$

In this way, individuals with comparatively few neighbors are relatively favored over individuals in crowded areas; however, no individual can get a worse modified rank than any individual with an inferior original rank. It is thus ensured that the current Pareto-front converges quickly to the true Pareto front, while the individuals are nevertheless well distributed.

Therefore, we define the fitness function F as follows: $F = F_o + F_p + r'(x)$.

Other operators and parameters in the implementation. We use the roulette wheel selection scheme in the selection operation and use single-point and two-point crossovers at the crossover rate of 0.25, which has been found to give good solutions. The mutation operation increases (or decreases) the number of lightpaths allocated on a randomly-selected primary path by 1, and at the same time decreases (or increases) the number of lightpaths on another primary path in the same set of alternate routes by 1. The number of reserved wavelengths on corresponding protection path should be also increased/decreased so that Constraints (18.3) through (18.5) can still be satisfied. The mutation rate is set to be 0.02.

18.3.3 An example of SAPP

Figure 18.5 gives an example result of SAPP by using SAR. Traffic requests, working lightpaths, reserved backup lightpaths and the wavelengths allocated for them are listed in Table 18.1. This example also demonstrates the advantage of SAPP. With SAPP, bandwidth can be reserved for protecting working path A-B on both backup paths, A-C-D-B and A-C-E-F-D-B, so that the traffic can be balanced better than that in single-path protection case (in this example, no solution can be found to

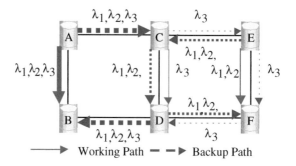

Figure 18.5 Alternate survivable routing. (B. Zhou and H.T. Mouftah, *IEEE Proceedings of International Symposium on Computers and Communications* (ISCC 2002), Taormina, Italy, July 2002, pp. 732–738. With permission.)

Table 18.1 An Example of Alternate Survivable Routing

s-d pair	Traffic	WP, λ_w	BP, λ_p
A → B	3	A-B, 1	A-C-D-B, 1
		A-B, 2	A-C-D-B, 2
		A-B, 3	A-C-E-F-D-B,3
C → D	1	C-D, 3	C-E-F-D, 3
E → F	2	E-F, 1	E-C-D-F, 1
		E-F, 2	E-C-D-F, 2

Note: WP, λ_w: working path and wavelength ID reserved for traffic along the working path; BP, λ_p: backup path and wavelength ID reserved along the protection path.

Source: B. Zhou and H.T. Mouftah, *IEEE Proceedings of International Symposium on Computers and Communications* (ISCC 2002), Taormina, Italy, July 2002, pp. 732–738. With permission.

satisfy all traffic requirements with single-path protection, if the maximum number of wavelengths in each link is three).

18.4 Simulations and numerical results

We carry out simulation experiments to show the effectiveness of SAR in solving the SAPP problem.

For different network topologies, the results obtained from the experiments are similar. In the following, we give the numerical results based on experiments over a 32-node network as shown in Figure 18.6. Each physical link in the topology represents a pair of unidirectional fiber links between two adjacent nodes with one in each direction (i.e., $p_{ij} = p_{ji} = 1$). In the simulation, the original traffic matrix is randomly generated within a specified range. The wavelength-continuity constraint is not considered (wavelength conversion at each node is assumed). For each experiment, we assume that $W = 64$, $K_w = 4$, and $K_p = 6$, if the parameters are not specified. We assume that the propagation delay on any link is the same (e.g., 50ms).

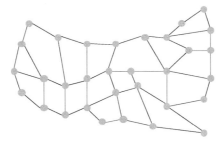

Figure 18.6 Network topologies used in the simulations.

Single-link failures are considered as the set of failure scenarios. Multi-failure scenarios are not considered in the simulations, because the survivability of traffic from multiple failures depends on the physical topology of the network and the locations of the failures; however, the SAR can handle the multi-failure scenarios virtually without any changes. Simulations are carried out over each network topology with a series of proportionately increased traffic demands (e.g., from 200 to 400 lightpath requests). The lightpath requests are randomly generated among all node pairs.

We compare WSGA and WPGA using the same set of weights for the multiple objectives and GA parameters. To facilitate the visualization of Pareto-front using WPGA, we only consider two contradictory objectives: minimizing average propagation delay on primary lightpaths (O_a) and minimizing network redundancy rate (O_c). A higher weight of an optimization objective reflects the higher preference of the objective. Using different preferences (i.e., O_a is more important than O_c) ($w_a = 10$ and $w_c = 2$), and O_a and O_c are equally important ($w_a = 5$ and $w_c = 5$), two different Pareto-fronts are presented in Figures 18.7a and 18.7b, respectively. The graphs show very clearly the separation of Pareto-front obtained using different preferences. The separation is caused by a combination of weighted objective function and Pareto-optimal rank, because the solutions in the preferred region of Pareto-front have higher fitness value and thus have better chance to be selected in reproduction than the other solutions outside the region in the Pareto-front. A single optimal solution can be obtained from WSGA without considering Pareto-optimal rank. We observed that after the same number generations, WPGA always performs better than WSGA, although the improvement may not be significant.

In the following paragraphs, we compare the protection design using SAPP with the results using another implementation of shared path protection, where primary lightpath and protection lightpath always go through the shortest link-disjoint route pair for each source and destination. We call the implementation scheme shortest-path based shared path protection (SSPP). We also test the performance of SAPP under different numbers of alternate primary and backup routes. In the following figures, we use SAPP-k_w-k_p to distinguish the implementations of SAPP with different maximum number of alternate routes of PRSs (k_w) and BRSs (k_p). For each node

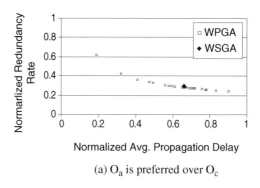

(a) O_a is preferred over O_c

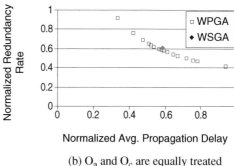

(b) O_a and O_c are equally treated

Figure 18.7 Pareto-optimal front and best solution obtained from WPGA and WSGA, respectively, after 500 generations.

pair, the k_w primary routes are selected from K_w candidate primary paths. The k_p backup routes are selected out of the K_p candidate backup routes for each primary route.

Figures 18.8 through 18.10 show the average propagation delay of primary lightpaths, redundancy rate, and maximum link load of SSPP,

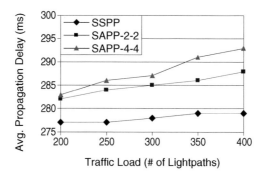

Figure 18.8 Impacts of traffic changes and protection schemes on the average propagation delay of primary lightpaths.

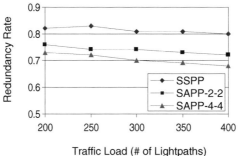

Figure 18.9 Impacts of traffic changes and protection schemes on the network redundancy rate.

SAPP-2–2, and SAPP-4–4, respectively. In fact, SSPP can be considered as a special case of SAPP-1–1. The weight of the optimization criterion O_a, O_b, and O_c are selected as 10, 2, and 5, respectively. We weight the average propagation delay of primary paths over other criteria because most of the time, the network is primarily under normal working condition without any failures. We simply use the same set of weights to select the best solution from the obtained Pareto-front.

Figure 18.8 is the average propagation delay (APD) on primary lightpaths vs. the changes of the traffic load with SSPP, SAPP-2–2, and SAPP-4–4. With the increase of traffic load, the APD on primary lightpaths are almost unchanged with SSPP, because the shortest path between a pair of nodes is always used as primary path. With SAPP-2–2 and SAPP-4–4, however, the ADPs of primary lightpaths increase with the increasing traffic load. This is because a larger amount of traffic load would result in more alternate routes used when numbers of alternate path are allowed.

Figure 18.9 is the network redundancy rate obtained from SSPP, SAPP-2–2, and SAPP-4–4. Clearly, SSPP produces solutions with highest redundancy rate. The redundancy rate obtained from SAPP-2–2 and SAPP-4–4 are about 10 and 15% lower than that from SSPP, respectively. With more numbers of alternate paths allowed, however, the redundancy rate cannot be reduced further. SAPP generates better protection solution in terms of redundancy rate than SSPP for two reasons. One is that the average primary path is longer in SAPP than that in SSPP. The other is that traffic is distributed over alternate routes in SAPP, which increases the share-ability of protection wavelengths among the lightpaths from different SRLG. Obviously, the latter reason is the main contribution to the reduction of network redundancy rate.

In Figure 18.10, we present the impacts of traffic changes and protection schemes on the maximum link load. Obviously, SAPP-4–4 generates solutions with the lowest maximum link load, because more numbers of alternate routes can provide more flexibility in lightpath routing and thus can achieve more balanced load distribution over the links in the network.

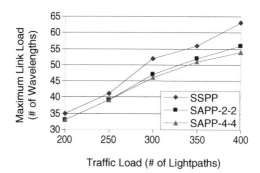

Figure 18.10 Impacts of traffic changes and protection schemes on the maximum link load.

18.5 Conclusions

The survivable optical network design is essential to the deployment of all-optical WDM networks. This chapter proposed a path protection scheme, which uses alternate-path routing for both primary and protection light-paths, deals with single failure as well as multiple simultaneous failures, and considers the shared-risk link group (SRLG) constraint and multiple practical optimization objectives. The problem has been formulated as an integer programming problem with linear constraints and multiple optimization criteria including minimizing average propagation delay of primary lightpaths, minimizing the network redundancy rate, and load balance. Because the problem is inherently NP-hard and practically impossible to obtain an exact optimal solution for the multiple contradictory criteria, we developed a survivable alternate routing algorithm using a k-disjoint path-routing algorithm to compute a set of candidate primary paths and backup paths for each node pair and a genetic algorithm (e.g., WPGA) to distribute the traffic among the primary paths and shared backup paths.

Through simulation experiments, we demonstrate the effectiveness of SAR in solving the SAPP problem. It is observed that the WPGA constantly outperforms the WSGA, which is used widely as a non-Pareto optimization GA. The WPGA guides the search to the interesting region of Pareto-front instead of the whole Pareto-front, and thus is more efficient than the corresponding non-bias, multi-objective optimization GAs. We also compared the performance of the SSPP with the SAPP under different traffic load. Although SSPP ensures the shorter average propagation delay than SAPP, it suffers higher network redundancy rate and higher traffic load on the most congested link. We also presented the impacts of the numbers of alternate routes in SAPP on the performance of the algorithm. With more numbers of alternate routes allowed, better performance can be achieved; however, longer computation times are expected. Through the simulations on the specific network, we observed that the improvement becomes insignificant when more than four alternate paths are allowed.

References

1. B.V. Caenegem, W.V. Parys, F.D. Turck, and P.M. Demeester, Dimensioning of survivable WDM networks, *IEEE J. Select. Areas Commun.*, vol. 16, no. 7, Sept. 1998, pp. 1146–1157.

2. O. Crochat and J.-Y.L. Boudec, Design protection for WDM optical networks, *IEEE J. Select. Areas Commun.*, vol. 16, no. 7, Sept. 1998, pp. 1158–1165.

3. Y. Miyao and H. Saito, Optimal design and evaluation of survivable WDM transport networks, *IEEE J. Selected Areas Commun.*, vol. 16, no. 7, Sept. 1998, pp. 1190–1198.

4. S. Ramamurthyt and B. Mukherjee, Survivable WDM mesh networks, part I — protection, *INFOCOM '99. Proc. IEEE*, vol. 2, 1999, pp. 744 –751.

5. R.D. Davis, K. Kumaran, G. Liu, and I. Saniee, Spider: a simple and flexible tool for design and provisioning of protected lightpaths in optical networks, vol. 6, no. 1, Jan.– Jun., 2001, pp. 82–97.

6. M. Alanyali, and E. Ayanoglu, Provisioning algorithms for WDM optical networks, *IEEE/ACM Trans. on Networking*, vol. 7, no. 5, Oct. 1999, pp. 767–778.

7. G. Shen et al., Designing WDM optical network for reliability: routing light paths efficiently for path protection, *Optical Fiber Commun. Conf.* 2000, vol. 3, pp.50 –52.

8. W.D. Grover et. al, New options and insights for survivable transport networks, *IEEE Commun. Mag.*, vol. 40, no. 1, Jan. 2002, pp. 34–41.

9. B.T. Doshi et. al., Optical network design and restoration, *Bell Labs Tech. J.*, vol. 4, no. 1, Jan. –Mar., 1999, pp. 58–84.

10. O. Gerstel and R. Ramaswami, Optical layer survivability — an implementation perspective, *IEEE J. Selected Areas Commun.*, vol. 18, no. 10, Oct. 2000, pp. 1885–1899.

11. W.D. Grover and D. Stamatelakis, Cycle-oriented distributed preconfiguration: ring-like speed with mesh-like capacity for self-planning network restoration, *IEEE ICC '98*, 1998, pp. 537–543.

12. R.R. Iraschko and W.D. Grover, A highly efficient path-restoration protocol for management of opical network transport integrity, *IEEE J. Select. Areas Commun*, vol. 18, no. 5, May 2000 pp. 779–794.

13. Y. Liu, D. Tipper, and P. Siripongwuikorn, Approximating optimal spare capacity allocation by successive survivable routing, *INFOCOM 2001, Proc. IEEE*, vol. 1, 2001.

14. B. Zhou and H.T. Mouftah, Balance alternate routing for WDM networks, *IASTED Int. Conf. on Wireless and Optical Commun. 2001*, Banff, Canada, July 17–19, 2001.

15. D. Saha, M.D. Purkayastha, and A. Mukherjee, An approach to wide area WDM optical network design using genetic algorithm, *Computer Commun.*, vol. 22, 1999, pp. 156–172.

16. S.W. Lee and C.S. Wu, K-best paths algorithm for highly reliable communication networks, *IEICE Trans. Commun.*, vol. E82-B, no.4, pp. 586–590, 1999.

17. N. Srinivas and K. Deb, Multiobjective optimization using nondominated sorting in genetic algorithms, *Evolutionary Computation*, vol. 2, no. 3, pp. 221–248, 1994.

18. C.A. Coello, A comprehensive survey of evolutionary-based multi-objective optimization techniques, *Knowledge and Info. Syst. — An Int. J.*, vol. 1, no. 3, Aug. 1999, pp. 269–308.

19. D. Cvetkovic and I.C. Parmee, Preferences and their application in evolution-ary multi-objective optimization, *IEEE Trans. on Evol. Comput.*, vol. 6, no. 1, Feb. 2002, pp. 42–57.

chapter nineteen

Optical transport networks: A physical layer perspective

M. Yasin Akhtar Raja
University of North Carolina, Charlotte
Mohammad Ilyas
Florida Atlantic University

Contents

19.1 Introduction

An ongoing effort is being made to improve the span of performance of optical transport networks that comprise traditional synchronous optical network (SONET)-based and recent high-capacity dense wavelength division multiplexing (DWDM) systems installed to increase the available bandwidth and reduce overall costs. Although extending the geographic reach of optical-transport equipment is an essential part of any cost-cutting strategy, it does not address some of the greatest cost factors in the network operations, such as those associated with optical cross-connects (OXCs)[1] and switching.[2]

In the "traditional" network architecture, with its back-to-back DWDM terminals and an OXC in between, one relies on the OXC to:

1. Speed up service delivery by routing the wavelengths involved in both dynamic and semi-permanent connections
2. Perform mesh-restoration functions; however, some architectures assign various other tasks to the OXC for which it is not well suited, leading to higher capital/operating costs

For example, when both the dynamic reconfiguration of all-optical trunks or "lightpaths," as well as mesh restoration is required, the OXC has a serious limitation in that it cannot scale readily to thousands of ports. Therefore, it hinders the equipment's ability to provide a broad range of new, dynamically provisioned services across the user base. It also creates "islands" of dynamically configured *lightpaths* with manual patching to connect between them in the optical transport networks landscape. In addition, the continuing growth of data applications in today's Internet-driven economy means that a significant portion of the communications traffic traversing the multi-node networks is express or pass-through traffic, especially at smaller, two-way junctions. The ideal solution is an alternative to the OXC — a more cost-effective way to create and manage the semi-permanent connections to handle this type of traffic. In the following sections, we review some selected aspects of optical transport networks pertinent to the contemporary architecture and future proliferation.

19.2 Introduction to the "photonic layer"

As mentioned in the beginning of Chapter 2, regardless of the type or category of optical networks, a common goal of and an important feature among all domains is high-speed, high-capacity, and error-free efficient transport of communication signals across the physical layer. With this in perspective, "optical transport networks" encompass all geographical areas from long-haul/core through regional/metro-core, and finally to edge/access and all the way to distribution network domains. Indeed, the integration of transport functionality of all subnets enables the implementation of *lightpath* concept[3–5] and constitutes optical/photonic-layer.[5] Various nodes containing optical switches/cross-connects (OCXs)/optical add/drop multiplexers (OADMs)[1,6,7] can set up lightpaths for various wavelengths but little or no quality of service monitoring. Network management, quality of service functions, and intelligence will become reality only in future all-optical networks. Among various enabling technologies for optical transport networks (also known as "photonic transport networks") are the signal generation equipment, various multiplexing technologies, optical amplification, switching and routing, and wavelength management with the capability of establishing lightpaths. This chapter briefly reviews some key enablers (in physical layer) in the optical transport networks (OTN). Obviously, the demand on optical equipment and component performance is the highest in long-haul and metro-core and becomes less stringent in the access networks/local-loops, with a proportional drop in cost as well. After optical transceivers, the multiplexing and de-multiplexing technologies hold the foremost importance with grooming, various amplification approaches, and subsequent switching, routing (i.e., OXCs) and wavelength management schemes, as well as some level of protection for reliability and quality of service (QoS). It is worth noting that as of today, except for limited OXCs nodes, almost all management, routing, and QoS functions are implemented in the electronic domain, leaving the all-optical signal processing as a futuristic technology.

19.3 Multiplexing technologies

In optical networking, there are different system considerations and design trade-offs because of different high-speed and optical implementation characteristics. Medium access control (MAC) and principles for using shared media (e.g., fiber, free space) are important from a practical perspective.[8] The three main approaches pertain to MAC; namely, time–domain medium access (TDoMA), wavelength–domain medium access (WDoMA), and code–domain medium access (CDoMA). The first two are physical entities and utilize two distinct approaches for MAC (i.e., multiple access and multiplexing). Therefore, two sub-categories are used for each (e.g., TDMA and TDM, and WDMA and WDM, respectively).[8] The last one is a logical entity and also uses multiple access and multiplexing for MAC, and OCDMA/OCDM is becoming a quite attractive approach. Both multiple-access and multiplexing have their merits and challenges. In terms of bandwidth usage, multiplexing is much more efficient than multiple-access.

Two major architectural categories of optical networks with respect to the multiplexing techniques/schemes are used. The first pertains to the time–domain multiplexing (TDM), which is further subdivided into electronic time–domain multiplexing (ETDM) and optical time–domain multiplexing (OTDM).[9] The second major category relates to wavelength–domain multiplexing (WDM)[10] where several wavelengths, each containing different ETDM data stream, are added to the same fiber strand. As mentioned previously, a third type of multiplexing, known as "optical code division multiplexing" (OCDM), has recently emerged.[11] The legacy optical networks using SONET/SDH[12,13] interfaces utilize ETDM (or simply known as TDM). On other hand, WDM[14,15] adds a third dimension to the time–bandwidth plane of data communication. That is, each single wavelength carries TDM data packets at the highest bit-rate, and multiple wavelengths are multiplexed using DWDM technology,[12,13] making an enormous fiber bandwidth available. A combination of these multiplexing technologies can be used to maximize the fiber bandwidth utilization. A high-bit TDM implemented together with DWDM in long-haul and metro-core increases the capacity of deployed fiber more than 100-fold, while OCDM,[11] an emergent technique, will further enhance the bandwidths of OTN.

19.3.1 ETDM (SONET/SDH)

The first-generation optical networks used electronic TDM and started with the definition of SONET/SDH interface where low-bit data streams are multiplexed into high-bit rate data streams using hierarchical TDM in an electronic domain. SONET/SDH employ quite sophisticated schemes,[16] which can be implemented using the contemporary very large-scale integration (VLSI) circuits. A low bit-rate data stream from asynchronous digital signals is mapped into the synchronous payload envelope (SPE) or in a "synchronous container" (SC) in SDH scheme. A set of overhead bytes is added to SPE/SC as a "path-overhead," and it remains unchanged until SPE/SC reaches its destination. The SPE, along with its path-overhead, defines a virtual tributary (VT or VC). The legacy optical networks utilize SONET/SDH interfaces that aggregate traffic from OC-3 (155Mb/s), OC-12 (622Mb/s), and OC-48 (2.5Gb/s) into OC-192 (10Gb/s) and into the emerging high bit-rate OC-768 that accommodates $4 \times$ OC-192 or $16 \times$ OC-48 signals. An interested reader should consult the references[12,17] to obtain details of SONET frame structure, as well as the definition involved in SPE and various overhead bits for line, path, and restoration and monitoring functions. Table 19.1 presents the equivalence of SONET and SDH and their capacity. There is no equivalence of OC-1 or SONET with a corresponding STM of SDH. Starting with OC-3 the equivalence of SDH standard STM-1 is established and a hierarchy uses quadruple multiplexing. Figure 19.1 summarizes the frame structure, while Table 19.2 lists the SONET overhead bytes. A smallest SONET frame known as *STS-1* is defined as a matrix with 9 (row) ∞ 90 (column) bytes. First, three columns (all rows) are reserved for "transport overhead," while the remaining 87 columns constitute SPE. In

Table 19.1 Equivalence of SONET and SDH and Corresponding Capacity

SONET	SDH	Capacity
OC-1	N/A	51.84 Mb/s
OC-3	STM-1	155.52 Mb/s
OC-12	STM-4	622.08 Mb/s
OC-48	STM-16	2.48832 Gb/s
OC-192	STM-64	9.953 Gb/s
OC-768	STM-256	39.812 Gb/s

Table 19.2 Listing of SONET "Overhead Bytes"

	Transport Overhead			Path Overhead
	Framing	Framing	Section Trace	Trace
	A1	A2	Jo/Zo	J1
Section Overhead	BIP-8	Orderwire	User Channel	BIP-8
	B1	E1	F1	B3
	Section Data	Communication	Channel	Signal Label
	D1	D2	D3	C2
	Pointer	Pointer	Pointer	Path Status
	H1	H2	H3	G1
	BP-8	APS	APS	User
	B2	K1	k1	F2
Line Overhead	Line Data	Communication	Channel	Indicator
	D4	D5	D6	H4
	D7	D8	D9	Growth
				Z3
	D10	D11	D12	Growth
				Z4
	Synch Status	REI-L	Orderwire	Tandem
	S1/Z1	Mo or M1/Z2	E2	Connection Z5

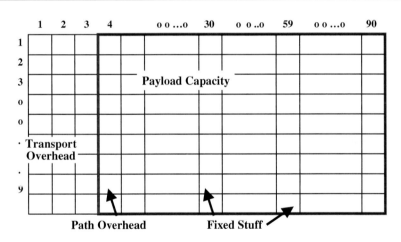

Figure 19.1 Structure of basic SONET frame, a 9 ×90 matrix with 84 data columns (i.e., 756 bytes for payload [48.384 Mb/s]).

SPE, the fourth column is used as the "path overhead" and two more columns, no. 30 and no. 59, are reserved for "fixed-stuff." The remaining 84 columns are all used for payload.

19.3.2 OTDM

Optical time–domain multiplexing (OTDM)[9,11] is functionally identical to ETDM. A major difference, however, is that the multiplexing and demultiplexing operations are performed completely in the optical domain at

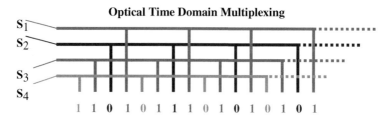

Figure 19.2 Concept of optical time division multiplexing (OTDM). It can be used in two modes: (a) bit-interleaving (as shown in the figure) and (b) optical packet-interleaving, where data-packets are substituted instead of bits in the diagram.

high OC-n level. A typical aggregate rate would exceed 100Gb/s and have the potential to reach 250Gb/s.[9] This in turn would require extremely short optical pulses in the order of few picoseconds, and such pulses have to be generated with modelocked lasers. From Fourier transform relations, it is well known that the narrower the pulse in time-domain, the greater the frequency content; thus such pulses would suffer a severe chromatic dispersion. Therefore, OTDM would require soliton pulses or would be feasible only with some special dispersion compensation provisioning. Two methods are used for OTDM: one involves bit interleaving and the second uses packet interleaving.[18] To date, OTDM remains a more futuristic technology because it has potentials and challenges for implementing extremely high bit rates. High-capacity transport reaching the Tb/s (terabits per second) is now a reality because of DWDM and ultra-DWDM (UDWDM); however, OTDM will become an enabling technology as the fiber permeates from metro-core to access space and ultimately reaches the desk in the office/home. The signal aggregation and de-aggregation would be handled in the optical domain at wavelength level, thus enabling the next generation of all-optical networks. Figure 19.2 illustrates the basic idea of OTDM technique.

19.3.3 Wavelength division multiplexing (WDM)

Point-to-point optical transport using WDM started in the early 1990s and forms the core of data-centric networks. The optical backbone networks really came as a cost-effective solution for bandwidth intensive Internet applications, given the phenomenal rate of growth in Internet traffic (i.e., more than 200-fold over the last decade). The WDM technology has evolved from the wideband-WDM (1310 and 1550 nm) and coarse-WDM, commonly known in literature as CDWM,[11] to a much denser channel spacing, adding hundreds of channels just 30 nm in wavelength (e.g., conventional band [C-band]). The advent of dense-WDM (DWDM) in 1996 caused an explosion in the bandwidth. Today, the DWDM is ubiquitous in long-haul and metro-core (data-centric) optical networks. From point-to-point open DWDM[14] to DWDM rings and mesh topologies, it

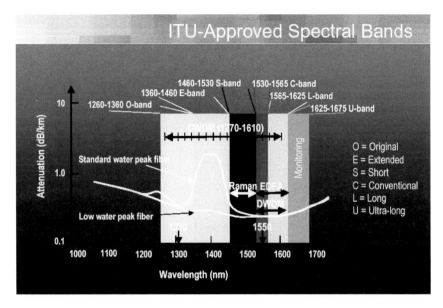

Figure 19.3 ITU-T-approved band assignment in the low attenuation window of the silica fibers; the wavelength range involves 1260–1360 nm = O-band; 1360–1460 nm = E-band; 1460–1530 nm = S-band; 1530–1565 nm = C-band; 1565–1625 nm = L-band; and 1625–1675 nm = U-band (used in monitoring). (Courtesy EXFO Electro-Optical Engineering Inc. Printed with permission.)

now spans the backbone/long-haul and metro core/edge domains. Various static DWDM rings now belong to legacy networks and dynamic routing/switching technologies are now permeating from long-haul to metro-core with a projected growth into metro-edge and optical access networks. In fact, DWDM networks deserve a detailed treatment with their ever-expanding growth in various geographical topologies and the addition of new wavelength bands around C-band (i.e., long- and short-wavelength bands [L- and S-band], respectively). Due to space limitation, however, we shall only review the salient features. Figure 19.3[19] illustrates the low attenuation band in the silica fibers and the band assignment as accepted in the literature (e.g., ITU-T Standards[20]). The hump (water absorption peak) in the old fiber has now been removed in the contemporary fiber using advanced fabrication technology. Table 19.3 gives the DWDM channel assignments for C-band with frequency spacing and central wavelengths as approved by ITU-T.[21]

In the late 1990s, various test beds were deployed for WDM-based optical networks, and one such well-known success was the multi-wavelength optical network (MONET).[4] A later report for further enhancements making MONET fully optical network can be found in the references.[22]

DWDM is slowly but steadily expanding in all geographic domains from network cores toward the end user, driven by inexhaustible demands for bandwidth. As mentioned earlier, DWDM has permeated fully long-haul

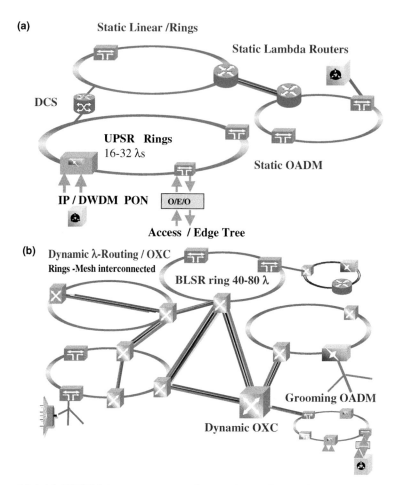

(a) Static Linear /Rings

Static Lambda Routers

DCS

UPSR Rings
16-32 λs

Static OADM

IP / DWDM PON   O/E/O

Access / Edge Tree

(b) Dynamic λ-Routing / OXC
Rings -Mesh interconnected

BLSR ring 40-80 λ

Grooming OADM

Dynamic OXC

Figure 19.4 (a) DWDM transport networks taxonomy from edge through core-networks. Second-generation, high-capacity (hundreds of Gb/s to Tb/s) DWDM transport networks with static/fixed OADMs and DCS. (b) Technology evolution continues from fixed to dynamic OADMs and OXCs. Third-generation "hundreds of Tb/s" DWDM dynamic λ-routing/OXC transport networks with optical access and tributaries.

(national/regional backbones) and is now migrating toward the metropolitan optical networks — in limited cases, even in the access networks infrastructure. Its role at the network core is quite straightforward: to provide as much capacity as possible by supporting multiple wavelengths on a single fiber. Improving the technology means increasing the number of wavelengths per fiber, the distance between optical amplifiers, and the capacity per wavelength. As already demonstrated, terabit DWDM transport networks will become commonplace after the year 2005. An almost universal understanding exists among backbone network providers that DWDM is the ultimate solution for raw capacity expansion. In future OTN, the vast majority of services will be carried on a common Internet protocol (IP) or

Table 19.3 Nominal Central Frequencies and Corresponding Wavelengths for C-Band with 100 and 50 GhZ Spacing

Nominal f_{50}(THz)	Nominal f_{100}(THz)	Nominal λ_1 (nm)	Nominal f_{50}(THz)	Nominal f_{100}(THz)	Nominal λ_1 (nm)
196.10	196.10	1528.77	194.05	—	1544.92
196.05	—	1529.16	194.00	194.00	1545.32
196.00	196.00	1529.55	193.95	—	1545.72
195.95	—	1529.94	193.90	193.90	1546.12
195.90	195.90	1530.33	193.85	—	1546.52
195.85	—	1530.72	193.80	193.80	1546.92
195.80	195.80	1531.12	193.75	—	1547.32
195.75	—	1531.51	193.70	193.70	1547.72
195.70	195.70	1531.90	193.65	—	1547.11
195.65	—	1532.29	193.60	193.60	1548.51
195.60	195.60	1532.68	193.55	—	1548.91
195.55	—	1533.07	193.50	193.50	1549.32
195.50	195.50	1533.47	193.45	—	1549.72
195.45	—	1533.86	193.40	193.40	1550.12
195.40	195.40	1534.25	193.35	—	1550.52
195.35	—	1534.64	193.30	193.30	1550.92
195.30	195.30	1535.04	193.25	—	1551.32
195.25	—	1535.43	193.20	193.20	1551.72
195.20	195.20	1535.82	193.15	—	1552.12
195.15	—	1536.22	193.10	193.10	1552.52
195.10	195.10	1536.61	193.05	—	1552.92
195.05	—	1537.00	193.00	193.00	1553.33
195.00	195.00	1537.40	192.95	—	1553.73
194.95	—	1537.79	192.90	192.90	1554.13
194.90	194.90	1538.19	192.85	—	1554.54
194.85	—	1538.58	192.80	192.80	1554.94
194.80	194.80	1538.98	192.75	—	1555.34
194.75	—	1539.37	192.70	192.70	1555.75
194.70	194.70	1539.77	192.65	—	1556 15
194.65	—	1540.16	192.60	192.60	1556.55
194.60	194.60	1540.56	192.55	—	1556.96
194.55	—	1540.95	192.50	192.50	1557.36
194.50	194.50	1541.35	192.45	—	1557.77
194.45	—	1541.75	192.40	192.40	1558.17
194.40	194.40	1542.14	192.35	—	1558.58
194.35	—	1542.54	192.30	192.30	1558.98
194.30	194.30	1542.94	192.25	—	1559.39
194.25	—	1543.33	192.20	192.20	1559.79
194.20	194.20	1543.73	192.15	—	1560.20
194.10	194.10	1544.53	192.10	192.10	1560.61

packet-based infrastructure. The new public network will require highly scalable solutions, with virtually limitless transport bandwidth, to drive data-rich applications. Yet, at the same time, the carriers need to deploy networks that provide lowest cost per bit at a reliability level better than

SONET/SDH based infrastructure.[12,17] DWDM[23] will become ubiquitous with its proliferation to access networks and will form the basis of the future all-optical networks.

19.4 WDM transmission capabilities

Figure 19.5 illustrates the typical loss of single mode dispersion-shifted fiber (DSF) as a function of wavelength, and a variety of optical amplifiers covering the low-loss region. The low-loss region (transmission loss < 0.4dB/km) has a very wide range; from 1250 to 1650nm (400nm). Recently, extension of the optical amplifier bandwidth has been intensively pursued, and wideband optical amplifiers with a bandwidth over 80nm have been reported.[24,25]

When designing wideband DWDM transmission systems, nonlinear effects of four-wave mixing (FWM), self-phase modulation (SPM), cross-phase modulation (XPM), and stimulated Raman scattering (SRS) in fibers must be taken into account. The following approach should address some of the issues:

- Wavelength channels arranged in zero-dispersion region of the fiber
- Wavelength channels arranged on either side of the zero-dispersion region, where they have almost the same group velocities and walk-off between channels is small

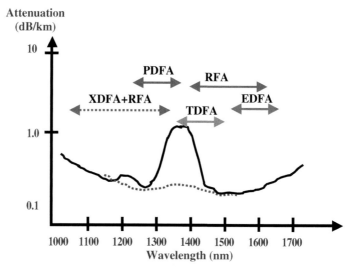

Figure 19.5 Shows the typical loss dB/km of single-mode fiber (silica) as a function of wavelength and a variety of optical amplifiers covering the low-loss region that constitutes the transmission window, which is subdivided into multiple bands. Rare-earth doped fiber amplifiers (XDFA) and Raman fiber amplifiers (RFA) provide gain over the entire 400 nm low attenuation window.

- Avoid the zero-dispersion wavelength region and place all channels to one side of the zero-dispersion wavelength region (not both sides)
- If simultaneous use of both sides of the zero-dispersion wavelength region is needed, employ band-by-band bidirectional transmission.[24] This ensures a very large walk-off between channels and results in phase mismatching in FWM, and averaged XPM and SRS over bits

The ideal carrier-generation system configuration uses simple and inexpensive lasers that independently generate optical channel frequencies with the desired optical frequency accuracy. If optical channel spacing is greatly decreased (e.g., UDWDM), the optical frequencies of the lasers must be carefully monitored. Although existing ideal frequency standards such as an iodine-stabilized He–Ne and a methane stabilized He–Ne are recommended by the International Metrology Committee, it is not easy to create an optical frequency chain from the output of these primary optical frequency standards for the optical frequency utilized by networks as approved by ITU-T. It is well known that atoms and molecules such as krypton, acetylene, hydrogen cyanide, and rubidium exhibit optical frequency resonant states that are very stable, and many of them match the optical channel band of the photonic transport systems.[24]

Although point-to-point WDM systems boost a network's capacity, they also provide the foundation of the optical networks supporting multiple "client overlay networks" and the transportation of various optical services. In support of legacy networks, "SONET/SDH over WDM" and "IP over WDM," as well as "Wavelength (λ) Services" are among the new services that are transported by WDM.

19.5 New multiplexing schemes

A variety of multiplexing schemes are emerging to make the use of the fiber bandwidth more efficient. Two of them are known as "multiband multiplexing" and "subrate multiplexing." In multiband multiplexing, DWDM traffic is carried, for instance, in the C-band (1528 to 1561nm), which is then overlaid onto a 1310nm SONET-based fiber network. A concept of a multi-band system layout is depicted in Figure 19.6.

Selectivity is obtained by using WDM couplers at each optical add/drop multiplexer (OADM)[1] to permit traffic to pass through the nodes where wavelength services are not required. The result is that the traffic "bypasses" the unnecessary "adds/drops" of wavelengths. In the Internet highway analogy, this would mean that communications traffic is routed around "stop-and-go traffic," much like trucks are routed around a city to avoid inner-city congestion.

Subrate multiplexing between the transponder and the customer interfaces make efficient use of wavelength capacity. At each node, this allows users to share a single wavelength that can carry multi-gigabit traffic, e.g., 10Gb/s (OC-192) and 40Gb/s (OC-768), respectively. The subrate-multiplexing features are illustrated in Figure 19.7.

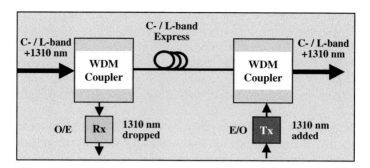

Figure 19.6 Multiband multiplexing carries 1310 nm SONET and C-band WDM traffic on the same fiber.

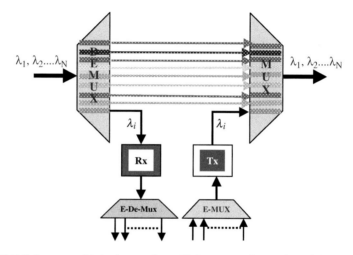

Figure 19.7 Subrate multiplexing makes efficient use of wavelength bandwidth by aggregating/grouping data at OADM nodes.

Although these techniques can help reduce costs somewhat to the service providers, their distance-dependent design considerations put limits on how much cost reduction is attainable. Service providers cannot justify the expenses in using DWDM terminals, OADMs, or state-of-the-art optical amplifiers in metro-edge and access networks where the number of potential customer-density is low, or where the customers are far from the service provider's/carrier's point-of-presence (POP).

19.6 Traffic grooming in optical transport networks

Grooming[26,27] in the telecom industry is a term that is related to an optimization of capacity utilization in the transport networks by means of multiplexing, cross-connections, or conversions between various different transport networks systems or layers of the same system. In various types of multiplexing, this may require synchronization in TDM signals and wavelength conversion in wavelength-routed services. Thus, grooming typically involves equipment for conversion of frequency (wavelength) or time-slots conversion by buffering the low-bit data and then converting it into high-bit rate frames.

From the previous definition, the grooming in optical networks can be analyzed through two aspects,[28] namely, multiplexing and routing.

19.6.1 Multiplexing and bundling

As evident from the foregoing sections, optical networks rely on three types of multiplexing and each can be viewed as partitioning of physical resources (space, time, wavelength) into distinct (separate) sets for independent assignments to users. For example, space-division multiplexing is based on multiple fibers bundled into a single cable, or subsequent use of several cables within a given network link. Simply put, space is divided into distinct fiber/cable channels.

WDM (Section 19.3.3) with varying densities (e.g., DWDM) enables a given fiber strand to carry multiple independent time-domain multiplexed (TDM) traffic on several independent wavelengths. Equivalently, the fiber bandwidth (low attenuation spectrum) is divided into distinct wavelength channels. Finally, TDM divides the bandwidth of a single channel (wavelength) into several time slots of well-defined durations. Using TDM (SONET standards), multiple signals share the same wavelength in nonoverlapping time slots.

19.6.2 Routing and channel assignments

Point-to-point source–destination connections are routed over optical networks via lightpaths, logical circuits from source node to destination node via single or multiple fibers in the networks. In wavelength-routed networks without conversion equipment, lightpaths are established by assigning a distinct channel to the circuit from end to end. This implies that wavelength continuity and distinct channel assignment necessitates that the signal be carried over the same wavelength on all links and domains in the lightpaths of the network. Figure 19.8 illustrates the basic concept. The network management and channel assignment becomes very easy if wavelength conversion equipment is present at the switching and routing nodes. This provides a total transparency to networks that become nonblocking at wavelength level. Therefore, grooming of optical networks involves multiplexing, switching, routing, and wavelength conversion. The grooming devices may include wavelength converters, optical cross-connects, signal-regenerators, E-O-E conversion, etc. These devices exist in various domains of optical networks, from access/edge networks to metro-core/regional and to ultra-long-haul domains. Consequently, grooming is extremely important in OTNs and provides traffic optimization in multiple transport systems (e.g., SONET and DWDM) with

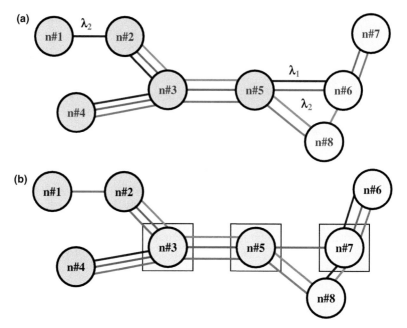

Figure 19.8 (a) Due to wavelength continuity constraints, certain traffic (e.g., λ_1, λ_2) (blue, green) is unroutable through certain nodes (no. 3 and no. 5) because those wavelengths are in use. (b) Wavelength conversion allows transparency throughout the network.

necessary x-connections and multiplexing and bundling issues.[29] Grooming also provides traffic optimization over multiple layers of a single system such as WDM and DWDM.[27,29]

According to Bala,[26] voice traffic today accounts for a little less than 50% of total telecom traffic and is growing at ~5% per year. Therefore, the grooming at STS-1 level granularity and below at the access/edge would continue. Currently, this traffic is groomed from 0.064Mb/s (DS0) to channels 1.54Mb/s (DS1/T1) and to 50Mb/s (STS-1) level using traditional TDM hierarchy. The networks are also becoming more data-centric with growth in the Internet and other bandwidth-intensive services. At the metro-edge a mix of traffic occurs, which includes OC-3 (155Mb/s), OC-12 (622Mb/s), and OC-48 (2.5Gb/s). Hence, traffic grooming becomes necessary for aggregation, synchronization, and multiplexing/demultiplexing the traffic to OC-192 (10Gb/s) for transport through the metro-core/regional optical networks.

As noted previously, the dominant traffic in today's network is data IP services that have maintained a growth rate of ~100% per year[26] over the past decade. Despite the current gloomy economy, an annual growth of > 30% would be more realistic in the upcoming decade. The data traffic is groomed at packet level by IP routing devices into OC-48 and OC-192 trunks. Such aggregated traffic is then transported to a single IP peer router at the destination location in the network. Any optical switches/routers that come across the transmission path can switch only the entire wavelength (OC-48 or OC-192) over the "lightpath."

OTNs are undergoing a revolution from access to metro and all the way to core/long-haul domains. The metro-core and long-haul networks are rapidly evolving from carrying tens of OC-48 wavelengths per fiber to carrying more than hundreds of wavelengths at OC-192 rate per fiber by using DWDM technology. Such a revolution drives the metro-core and long-haul optical transport networks grooming at STS-12 and OC-48 granularity. As the TDM hierarchy has grown to 40Gb/s (STS-768), the networks will migrate to OC-768. Grooming at STS-48 granularity, this would comprise of 16 blocks per 40Gb/s (STS-768) as compared to 768 blocks at STS-1 granularity. Therefore, the complexity of grooming at STS-1 level is 48 times more than that of grooming and managing at STS-48 level. With DWDM and OC-768 as well as permeation of OC-192 and OC-48 into metro-edge/access networks, grooming and traffic management present a greater challenge. The revolutionary trends in fiber-bandwidth utilization now require an expansion of logical networks' topologies from prevalent optical rings to optical mesh for scalability and manageability.

Optical mesh has tremendous advantages for fast provision and restoration at the core, and it outperforms the traditional ring networks topology. Ultimately switching routing at the wavelength level would be needed not only in the metro-core but also at metro-edge and in future access optical networks.

19.7 Switching and routing in the OTN

The success of present and future optical transport networks hinges on the efficient optical signal switching and routing.[2] Without reliable and efficient switching (electrical/optical) optical transport networks simply cannot function. With respect to switching and routing, OTNs are grouped into two major categories:

1. *Opaque/Translucent Networks* — The routing and switching is performed electronically, thereby requiring signal regeneration, which in turn involves multiplexing/de-multiplexing and amplification before the signal is launched to the next node.
2. *Transparent Network* — This does not need signal regeneration, and all optical switching, routing, and amplification is implemented and transport is achieved between the nodes as if the nodes were transparent. Signaling protocols for the establishment of connections in optical networks — setting up the lightpaths using specific control messages (dynamic path).

At the switching and routing nodes, several other functions, such as grooming and dispersion management, are also performed. Figures 19.9a and 19.9b illustrate "opaque/translucent" and "transparent" optical switching architec-

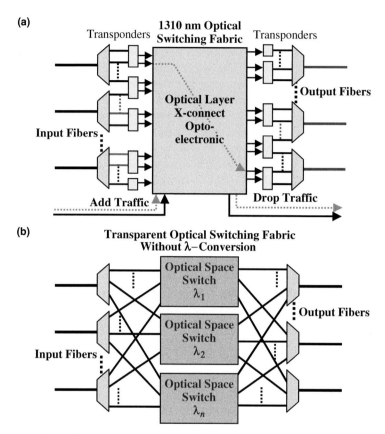

Figure 19.9 Optical cross-connects for optical transport networks: (a) opaque OXC schematic; (b) transparent OXC without wavelength conversion.

tures, respectively. Studies also indicate that at nodes where 10Gb/s express pipes bypass switching and grooming elements, one can achieve a savings of over 50% from an all-optical bypass within an OADM, compared with that of traditional architecture. This is because the OADM within the transparent network architecture can drop selected wavelengths; its all-optical switching element is remotely configurable. The traditional architecture uses a full-function OXC to cross-connect wavelengths[6] that pass through or drop at a site. The savings stem from the cost difference between the OADM's switching matrix and the full-function OXC. Indeed, the switching and routing defines several categories of optical networks (e.g., opaque, transparent, translucent, intelligent routing, fixed/dynamic, and wavelength/packet-routed networks, etc.), but due to space limitation we will omit any further discussion in this chapter.

19.8 Optical amplifiers — an enabling technology

The advent of optical amplifiers[25,30,31] not only extended the reach of TDM (SONET/SDH)-based optical signals on a specific single wavelength, but also made WDM technology possible. Today, four main types of optical amplifiers provide optical amplification functions in different segments of the optical transport networks. Erbium-doped fiber amplifiers (EDFAs) are used in metro-core and long-haul transport networks. Typical gains of ~25 to 30dB can be achieved with commonly available semiconductor pump lasers. EDFA gain is independent of signal polarization, but gain-flattening filters are required to keep DWDM signals at equal strength.[30,31]

Erbium-doped waveguide amplifiers (EDWA) constitute a new class that is also based on the EDFA principles, but use integrated optical circuitry in order to create a much smaller footprint. The difference between the two is that EDFAs use several meters of erbium-doped fiber, whereas EDWAs consist of only a few centimeters-long waveguides. The other components are pretty much the same for both; however, the EDWA integrates its optical components into a compact module. A tradeoff in manufacturing EDWAs is between the erbium-doping densities used vs. the lower efficiency of the Er-ion as it becomes more concentrated. The low-gain but cost-effective EDWAs may be placed only at the points in the metro- and access-networks requiring relatively lower gain. This would reduce the overall cost of the network's infrastructure.

A third class of optical amplifiers is known as "the Raman amplifiers," which allow a continuous low-gain amplification in the entire transport fiber of the network. The success of the amplification depends on the power of the pump and the ability to separate the signal from the pump. The Raman pump is used in counter-propagation to the signal.

A fourth class, and the oldest optical amplifier technology, is based on semiconductor optical amplifiers (SOAs). These are very compact and lend themselves readily to be integrated as optical circuits. SOAs have an inherently high gain that is polarization-dependent in general. Such amplifiers are very suitable for compensating the optical losses at the switching and multiplexing locations. SOAs also have applications in wavelength converters at the switching and routing nodes.

A variation of EDFA or Raman amplifiers allow a "distributed amplification" if necessary when a limited amount of gain is to be incorporated at various points in the SONET/SDH or WDM rings. Though not commercially available today, distributed amplification is promising in meeting the requirements for service providers who desire to extend fiber rings and trees at a low cost. Although the EDFA provided high-gain optical amplification for the long-haul networks, the new class of low-gain amplifiers (i.e., EDWA) is emerging specifically for the metro-edge and access networks. Such amplifiers must have lower costs, take up a small portion of real estate, use existing fiber, and have flexibility for the evolving metro network market. Such amplifiers will deliver only the amount of gain needed at a specific point in the metro network.

A desired cost reduction would be achieved by developing new optical components based on the relaxed requirements. Service providers can therefore gain a competitive advantage by using a new class of amplifiers to deliver bandwidth at a low cost. Such amplifiers would combine optical components such as circulators, combiners/splitters, wavelength-converters, and dispersion-compensating filters to create agile OADMs and DWDM terminals.

19.9 Five pillars of OTNs

From various test beds for OTNs such as MONET,[22] the demand of new flexible services, and the QoS guideline from standard bodies have emerged five major points that have been recognized as "five pillars" of WDM networks. These are massive aggregation, reconfigurability, transparency, scalability, and provisioning and interoperability.

1. *Massive aggregation* — This is the ability to enhance and utilize the available bandwidth of the fiber. It requires various multiplexing technologies that enable the simultaneous transmission of various traffic type and bit-rates.
2. *Reconfigurability* — A mechanism must be fully built in the network infrastructure so that when an optical signal/data enters on a particular wavelength in the *lightpath,* then that signal/wavelength could be routed through the network anywhere and to any drop point and any time at a node irrespective of data rate or format.
3. *Transparency* — The technology can support SONET/SDH and "native" data signal formats such as asynchronous transfer mode (ATM), gigabit ethernet, enterprise systems connection (ESCON), and fiber channel. Also wavelength-routed transport networks (e.g., DWDM) require static/reconfigurable optical cross-connects and wavelength converters at switching and routing nodes.
4. *Scalability* — DWDM can quickly accommodate demand for capacity on point-to-point links, and more importantly on individual spans of existing SONET/SDH rings.
5. *Dynamic provisioning and interoperability* — DWDM offers the ability to provide high-bandwidth transport and services such as virtual-private-network and broadband private line in days instead of months. Its prerequisite, of course, is the interoperability across all domains of optical transport networks. The protocol independence and wavelength transparency are not just highly desired features but are essential requirements of future OTN.[32]

Simply put, "optical networking" is first the demarcation of equipment protection, switching, and facility restoration as well as the implementation of facility-based restoration completely and solely in the "photonic layer" of the OTN. It should offer standardized interfaces to the other network layers to ensure compatibility, interoperability, performance monitoring, fault detection, and subsequent recovery. In addition, optical-networking applications should include dynamic reconfiguration, facility-based (not equipment-based) protection of capacity ratio optimization, and centralized restoration. From an architectural perspective, the most valuable first step is to implement optical self-healing rings. An optical ring, compared to architectures used in current SONET/SDH ring configurations, provides several advantages, including the ability to support survivable multiple wavelengths without 1+1 high-speed equipment protection. Facility restoration is performed at the photonic layer, so equipment protection can be implemented without any single point of failure.

According to prevailing trends in the industry, there is a dramatic shift in thinking from point-to-point solutions that relieve fiber exhaustion to seriously considering how such an all-optical network will be built. Optical-networking components that already reside in the network include DWDM, optical amplifiers, and wavelength converters. To increase momentum, the more challenging elements of optical cross-connects and particularly all-optical switching must mature. Although DWDM is one of the optical-networking pillars, there is a great deal of functionality that must be built into the network. Optical switching/routing are key aspects, as well as survivability and the ability to provision services and manage the OTN. An interested reader should refer to recent ITU-T's G-series recommendations G.870–G.879, where OTN architecture and standards have been specified.[33]

19.10 Future photonic transport layer and concluding remarks

The current and future growth in Internet traffic demands a paradigm shift from a voice network to data-centric networks. The first and most important requirement of future optical transport networks is a large bandwidth transport capability with a high degree of transparency. In addition, not only is the traffic volume of data communication increasing, but also requirements for new service attributes are becoming more tangible. We utilize Internet services every day, such as e-mail and file transfers, etc. These services are relatively tolerant of delay and loss. In the future, however, services that require intensive bandwidth and much smaller transfer delay will become prevalent. The desire for quick content transmission, say 1 megabyte/mouse click, and for cost-effective transmission of MPEG-2-level stream data (5 to 10Mb/s) are two examples. The increasing heterogeneity with regard to traffic conditions, a level of QoS requirements, mobility requirements, and protocols needs to be readily accommodated.[33]

High reliability will also be indispensable for most services, since the multimedia network will be the basis for the information society. Optical network technologies (i.e., photonic layer) will provide the solution.[5] The optical/photonic layer[31] enables quantum leaps in both transmission capacity and transport node throughput simultaneously by exploiting DWDM and OXC and wavelength routing. It will also offer flexibility to the client electrical signals (grooming), which will be effective in accommodating different signal formats and protocols.[33] Furthermore, DWDM transmission systems with a capacity of 40Gb/s or more will be introduced in the near future, particularly in North America. The transparency of the network is critical, since new protocol and transport formats will be developed incessantly, leading to new interface structures. The increased heterogeneity mentioned previously must be accommodated in all-optical (photonics) networks.

The core OTN[33] should provide abundant transmission capacity with large-throughput optical nodes with high transparency. Electrical processing, if used, is performed only in the edge/access networks; the nodes or node systems in the edge network will be connected via the core network with direct lightpaths. In existing ATM networks, for example, cell-by-cell routing needs to be done at every transport node, as mentioned earlier, which increases delay and limits transport node throughput due to the processing burden. In photonic networks, cell-by-cell or packet-by-packet electrical processing will not be performed in transit nodes (optical cross-connect nodes), which minimize transfer delay. The removal of electrical processing and the introduction of wavelength routing will yield a core transport with vast throughputs and future expandability.

References

1. P. Jaggi, H. Onaka, and S. Kuroyanagi, Optical ADM and cross-connects: recent technical advancements and future architectures, *National Fiber Optic Engineers Conf. (NFOEC) 1998,* Orlando, FL, Sept. 1998.
2. Special feature issue on optical switching, *IEEE Commun. Mag.,* vol. 40, no. 3, March 2002, pp.72–101.
3. B. Mukherjee, *Optical Communication Networks,* McGraw-Hill, New York, 1997.
4. R. Ramaswami and K.N. Sivarajan, *Optical Networks: A Practical Perspective,* Morgan Kaufmann Publishers, San Francisco, 1998.
5. K. Sato, Introduction strategy of photonic network technologies to create bandwidth abundant multimedia networks, *Trends in Optics and Photonics,* vol. 20, June 1998.
6. Y.K. Chen and C.C. Lee, Fiber Bragg grating-based large non-blocking multiwavelength cross-connects, *J. Lightwave Technol.,* vol. 16, no. 10, 1998, pp. 1746–1756.
7. J.E. Ford, V.A. Aksyuk, D.J. Bishop, and J.A. Walker, Wavelength add/drop switching using tilting micro-mirrors, *J. Lightwave Technol.,* vol. 17, no. 5, May 1999, pp. 904–911.
8. M.M-K. Liu, *Principal and Applications of Optical Communications,* Irwin, Inc., Boston, 1996.
9. U. Feiste et al., 160Gb/s transmission over 116 km field-installed fiber using 160Gb/s OTDM and 40Gb/s ETDM — changing at speed of light, *OSA Technical Digest, OFC 2001,* March 22, 2001, pp. ThF3–1–3, and reference therein.
10. B. Mikkelson et al., High spectral efficiency (0.53 bit/s/Hz) WDM transmission of 160Gb/s per wavelength over 400 km of fiber — changing at speed of light, *OSA Technical Digest, OFC' 2001,* March 22, 2001, pp. ThF2–1–3.
11. Technical Digest, *Optical Fiber Communication OFC 2002,* Anaheim CA, March 17–22, 2002, see for example, *ThEE sessions,* pp. 600–613.
12. S.V. Kartalopoulos, *Understanding SONET/SDH and ATM, Communications Networks for the Next Millennium,* IEEE Press, Piscataway, NJ, 2002.

13. U. Black, *Optical Networks: Third-Generation Transport Systems*, Prentice Hall, Upper Saddle River, NJ, 2002.

14. K. Bala, Multi-wavelength optical networks (WDM), in *Handbook of Emerging Communications Technologies: The Next Decade,* Saba Zamir, Ed., CRC Press LLC, Boca Raton, FL 2000.

15. J.M.H. Elmirghani and H.T. Mouftah, Technologies and architectures for scalable dense WDM networks, *IEEE Commun. Mag.,* vol. 38, no. 2, Feb. 2000, pp. 58–66.

16. A. Shivji, SONET gear for the optical generation, *National Fiber Optic Engineers Conf. (NFOEC) 2000,* Denver, CO, August 2000.

17. M. Chow, *Understanding SONET/SDH, Standards and Applications,* Andan Publisher, Holmdel, NJ, 1995.

18. J.E. Midwater, Ed., *Photonics in Switching: Vol. II, Systems,* Academic Press, San Diego, CA, 1993.

19. © EXFO Electro-optical Engineering, Inc. Printed with permission.

20. ITU-T Draft Recommendation. Extension of the nominal central frequencies grid included in Annex A of the present G92 draft recommendation for future version of G.692; http://www.ITU.Int/ITUDOC/ITU-T/COM15/DCON-TR/DC-OCT98/306.html.

21. 21.© EXFO Electro-optical Engineering, Inc. (table extracted from the poster). Printed with permission.

22. R. Pease, Breakthrough technologies to make MONET fully optical networks, *Lightwave,* Sept. 1999, pp 44–46.

23. S. Clavenna, Access networks: DWDM's newly charted territory is metropolitan access, *Lightwave Mag.,* Special Reports, Sept. 1999, pp. 49–59.

24. H. Yoshimura, K. Sato, and N. Takachio, Future photonic transport networks based on WDM technologies, *IEEE Commun. Mag.,* vol. 37, no. 2, Feb. 1999, pp. 74–81.

25. A.M. Samara, M.Y.A. Raja, and S.K. Arabasi, High-gain short-length phosphate glass Er-Yb-doped fiber amplifier, *SPIE Symp. on Optical Devices, Components and Systems,* OPTO South East, Clemson, SC, Oct. 4–5, 2001.

26. K. Bala, Grooming in the optical network, *SPIE Optical Network Mag.,* vol. 3, no. 1, Jan/Feb 2002.

27. T. Cox, and J. Sanchez, Cost savings from optimized packing and grooming of optical circuits: mesh vs. ring comparisons, *Optical Network Mag.,* vol. 2, no. 3, pp. 72–90, 2001.

28. R. Barr and R.A. Patterson, Grooming telecommunications networks, *Optical Networks Mag.,* vol. 3, no. 2, May/June 2001, pp. 20–23.

29. A. Lardies, R. Gupta, and R. Patterson, Traffic grooming in a multi-layer network, *Optical Network Mag.,* vol. 2, no. 3, pp. 91–99, 2001.

30. P.C. Becher, N.A. Olsson, and J.R. Simpson, *Erbium-Doped Fiber Amplifiers: Fundamentals and Technology,* Academic Press, San Diego, CA, 1999.

31. Y. Sun, A.K. Srivastava, J. Zhou, and J.W. Sulhoff, Optical fiber amplifiers for WDM optical networks, *Bell Labs Tech. J.,* vol. 4, no. 1, 1999, pp. 187–206.

32. S. Okamoto, M. Koga, H. Suzuki, and K. Kawai, Robust photonic transport network, implementation with optical cross-connected systems, *IEEE Commun. Mag.,* vol. 38, no. 3, March 2000, pp 94–103.

33. ITU-T Recommendation, optical transport network (OTN), recommendation G870–G.879, e.g., Architecture of optical transport networks, G.872, Nov. 2001.

chapter twenty

Fiber optic sensors

Tariq Iqbal
City of Riviera Beach

Contents

20.1 Introduction — BASIC principal

In its simplest form, a fiber optic sensor consists of a light source, an optical fiber, a sensing element, and a detector as illustrated in the Figure 20.1. The light source may be a broadband, a light-emitting diode, or a laser, depending upon the nature of the sensor. When the sensing element is an essential part of the fiber, the sensor is called an intrinsic sensor. When the fiber is only used to guide light to and from the sensing element the arrangement is known as an extrinsic sensor, as depicted in Figure 20.2a and Figure 20.2b. Physical properties of light such as amplitude, frequency, phase, and polarization passing through the sensing element can be affected by the change

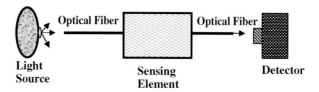

Figure 20.1 Basic arrangement of a fiber optic sensor.

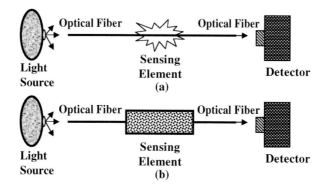

Figure 20.2 (a) Intrinsic sensor: a part of the fiber is being used as a sensing element. (b) Extrinsic sensor: the sensing element is not an integral part of the fiber.

in the environment surrounding the sensing part of the sensor. These changes can easily be detected and are used to design a variety of sensors such as pressure, temperature, chemical, and biological.

20.2 Distributed sensors

Many intrinsic or extrinsic sensors operating at different frequencies can be connected in series or parallel combinations, as illustrated in Figures 20.3a, 20.3b, and 20.4a. A single fiber is used to send a broadband light and to perform wavelength-division multiplexing in a series-distributed fiber optic sensor. Various parameters can also be detected simultaneously by using

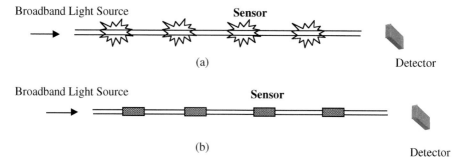

Figure 20.3 Serial distribution of sensors: (a) intrinsic sensors; (b) extrinsic sensors.

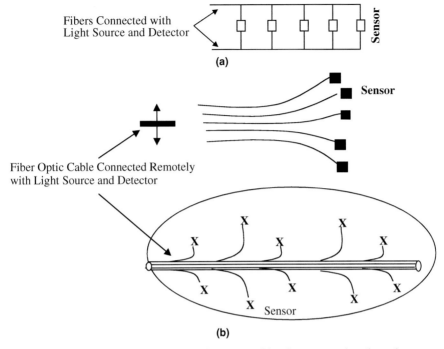

Fibers Connected with Light Source and Detector

Sensor

(a)

Sensor

Fiber Optic Cable Connected Remotely with Light Source and Detector

Sensor

(b)

Figure 20.4 (a) Parallel distribution of sensors. (b) Fiber optic distributed sensors mounted over a large area and connected remotely with a light source and a detector.

this arrangement. Fiber optic sensors connected with various fibers can also be distributed over a large area to remotely monitor the pressure, temperature, humidity, etc., as presented in Figure 20.4b.

20.2.1 Advantages

Fiber optic sensor technology offers significant advantages over other conventional sensors. Some of the attractive features of the fiber optic sensors include their small size, real-time monitoring, fast response, stability, large dynamic range, and remote access. Moreover, light consists of photons, which do not carry any charge and are immune to the electromagnetic fields. Therefore fiber optic sensors are more reliable and safe when deployed in adverse environmental conditions.

20.3 Optical fibers

Every optical fiber optic sensor is designed to work in a specific frequency range and requires optical fibers that can transmit these frequencies without affecting the sensitivity and performance of the sensor. The transparency of the optical fibers can be divided into three ranges of the electromagnetic spectrum; namely, ultraviolet (UV)(250 to 350 nm), visible (400 to 800 nm),

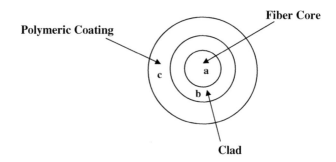

Figure 20.5 A cross-sectional view of an optical fiber: (a) fiber core; (b) clad polymeric coating.

and infrared (IR)(800 to 2500 nm) for sensor applications. Optical fibers made from the silica glass can transmit some UV, visible, and near-IR signals. These fibers are mainly used for communications but have also been used in the development of stress/strain and interferometric sensors.[1-5] In recent years, there has been growing interest in fluoride glass fibers.[6-8] Due to their middle IR range transparency of 2 to 5 microns, these fibers have been used in numerous chemical sensors.[9-12] Chalcogenide fibers can transmit from 5 to 20 microns.

Typical optical fiber consists of core, clad, and protective polymer coating. A cross-sectional view of an optical fiber is presented in Figure 20.5. Fiber core and clad are made of glass. The refractive index of the core is kept slightly higher than that of the clad, so that the light can propagate through the core by total internal reflection. The diameter of the core plays an important role in the design and application of the fiber. Optical fibers with core diameters greater than 10 microns are known as multi-mode fibers, while fibers with core diameters less than 10 microns are known as single mode. The relative refractive index difference between core and clad Δ, the core radius a, and the operating frequency λ, are related with the "V" number value of the fiber by Equation (20.1).[13]

$$V = 2\Pi/\lambda \ an_1(2\Delta)^{j}$$ (20.1)

for single-mode fibers $0 < V < 2.405$, where n_1 is the refractive index of the core.

Figure 20.6 compares the transmission spectra of various glass compositions. Typical components of various glass systems are listed in Table 20.1.

20.4 Sensors

20.4.1 Interferometric sensors

The variation of temperature and pressure around an optical fiber can change the phase of light passing through the fiber. The interferometric sensors are based on the detection of changes in the phase of light emerging mostly out

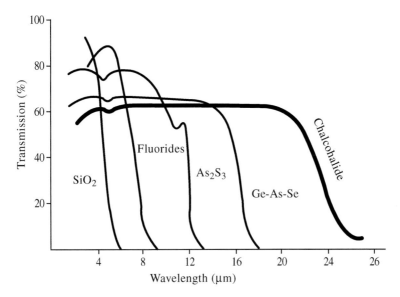

Figure 20.6 Transmission spectra of various glass systems.

Table 20.1 Typical Components of Various Glass Systems

Glass Type	Glass-Forming Systems
Silica glass	Na_2O - Ba_2O_3 -SiO_2
Fluoride glass (ZrF4-based)	$ZrF4_2$-BaF_2-LaF_3-$AlF3_2$-NaF
Fluoride glass (AlF3-based)	AlF_3-BaF_2-CaF_2-YF3-SrF_2-NaF-ZrF_4
Chalcogenide	Ge-As-Se
Chalcohalide	As-S-Cl
Sulphide glass	As_2S_3-La_2S_3

of a single mode fiber. Fabry–Perot[14] (FP) and Mach-Zehnder, Michelson[15] interferometers have been used for the measurement of strain, temperature variations, and health monitoring of composite materials and civil struc-tures.[16] FP sensors are based on multiple beam interference and do not need a reference arm for stabilization as required in the case of Mach-Zehnder, Michelson interferometers.

As illustrated in Figure 20.7, the Fabry–Perot cavity has air gap between the fiber ends, and the sensor output is not affected by the perturbation in the input/output fibers. Moreover, this sensor is classified as an extrinsic sensor. The Fabry–Perot sensor consists of a single mode and multimode fibers with their end face, cleaved perpendicular to the long axis of the fiber. These fibers are held parallel and are fixed at a predetermined dis-tance inside a hollow fiber with an inner diameter slightly greater than the outside diameter of the single and multi-mode fibers. Light is injected from a semiconductor laser diode into the single mode fiber. Input laser light is partially reflected into the fiber at the glass-air interface of the single-mode

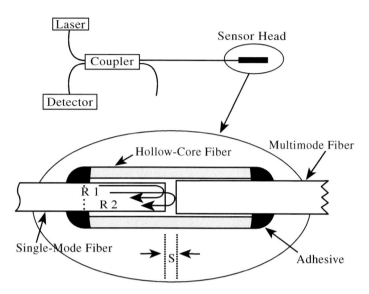

Figure 20.7 Fabry–Perot sensor system with detail of sensor. (From K.A. Murphy et al., *SPIE Proc.*, 1588, 134, 1991. With permission.)

fiber end face, as depicted by R_1 in the figure. The transmitted part of the light travels through the air gap between fibers, is reflected by the glass-air interface of the multi-mode fiber, and then travels back into the input fibers, as depicted by R_2 in Figure 20.7. These two reflected signals interfere in the single mode fiber and travel back toward the optical detector through a coupler.

The output signal intensity at the optical detector is proportional to the sinusoidal function of the phase difference between the fields reflected at the two fiber end faces. This phase difference is also proportional to the relative separation or air gap between the two end faces within the hollow tube. Any external stress/strain applied on the sensor can change the air gap and hence the phase difference and the intensity of the output signal. These changes can easily be detected and used to monitor any perturbation in external stress/strain or the temperature. Extrinsic Fabry–Perot fiber optic sensors have been used in the aerospace industry[14] and in monitoring the health of the civil structure.[17] Figure 20.8 shows the variation of the phase of output light with strain for a Fabry–Perot sensor embedded in graphite/PEEK 9 (gauge length = 6.04 mm).

20.4.2 *Fiber optic Bragg grating sensors*

Fiber optic Bragg grating sensors are produced by exposing a short length of the optical fiber that has germanium-doped core to the UV laser beam.[18,19] The protective coating is removed from that part of the fiber on which gratings are written. Figure 20.9 shows an optical fiber with Bragg grating. The gratings

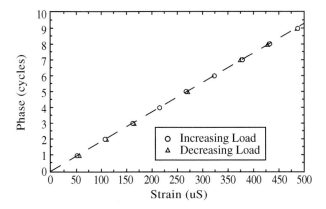

Figure 20.8 The phase-strain response of a typical uniaxial FTP embedded in graphite/PEEK (gauge length = 6.04 mm). (From T. Valis et al., *SPIE Proc.*, 1370, 154, 1990. With permission.)

form a narrow reflection band at the wavelength, which matches the Bragg conditions.

$$\lambda b = 2\,nd \qquad (20.2)$$

where n is the refractive index of the core, and d is the spacing between gratings.

An optical signal with a wavelength that is twice the spacing of the Bragg grating is reflected back. Figure 20.9 shows an experimental setup of producing fiber optic Bragg gratings.

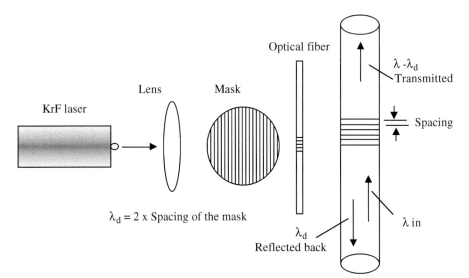

Figure 20.9 Experimental setup for making fiber gratings.

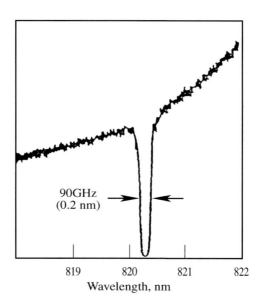

Figure 20.10 Typical fiber grating transmission spectrum. (From W.W. Morey et al., *SPIE Proc.*, 1586, 1, 1991. With permission.)

By changing the angle of the exposing UV beam on the side of the fiber, the grating spacing and hence the Bragg wavelength can be controlled, from the visible spectrum to 1650 nm. Figure 20.10 shows the transmission spectra of an optical fiber with Bragg gratings.[51] When the fiber containing the Bragg grating is subjected to external stress/strain or to temperature variations, the spacing between the gratings and the index of reflection are modified, which in turn changes the wavelength of the reflected light. Bragg gratings have been used in civil structures[20,21] to monitor their health due to temperature changes, stress/strain, and leak detection of petroleum hydrocarbons[22,23] and hydrogen gas.[24-25] Figure 20.11 shows the shift in Bragg grating spectrum when the fiber is under a stress of 32 Kpsi.[19]

Various parts of a single fiber can be exposed to UV light under different conditions so that each portion can reflect a different wavelength as illustrated in Figure 20.12. Therefore, Bragg grating can also be used for wavelength-division multiplex distributed fiber optic sensor for simultaneous detection of any change in stress/strain or temperature.

20.4.3 Microbend sensors

When external pressure is applied on the fiber, a small amount of light is lost from the core into the cladding of the fiber due to the phenomena known as micro-bending. Weakly guided optical fibers with a relatively low value of numerical aperture are used to fabricate the micro-bend sensors. In order to enhance the losses due to the micro-bending, the optical fiber may be passed through a specially designed set of parallel plates having a built-in

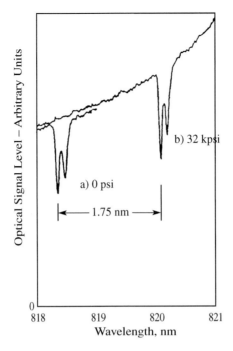

Figure 20.11 Shift in Bragg grating spectrum with tension on the fiber, producing a stress of 32 Kpsi.

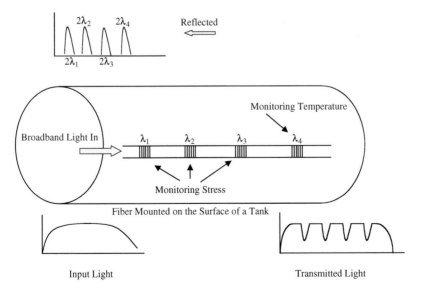

Figure 20.12 Wavelength-division-multiplexed fiber-optic Bragg grating sensor for simultaneous detection of stress and temperature.

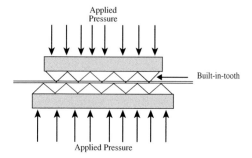

Figure 20.13 Optical fiber placed between two parallel plates with a built-in tooth.

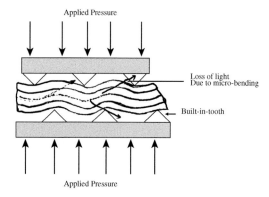

Figure 20.14 Mechanism of light loss due to micro-bending in an optical fiber.

tooth, as depicted in Figure 20.13. The mechanism of light loss by micro-bending is given in Figure 20.14. Micro-bend sensors have been used for the optimization of composite cure process,[26,27] monitoring of structural strain,[28,29] dynamic fluid pressure,[30] and electric current.[31]

20.4.4 Chemical sensors

When light passes through an unclad optical fiber, some modes leak out of the fiber and interact with the environment around it, as illustrated in Figure 20.15. These modes are known as evanescent modes.[32,39] This phenomenon has been used in intrinsic evanescent sensors. Glass rods that are 10 mm in diameter and 100 mm in length with nominal composition; mol%: 30.2 AlF_3-10.2 BaF_2- 20.2 CaF_2- 8.3 YF_3- 13.2 SrF_2- 3.5 MgF_2-3.8 NaF-10.2 ZrF_4 are prepared by using techniques explained in Shahriari et al. and Iqbal et al.[33,34] Optical fibers are drawn from these rods are by fixing them in a drawing tower enclosed by a unique three-stage vertical glove box in which the moisture levels are maintained below 1 part in 10^6.[35] These fibers can transmit up to 4.5 microns.

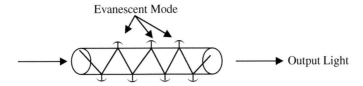

Figure 20.15 Light passing through unclad fibers and the evanescent mode.

Fiber optic chemical sensor technology offers significant advantages for industrial and environmental monitoring.[36] These devices can be employed efficiently for online measurement of chemical changes to control and guide a basic manufacturing process. This eliminates the cost of sample collection, transportation laboratory analysis, handling, and storage.

Remote sensing with optical fibers can be accomplished by:

1. Direct evanescent technique
2. Direct absorption/luminescent
3. Tip-/side-coated fibers
4. Porous glass fibers and sol-gel techniques

The IR region of the electromagnetic spectrum contains the "fingerprint" of molecular species and can be used for their quantitative analysis. Figure 20.16 shows a list of such chemical species and their respective IR signatures in the frequency range between 2 to 11 microns. Conventional silica fibers cannot be used to detect these chemicals because they do not transmit beyond 2 microns. Therefore, IR transmitting halide and chalcogenide glass fibers are used for such applications.

20.4.4.1 Direct evanescent technique

Unclad AlF_3-based glass fibers, which are 250 microns in diameter and have enhanced chemical durability, have been used as intrinsic evanescent IR sensors

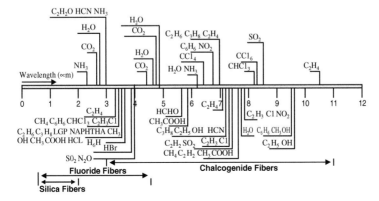

Figure 20.16 IR signatures of food products, fertilizers, drugs, plastics, petrochemicals, and gases, as well as the transmission range of various fibers.

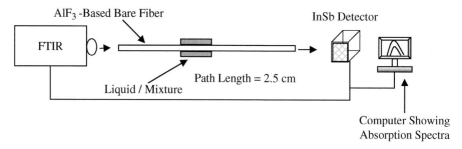

Figure 20.17 Experimental setup for the detection of chemical species by evanescent absorption of fundamental modes by using AIF₃-based fibers.

for monitoring liquid chemicals[9] with absorption band between 1 and 4.5 microns. These fibers are coupled to a Perkin-Elmer FTIR spectrophotometer to remotely monitor the concentration of acetinitrile mixed with alcohol. It is possible to monitor 1 vol% of acetonitrile in alcohol by immersing 2.5 cm of fiber in liquid mixture. The experimental arrangement is presented in Figure 20.17.

Figure 20.18 shows the absorption spectra of different concentrations of actonotrile in alcohol.[9] It can be seen that as the concentration of acetonitrile is decreasing in the mixture, so is the absorption peak around 4.4 microns, which is the characteristic of C ≡ N bond in acetonitrile. It is possible to sense 0.5% of acetonitrile in alcohol using this arrangement. Figure 20.19 shows the calibration curve for various concentrations of acetonitrile in alcohol.

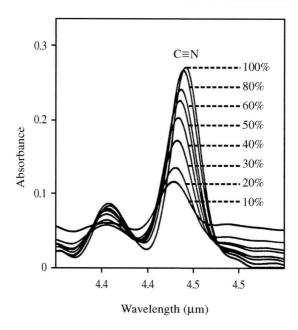

Figure 20.18 Evanescent wave absorption spectra for different concentrations of acetonitrile in alcohol. (From T. Iqbal et al., *Optics Lett.*, 16, 1611, 1991. With permission.)

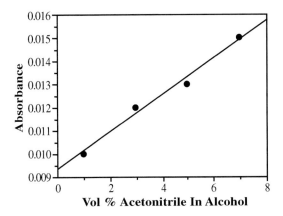

Figure 20.19 Calibration curve for acetonitrile in alcohol. (From T. Iqbal et al., *Optics Lett.*, 16, 1611, 1991. With permission.)

Wavelength division multiplexing can be achieved by immersing various parts of the fiber in different chemicals with distinct and nonoverlapping absorption bands. Figure 20.20 shows an experimental arrangement for simultaneous detection of alcohol and acetonitrile. Figure 20.21 gives the results of wavelength division multiplexing[9] when two segments of the fibers are used to sense two different chemicals simultaneously, such as alcohol and acetonitrile. The main absorption corresponding to C-H in alcohol appears at 3.4 microns, while $C \equiv N$ absorption in acetonitrile appears at 4.4 microns.

20.4.4.2 Direct absorption/luminescent

In this technique, multi-mode optical fibers with core/clad/coating structure are used to guide the light to and from the test chamber containing the analyte. Figures 20.22a and 20.22b show the two different arrangements of direct absorption sensors. These are also examples of the extrinsic sensors. Multi-mode IR-transmitting optical fibers have been used for remote sensing

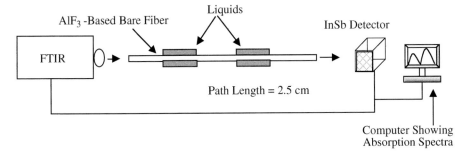

Figure 20.20 Experimental setup for a wavelenth-division-multiplexed distributed fiber optic chemical sensor for simultaneous detection of a variety of chemical species.

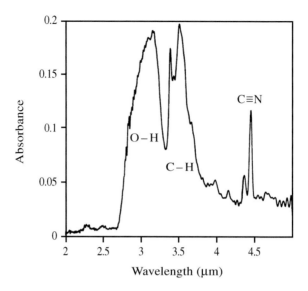

Figure 20.21 Remote absorption spectra of alcohol and acetonitrile taken by a wavelenth-division-multiplexed distributed AIF₃-based fiber-optic chemical sensor. (From T. Iqbal et al., *Optics Lett.*, 16, 1611, 1991. With permission.)

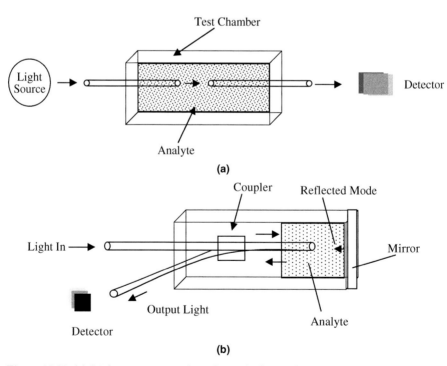

Figure 20.22 (a) Light interacting directly with the analyte via an optical fiber. (b) Light interacting indirectly with the analyte via an optical fiber.

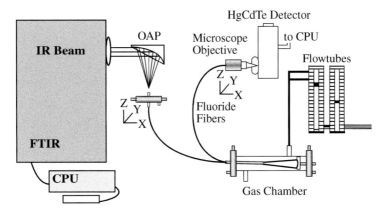

Figure 20.23 Fluoride glass fibers coupled with FTIR for sensing different gases. (From S.J. Saggese et al., *SPIE Proc.*, 1172, 2, 1990. With permission.)

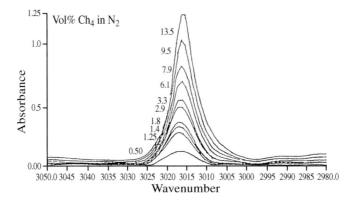

Figure 20.24 Absorption spectra for various concentrations of methane in nitrogen. (From S.J. Saggese et al., *SPIE Proc.*, 1172, 2, 1990. With permission.)

of combustion products mixed with N_2 gas[10] by the direct adsorption technique. A fiber is coupled with a spectrophotometer to transmit light into a glass chamber, and a second fiber is used to transmit reflected light by a concave mirror mounted at the end of the chamber and to a detector, as illustrated in Figure 20.23. It is possible to detect 0.2 to 15 vol% of methane (CH_4), 0.3 to 100 vol% of carbon monooxide (CO), and 0.05 to 2.5 vol% of carbon dioxide (CO_2) in nitrogen gas by using this arrangement. The overall absorption path inside the gas chamber is 10 cm. Figure 20.24 shows the absorption spectra for various concentrations of methane in nitrogen gas, while Figure 20.25 shows the calibration curve.[10]

20.4.4.3 Tip- and side-coated fibers

The sides of an unclad fiber can be coated with indicators so that the evanescent mode of the light passing through the fiber interacts with the coating,

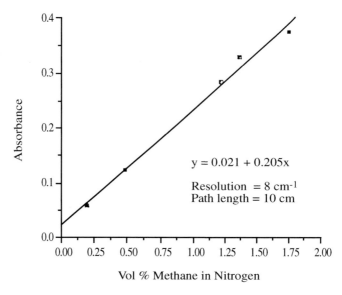

Figure 20.25 Calibration curve for methane in nitrogen. (From S.J. Saggese et al., *SPIE Proc.*, 1172, 2, 1990. With permission.)

and the change in the absorption spectra of output light can be used to detect the changes in the environment around the fiber. The tip of a fiber can also be coated with an indicator for such purposes.

20.4.4.4 Porous glass fibers and sol-gel techniques

Porous glass fiber optic sensors have been used for monitoring PH,[37,39,40] hydrogen,[41,42] HCl gas,[43,44] oxygen,[45,46] carbon monooxide,[47,48] and moisture.[49] Figure 20.26 shows the scheme of producing porous glass fiber optic sensors.

Optical glass fibers are produced from an alkali-borosilicate glass preform of selected composition. These solid glass fibers, about 3 to 5 inches in length, are heat treated in a conventional oven with a defined heat zone, between 550 and 600°C, to induce the phase separation. After the heat treatment two phases, one rich in alkali-borate and other in silica, are formed within the fibers. Now, diluted mineral acid is used to leach the borate-rich phase, leaving behind a silica-rich porous fiber skeleton.[50]

Figure 20.27 shows the cross-sectional view of a porous fiber. As depicted in this fiber, only the outer layer of the phase-separated glass fiber is leached and the inner part of the fiber remains solid. Various optically sensitive chemical indicators can be introduced into the porous section of the fiber. The light passing through the solid core of the sensor interacts with the indicator's molecules trapped inside the core. The sensitivity and response time of the fiber optic sensor can be improved by controlling the depth of the porous layer. Due to the high surface area of the porous fiber the light interaction is increased, which enhances the sensitivity of the device. Monitoring of humidity levels in

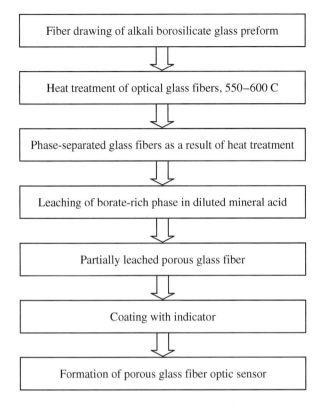

Figure 20.26 Flow chart for the production of porous fiber optic sensors.

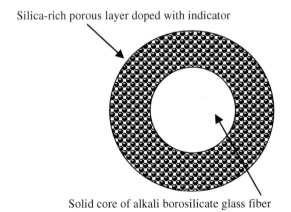

Figure 20.27 Cross section of porous glass fiber.

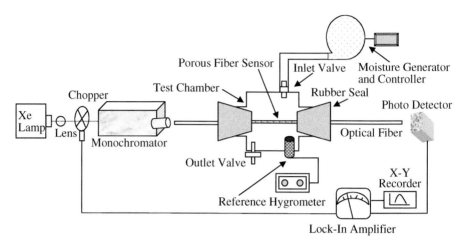

Figure 20.28 Schematic diagram of experimental setup for porous glass fiber optic humidity sensor.

a gaseous environment at elevated temperature is very important in numerous industries such as device fabrication, food processing, and storage. Porous fibers have been used successfully as humidity sensors at room as well as high temperatures. The porous section of the glass fiber with length about 0.5 cm is immersed into an aqueous solution of $CoSO_4.7H_2O$.[49] The sensitivity and the dynamic range of the sensor depends upon the concentration of the solution. Figure 20.28 shows the experimental setup for the porous fiber optic humidity sensor. It has been demonstrated that the transmission spectra of porous fiber immersed in an aqueous solution of $CoSO_4.7H_2O$ changes with humidity around 670 nm. Cobalt ll Sulphate exhibits blue color, but its color changes from pink to red,[49] as it goes under hydration process in the presence of humidity.

Figure 20.29 shows the transmission spectra of porous fiber[49] coated with Cobalt ll Sulphate in the temperature range 50 to 125°C when exposed to 0.05 mole/l moisture concentration. The optical intensity of the output signal decreases between 500 and 700 nm with the increase of temperature. Figure 20.30 shows a calibration curve[49] of the porous fiber optic sensor at 160°C. In this figure, the optical transmission at 670 nm is plotted vs. the moisture concentration from 0.005 to 0.35 mole/l.

The response time of the porous fiber is affected by the pore size, concentration of the indicator, moisture, temperature, and the flow rate of the introduced air. Response time between 1 and 3 minutes has been reported for this sensor. Porous glass fiber immersed in $CoCl_2$ has been used as a humidity sensor at temperatures below 100°C. $CoCl_2$ forms a hydrated salt having 6 molecules of bound water with an octahedral structure and changes its color from pink to blue as it undergoes the hydration process in the presence of humidity. The best performance of porous fiber sensor coated with $CoCl_2$ has been reported around 690 nm.

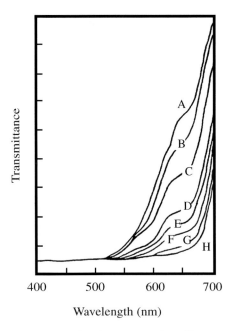

Wavelength (nm)

Figure 20.29 Transmission spectra for the porous glass fiber optic sensor doped with hydrated cobalt sulphate at 0.05 mole/l moisture environment: (A) 50°C, (B) 55°C, (C) 60°C, (D) 70°C, (E) 75°C, (F) 85°C, (G) 105°C, and (H) 125°C. Spectra were measured using the porous glass fiber system. (From J.Y. Ding et al., *Int. J. Optoelect.,* 6, 385, 1991. With permission.)

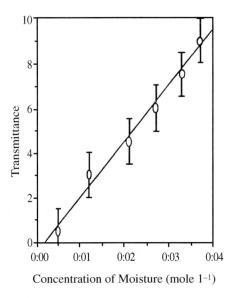

Concentration of Moisture (mole 1^{-1})

Figure 20.30 The calibration curve for the porous glass fiber optic moisture sensor doped with 0.03 b/ml hydrated cobalt (II) sulphate at the temperature of 160°C. (From J.Y. Ding et al., *Int. J. Optoelect.,* 6, 385, 1991. With permission.)

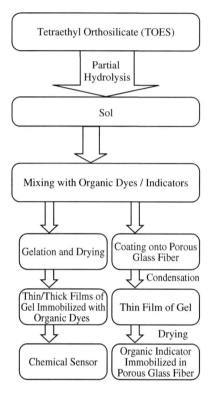

Figure 20.31 Flow chart for the preparation of sol-gel immobilization techniques for porous glass fiber and thin film fiber optic chemical sensors.

Sol-gel immobilization technique has been used for the immobilization of organic photochemical indicators to an inorganic substrate for fiber optic application. It is a low temperature process, in which photochemical indicators are entrapped into the pores without changes in their molecular structures.[52] A complete procedure for the fabrication of a porous fiber optic PH sensor by the sol-gel technique is given in Figure 20.31. Tetraethyl orthosilicate (TEOS) and Methyl tri-methoxy-silane (MTMS) dissolved in alcohol are partially hydrolyzed by water under ambient conditions. The partially hydrolyzed TEOS is then mixed with indicator,[53] fluorecent Ru-(di phenyl phenanthroline), and phenol red. Ethanol is used as a mutual solvent to dissolve all reactants. Partially hydrolyzed TEOS is obtained by the evaporation of alcohol. Coating of the sol solution on the porous glass fiber is performed when the solution becomes viscous due to the condensation of TEOS and MTMS.

Sol-gel and porous fiber techniques have been used for the monitoring of pH and O_2, and dissolved oxygen (DO). The sol penetrates into the fiber via the interconnected porosity, and the light directly interacts with the organic molecules.

During the fabrication of the oxygen sensors, $Ru^{ll}(Ph2Phen)_3^{2+}$ is used as an indicator and is immobilized into the gel,[54] which is then coated onto

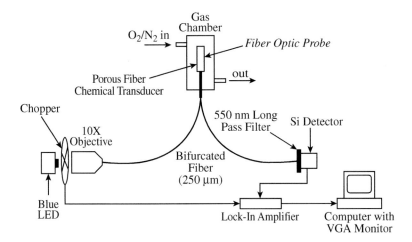

Figure 20.32 A schematic of the oxygen sensor measurement. (From M. Krihak and M.R. Shahriari, *SPIE Proc.*, 2508, 358, 1995. With permission.)

the porous fibers. After the fiber preparation, the oxygen transducer is placed into the sensor system as illustrated in Figure 30.32. The dissolved oxygen measurement is based upon the principle of fluorescence quenching by oxygen molecules.[55,56] The excitation wavelength for Ru(diphenyl phenanthroline) is 470 nm and the fluorescence wavelength is 610 nm. A blue LED is launched into one leg of a bifurcated fiber, which excites the sol-gel coating on the porous fiber. The emitted as well as scattered excitation light is collected through the second leg of the bifurcated fiber, which guides the light into the silicon detector. It is possible to detect very low concentrations of oxygen ranging from 0.5 to 100 ppm, as presented in Figure 30.33.

Table 20.2 lists the various techniques, indicators, and the principal of operation of the fiber optic chemical sensors.

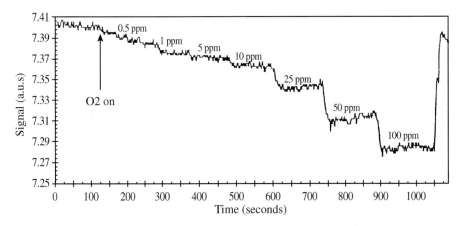

Figure 20.33 PPM level oxygen detection by the sol-gel coated porous fiber. (From M. Krihak and M.R. Shahriari, *SPIE Proc.*, 2508, 358, 1995. With permission.)

Table 20.2 Fiber Optic Chemical Sensors

Sensor	Technique	Indicator	Principle	Wavelength (nm)			Detection limits	Sensitivity	Response Time	Ref.
				Excitation	Emission	Absorption				
Hydrogen sulphide	Porous fiber and sol-gel coating	Thionine	Florescence quenching	580	630	—	> 50 ppb in H2S in water	~5 ppb	~5 S	57
pH	Porous fiber and sol-gel coating	Bromocresol green	Transmission/absorption	—	615	—	3–6	–0.01 pH	~2 S	37
pH	Porous fiber and sol-gel coating	Bromocresol purple	Transmission/absorption	—	—	580	6–9	–0.01 pH	~2 S	37
Hydrogen	Palladium coated fiber	Palladium thin films	Transmission/absorption	—	—	650	0.2–0.6%	—	20–30 S	41
Carbon monoxide	Porous fiber	Organometallic complex in chloroform solution	Transmission/absorption	—	—	450	9–28 vol.% in N_2 gas	–0.5 vol%	~100 S	27
Carbon dioxide dissolved in sea water	Fiber dipped in HPTS solution	Aqueous solution of 8-hydroxy-1,3,6-pyrenetrisulfonic acid-trisodium salt	Fluorescence	450	530	—	0–600 ppm	—	5 min	58
Oxygen dissolved oxygen	Porous fiber and sol-gel coating	Ruthenium complex	Fluorescence quenching	450	610	—	0–10% Oxygen 0.25–9ppm dissolved oxygen	—	—	46
pH	Porous cellulose tricaetate fiber	Congo-red	Transmission	610	—	—	—	—	1–2 min	59

References

1. Wang, A., High temperature strain, temperature and damage sensors for advanced aerospace materials, *1st European Conf. on Smart Structures and Materials,* 127, 1992.
2. Berthold, J.W., Measurement of axial and bending strain in pipelines using Bragg grating sensors, *SPIE Proc.,* 4204, 11, 2001.
3. Kin-Tak, L. et al., Strain monitoring in composite-strengthened concrete structures using optical fiber sensors, *Composites, Part B,* 32B, 33, 2001.
4. Kalinowski, H.J. et al., Multiplexed fiber optic Bragg grating sensors for strain and temperature monitoring, *SPIE Proc.,* 3666, 544, 1999.
5. Tapanes, E., Real-time structural integrity monitoring using a passive quadrature demodulated, localized Michelson optical fiber interferometer capable of simultaneous strain and acoustic emission sensing, *SPIE Proc.,* 1588, 356, 1991.
6. Tran, D.C. et al., Heavy metal fluoride glass fibers, *J. Lightwave Technol.,* 5, 566, 1984.
7. Iqbal, T. et al., New highly stabilized AlF_3-based glasses, *J. Non. Cryst. Solids,* 184, 190, 1995.
8. Iqbal, T. et al., Preliminary study of fiber drawing of AlF_3-based glass fibers, *J. Mat. Res. Eng.,* B12, 299, 1992.
9. Iqbal, T. et al., Remote infrared chemical sensing using highly durable AlF_3-based glass fibers, *Optics Lett.,* 16, 1611, 1991.
10. Saggese, S.J. et al., Evaluation of an FTIR/fluoride optical fiber system for remote sensing of combustion products, *SPIE Proc.,* 1172, 2, 1990.
11. Jung, C.C. et al., Fiber optic chemical sensors, *SPIE Proc.,* 3489, 2, 1998.
12. Wolfbeis, O.S., Fiber-optic chemical sensors and biosensors, *Analytical Chem.,* 72, 12, 81R, 2000.
13. Senior, J.M., *Optical Communications Principles and Practice,* Prentice Hall, 1992.
14. Murphy, K.A. et al., Fabry–Perot fiber optic sensors in full fatigue testing on an F-15 aircraft, *SPIE Proc.,* 1588, 134, 1991
15. Yoshino, T. et.al, Fiber optic Fabry–Perot interfereometer and its sensor applications, *IEEE J. Quantum Electron.,* QE-18, 1624, 1982.
16. Valis, T. et al., Composite material embedded fiber-optic Fabry–Perot strain sensor, *SPIE Proc.,* 1370, 154, 1990.
17. Kruschwitz, B. et al., Optical fiber sensors for the quantitative measurement of strain in concrete structures, *1st European Conf. on Smart Structures and Materials,* 241, 1992.
18. Meltz, G. et al., Formation of Bragg gratings in optical fibers by transverse holographic method, *Optics Lett.,* 14, 823, 1989.
19. Morey, W.W. et al., Fiber optic Bragg grating sensors, *SPIE Proc.,* 1169, 98, 1989.
20. Storoy, H. et al., Fiber optic condition monitoring during a full-scale destructive bridge test, *J. Intelligent Mat. Syst. Structures,* 8, 633, 1997.
21. Maaskant, R. et al., Fiber optic Bragg sensors for bridge monitoring, *Cement and Concerete Composite,* 19, 21, 1997.
22. Spirin, V.V. et al., Fiber Bragg grating sensor for petroleum hydrocarbon leak detection, *Optics and Laser Eng.,* 32, 497, 1999.
23. Shlyagin, M.G. et al., Fiber optic Bragg gratings for liquid hydrocarbon detection, *SPIE Proc.,* 4074, 108, 2000.

24. Sutapun, B. et al., Fiber-optic Bragg grating sensors for hydrogen gas sensing, *SPIE Proc.*, 3740, 278, 1999.

25. Sutapun, B. et al., Pd-coated elastooptic fiber optic Bragg grating sensors for multiplexed hydrogen sensing, *Sensors and Actuators B: Chemical*, 60, 27, 1999.

26. Bocquet, J.C. et al., Optical sensors embedded in composite materials, *SPIE Proc.*, 1588, 210, 1991.

27. Zhang, B. et al., Composite cure process optimization with the aid of fiber optic sensors, *SPIE Proc.*, 4202, 15, 2000.

28. Luo, F. et al., A fiber optic micro bend sensor for distributed sensing applications in the structural strain monitoring, *Sensors and Actuators A: Physical*, 75, 41, 1999.

29. Donlabic, D. and Zavrsnik, M., Fiber-optic microbend sensor structure, *Optics Lett.*, 22, 837, 1997.

30. McColluma, T. and Spector, G.B., Fiber optic microbend sensor for detection of dynamic fluid pressure at gear interfaces, *Review of Scientific Instruments*, 65, 724, 1994.

31. Petrik, S., A microbend fiber optic sensor for electric current, *Elektrotechnicky Casopis*, 42, 103, 1991.

32. Messica, A. et al., Theory of fiber-optic evanescent-wave spectroscopy and sensors, *Applied Optic*, 35, 2274, 1996.

33. Shahriari, M.R. et al., Synthesis and characterization of aluminum fluoride-based glasses and optical fibers, *Mat. Sci. Forum*, 32–33, 99, 1989.

34. Iqbal, T. et al., Synthesis, characterization and potential applications of highly chemical durable glasses based on AlF_3, *J. Mat. Res.*, 6, 401, 1991.

35. Iqbal, T. et al., AlF_3-based glass fibers with enhanced optical transmission, *Electr. Lett.*, 27, 110, 1991.

36. Seitz, W.R., Chemical sensors with fiber optics, *Crit. Rev. Anal. Chem.*, 19, 135, 1988.

37. Ding, J.Y. et al., Porous fiber optical sensor for pH measurement, *Proc. 8th Optical Fiber Sensor Conf (OFS)*, 321, 1992.

38. Goswami, K. et al., Evanescent wave sensor for detecting volatile organic compounds, *SPIE Proc.*, 3540, 115, 1999.

39. Wroblewski, W. et al., Cellulose-based pH optomembranes, *Sensors and Actuators B: Chemical*, 1, 471, 1998.

40. Maher, M.H. and Shahriari, M.R., A fiber optic chemical sensor for the measurement of groundwater pH, *J. Testing Eval.*, 21, 448, 1993

41. Tabib-Azar, M. et al., Highly sensitive hydrogen sensors using palladium coated fiber optics with exposed core and evanescent field interactions., *Sensors and Actuators B: Chemical*, B56, 1, 158, 1999.

42. Jung, C.C. et al., Fiber optic hydrogen sensor, *SPIE Proc.*, 3489, 9, 1998.

43. Supriyatno, H. et al., Optochemical sensor for HCL gas based on tetraalkoxyphenylporphyrin dispersed in an acrylate polymer matrix, *Sensors and Materials*, 13, 359, 2001.

44. Nakagawa, K. et al., HCL gas sensing properties of TPPH$_2$ dispersed in copolymers, *Sensors and Actuators B: Chemical*, 65, 138, 2000.

45. Murtagh, M.T. et al., Development of a highly sensitive fiber optic O2/D) sensor based on a phase modulation technique, *Electr. Lett.*, 32, 477, 1996.

46. Campo, J.C. et al., Measurement of low oxygen concentrations by phosphorescence lifetime using fiber optics, *IMTC/98 Conf. Proc.*, 2, 1162, 1998.

47. Dulebohn, J.I. et al., Reversible carbon monoxide additions to sol-gel derived composite films containing a cationic rhodium (ll) complex: toward the development of a new class of molecule-based CO sensor, *Chem. Mater.*, 4, 506, 1992.

48. Pattison, R. et al., Measurement and control of dissolved carbon dioxide in mammalian cell culture processes using an *in situ* fiber optic chemical sensor, *Biotechnol. Progess*, 16, 769, 2000.

49. Ding, J. et al. Development of high-temperature humidity sensors based on porous optical fibers, *Int. J. Optoelectronics*, 6, 385, 1991.

50. Ding, J. et al., Fiber optic moisture sensors for high temperatures, *Ceramic Bull.*, 70, 1513, 1991.

51. Morey, W.W. et al., Multiplexing fiber Bragg grating sensors, *SPIE Proc.*, 1586, 1991.

52. Hartmeier, W., *Immobilized Biocatalysts*, Spring-Verlag, NY, 22–40, 1986.

53. Krihak, M. and Shahriari, M.R., Highly sensitive, all solid-state fiber optic oxygen sensor based on the sol-gel coating technique, *Electr. Lett.*, 32, 240, 1996.

54. Krihak, M. and Shahriari, M.R., Fiber optic chemical/environmental sensors based on active coatings, *SPIE Proc.*, 2508, 358, 1995.

55. Ding, J. et al., Fiber optic pH sensors prepared by sol-gel immobilization technique, *Electr. Lett.*, 27, 1560, 1991.

56. MacCraith, B.D. et al., Fiber optic oxygen sensor based on fluorescence quenching of evanescent-wave ruthenium complexes in sol-gel derived porous coatings, *Analyst*, 118, 385, 1993.

57. Shahriari, M.R. and Ding, J., Active silica-gel films for hydrogen sulphide optical sensor application, *Optics Lett.*, 19, 1, 1994.

58. Goswami, K. et al., A fiber optic chemical sensor for carbon dioxide dissolved in seawater, *SPIE Proc.*, 1172, 225, 1989.

59. Shahriari, M.R. et al., Development of a porous polymer pH optrode, *Optics Lett.*, 17, 1815, 1992.

chapter twenty-one

Wavelength converters

Mohammed A. Alhaider
King Saud University

Contents

21.1 Introduction

Demands on communications are increasing rapidly worldwide, particularly in multimedia and Internet applications. This has led researchers to explore

0-8493-1333-3/03/$0.00+$1.50
© 2003 by CRC Press LLC

new possibilities of accessing a large fraction of the approximately 50 THz theoretical bandwidth of a single mode fiber.[1]

Dense wavelength division multiplexing (DWDM) and optical amplification were utilized to get transmission capacity in the Terabit/second.[2–6] Systems with up to 128 wavelengths per fiber have been designed and are commercially available, although theoretically more than 1000 channels may be multiplexed in one fiber.[7] The use of wavelengths is not limited to increasing capacity; they can also be used to perform functions such as switching and routing.[8] Different enabling technologies have been employed to address various issues in deploying optical networks, which include routing in the optical domain, interoperability, scalability, and transparency.[9–12] Wavelength converters (WCs) add new dimension in this regard, as they enable conversion of data from one wavelength channel to another. This resolves wavelength contention, increases flexibility and capacity of the network, and enables distribution of control and management into smaller subnetworks.[11]

Wavelength converters have many classifications, which depend on the mapping functions and the form of control signal, signal routing, or signal properties.[11] No matter what classification is chosen, important features that one should look for include:[9,13]

- Bit rate transparency (up to 10Gb/s or more)
- No extinction ratio degradation
- High signal-to-noise ratio (SNR) (to ensure cascadability)
- Moderate input power (~ 0 dBm)
- Large wavelength span for input and output signals (at least for erbium doped fiber amplifier [EDFA], 30 to 34 nm range)
- Possibility of same input and output wavelengths (no conversion) to avail usage in WDM switch blocks
- Low chirp
- Fast setup time of output wavelengths
- Insensitive input signal polarization
- Good conversion efficiency
- Cost issues

It may be difficult to find a wavelength converter that meets all the above features, but the choice will be system dependent.

Because semiconductor optical amplifiers (SOAs) and interferometers are frequently used in wavelength conversion techniques, this chapter begins with a tutorial on SOA in Section 21.2, and one on interferometers in Section 21.3. Section 21.4 presents optoelectronic WCs. Optical gating WCs are presented in Section 21.5 followed by four-wave mixing (FWM) WCs in Section 21.6. The conclusion is presented in Section 21.7. The references listed at the end of the chapter do not represent a full survey of the literature in this dynamic area of research. Instead, they are a guide for further reading, for those interested in this topic.

21.2 Semiconductor optical amplifier (SOA)

SOAs were originally developed to amplify signals in the optical domain of the wavelength band of interest of optical networks; however, their applications are not limited to amplification but are extended to other functions such as regeneration, switching, and wavelength conversion in the optical domain, where the nonlinearity of the SOA is utilized.

SOAs are made of alloys from Group III–V materials such as gallium and indium from Group III elements, and arsenic and phosphorus from Group V elements. At the wavelengths of interest, namely at 1300 and 1550 nm, an alloy of InGaAsP is used in the active region. The III-V compounds have a direct energy gap in which electrons and holes recombine directly across the band gap. The active layer is surrounded by p-type and n-type layers as depicted in Figure 21.1. Oscillation occurs in the active region, which has an energy gap that corresponds to the desired frequency. The operation of the SOA starts with injection of external current in the active region, which causes the population density in the excited states to increase until it exceeds that of the ground state. This condition is referred to as the population inversion. When population inversion occurs, the active region acts as an optical amplification medium. The electron at the excited state can make a downward transition to the ground state through external stimulation. If a photon with proper energy collides with an electron in the excited state, it is stimulated to drop to the ground state, giving off a photon with the same energy of the incident one. The emitted and incident photons are in phase, and the emission is called stimulated emission.

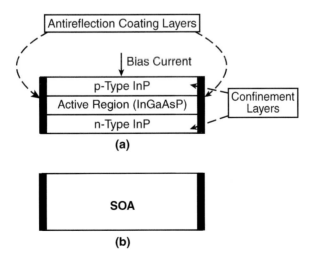

Figure 21.1 (a) Cross-sectional view of SOA showing the active region surrounded by p-type and n-type; (b) symbol of an SOA.

SOA is based on the Fabry–Perot (FP) laser in which light is confined to a cavity with two facets, of appropriate Fresnel reflection (reflectivity), at both ends of the cavity. Pumping of the FP for population inversion is done through external electrical current. To get amplification only out of this amplifier, it is important to prevent lasing. This is achieved by limiting the bias current to just below threshold, and the two ends of the semiconductor crystal are cleaved at an angle. The two cleaved facets act as partially reflective end mirrors. The natural reflectivity of the facets is approximately 32%, which can be enhanced by the deposition of antireflection coatings on the facets.[14] When an optical input signal is applied to the active region of the FP amplifier (FPA), it goes back and forth between the facets and is amplified during this process until it is emitted at a high intensity. The FPA is easy to fabricate, but as a resonant device its gain suffers from variations in temperature and it requires good temperature stability and well-controlled bias current. FPAs with low reflectivity of 3% have a gain of 30 dB and 3 dB bandwidth of 10 GHz for a 250 μm-long device.[15]

Another type of SOA is available. It is known as the traveling-wave amplifier (TWA), which is nonresonant and has the same structure of the FPA except that the end facets are coated or cleaved so that reflectivity is around 10^{-4} and internal reflection does not occur. Thus, the input optical signal gets amplified once during its single passage through the TWA.

TWAs are favored over the FPAs because they have low noise, high output saturation power (defined as the output power for which the gain is reduced by 3 dB), low polarization sensitivity, and a bandwidth of more than 5 THz (e.g., 40 nm in the 1550 nm window),[16] which is about 3 orders of magnitude greater than FPA. To get a high-gain TWA, the following conditions have to be satisfied:[15]

1. The design structure of the device must provide large small-signal gain.
2. A very low reflectivity antireflection coating should be used.
3. Gain saturation by amplified spontaneous emission (ASE) should be avoided. In most of the published work on the use of SOA as a WC, TWA is implied.

Population inversion in the TWA SOA is obtained by applying a bias current. When an input optical power signal is applied to the active region of the SOA, it will be amplified and the single-pass gain over a length L of the amplifier is given as:[14]

$$G = \exp\,[\Gamma\,(g_m - \alpha)L] = \exp\,g(z)\,L$$

where Γ = optical confinement factor in the cavity, g_m = material gain coefficient, α = effective absorption coefficient, and $g(z)$ = overall gain per unit length.

Typical gain of a TWA is around 25 dB with reflectivity of 10^{-4}.

21.3 Interferometers

An interferometer is an optical instrument that splits a wave in two. It delays them by unequal distances, redirects and recombines them, and detects the intensity of their superposition. Three types of interferometers are used in wavelength conversion: Mach–Zehnder, Michelson, and nonlinear-optical loop mirror (NOLM).

21.3.1 Mach–Zehnder interferometer (MZI)

A schematic diagram of a planar Mach–Zehnder interferometer is given in Figure 21.2a. The optical input signal is divided between the two branches of the interferometer, and after traveling through the two branches they recombine again at the output. If the two branches are symmetrical, no phase difference occurs between the two guided waves and they recombine constructively. If a phase difference occurs between the signals in arm 1 and arm 2, however, the output optical power will depend on the phase difference between them. They will cancel each other at the output if the phase difference between them is 180°. Many methods are used to change the phase difference between the two arms, such as using an interferometer with asymmetrical arms or changing the index of refraction on one of the arms, thus changing the optical path. The Michelson interferometer (MI) in Figure 21.2b has reflective facets which make it look like a folded version of MZI, and it works similarly.

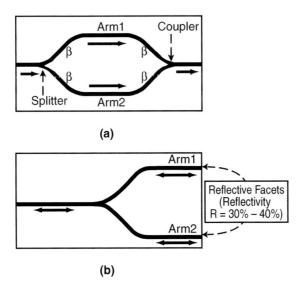

(a)

(b)

Figure 21.2 (a) Mach–Zehnder interferometer; (b) Michelson interferometer.

21.3.2 Nonlinear-optical loop mirror (NOLM)[11,17]

The operation of the NOLM interferometer resembles that of the classical Sagnac interferometer in Figure 21.3, where a beam of light is split by a beam splitter M into two beams. These beams are directed around a set of mirrors ($M_1 - M_3$) in opposite directions, and travel equal paths and arrive at the detector simultaneously and in phase. If the entire device, including source and detector, is rotating, then the beam traveling around the loop in the direction of the rotation will have to go further than that traveling counter to the rotation. Thus, the two beams will reach the detector at different times and will not be in phase, producing interference fringes that can be measured. The NOLM is a fiber Sagnac interferometer with an optical fiber as the nonlinear medium and two 3 dB couplers as illustrated in Figure 21.4.

A probe beam is applied to the 3 dB coupler 1, which splits it into two equal parts; one propagates clockwise and the other counterclockwise. In the absence of nonlinear interaction, they acquire identical phase shift. They interfere constructively at coupler 1 and exit back through input of coupler

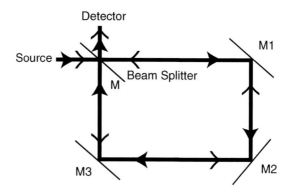

Figure 21.3 Classical sagnac interferometer.

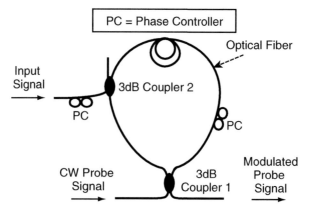

Figure 21.4 Nonlinear-optical loop mirror using optical fiber.

1. Thus, the output port at coupler 1 sees no probe beam. An input signal is coupled by coupler 2, which co-propagates with one of the two probe beams and modulates the optical index of the fiber, thus shifting its phase. When the two probe beams interfere at coupler 1, the output port sees the probe beam. A length of 1 to 10 km of fiber is required to get enough nonlinearity.

Despite the above, nonlinearity can be enhanced, and thus reduce the length of fiber through insertion of SOA in the loop, or by the use of highly nonlinear dispersion shifted fiber (HNL-DSF).[18–20]

21.4 Optoelectronic wavelength converters

Most of the current research is directed toward all-optical WCs. However, for completeness, optoelectronic WC will be briefly discussed. The simplest form of optoelectronic wavelength converter uses a detector, which converts the signal from the optical domain to the electrical domain. The signal is then used to drive a source at the desired wavelength. Prior to commercial applications of the optical fiber amplifiers, regenerators that consisted of three major components, namely optical receiver, radio frequency (RF) amplifier, and optical transmitter, were typically placed a distance of 40 km in an optical link. Regenerators are classified as 2R if they amplify and reshape and 3R if they amplify, reshape, and retime; the optical amplifier is classified as 1R (amplify only). A regenerator handles one wavelength only and needs a great deal of maintenance. Thus, their use in multiwavelength links will yield a high cost, low reliability, and poor upgradability, and they are used only where absolutely necessary.

In order to have a fair comparison, one should look at the optoelectronic conversion without regeneration. The optoelectronic wavelength converters have limited transparency where information in the form of phase, frequency, and amplitude is lost during the conversion process but have high degree of transparency for the digital signals up to a certain bit rate.[11] The RF amplifier used in the optoelectronic WC increases the conversion efficiency (defined as the ratio between the converted signal and input signal) at the expense of signal-to-noise ratio (SNR) and the noise figure (NF). NF for EDFAs is around 3.1 dB, whereas those for RF amplifiers and SOAs are in the range of 7 to 8 dB. This leads to the fact that using EDFA with SNR of 4 to 5 dB, along with all-optical WC with no excess noise, should outperform optoelectronic WC in many respects; however, optoelectronic WCs have good conversion efficiencies, and their technology is mature. Yet they are complex and have higher power consumption, especially when operating at 10Gb/s or above.[11]

21.5 Optical gating wavelength converters

This category of all-optical wavelength converters has received much attention in the past few years. Wavelength conversion is achieved through a device that transfers information from one wavelength to another without entering the electrical domain. This category includes many wavelength conversion methods due to the availability of various optical gating mechanisms.

The SOA is the main device of this category. It has been investigated by various research groups and used in cross-gain modulation (XGM)[9,11–13,21–29] and cross-phase modulation (XPM).[9,11–13, 7–39] Gain-suppression in semiconductor lasers is also utilized to get wavelength conversion, which includes distributed feedback (DFB) lasers,[40] T-gate lasers,[41] and Y-lasers.[42] Other devices include semiconductor lasers with saturable absorption[43,44] and nonlinear-optical loop mirrors (NOLM).[17,19]

21.5.1 Cross-gain in semiconductor optical amplifiers

The rate of stimulated emission in SOA depends on the optical input power. When high optical input power gets into the active region of the SOA, the carrier concentration is depleted through stimulated emission such that its gain is reduced. This is known as gain saturation and typically occurs for input powers of 100 μw or more. Gain saturation can be used to convert data from one wavelength to another.[12] Figure 21.5a is an SOA with two optical signals at the input of the SOA. One is an intensity-modulated signal with wavelength λs, and the other is a continuous wave (CW) probe signal at the target wavelength λc. If the peak optical power of the modulated signal is near the saturation power of the SOA, the gain will be modulated. Thus, the gain modulation is imposed on the CW input probe signal.

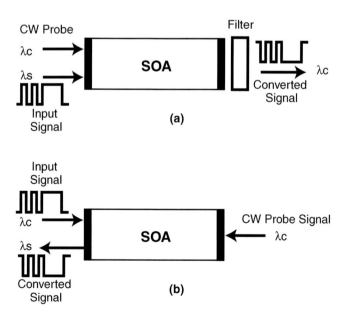

Figure 21.5 XGM wavelength conversion principle: (a) co-propagating signals; (b) counter-propagating signals.

When the information signal is 1 (high), the probe signal receives no amplification, and is amplified when the input signal is zero (low). Thus, transfer of information from λs to λc has occurred except for the inversion of the bit stream as depicted in Figures 21.5a and 21.5b. The information signal and the CW probe signals can be launched either co-propagating as depicted in Figure 21.5a or counter-propagating as depicted in Figure 21.5b. In the co-propagating mode, an optical band pass filter is required to isolate the target wavelength λc. In the counter-propagating mode the filter is not required, provided the antireflection coating of the facet is high and could suffer from speed limitation because input and probe signals are not traveling together.

In XGM, the speed of wavelength conversion using SOA gain saturation was thought to be limited by the intrinsic carrier lifetime of around 0.5 ns; however, the speed of such devices, as discussed in the literature, is more than what this lifetime implies. It was found that high power injection increases the stimulated emission and thus reduces the lifetime to around 10ps, allowing bit rates of 100Gb/s with a small bit ratio penalty to be achieved.[12] Theoretical and experimental research continued on finding ways and means of having better performance out of the SOAs, particularly modulation bandwidth. Parameters that influence the SOA bandwidth were identified as:[13,29]

1. Large injection current
2. High optical input power
3. Large confinement factor
4. Large differential gain

With these parameters in mind, experimental work was pursued to get high-speed converters. The dimensions of the SOA on the performance of the converter are important, as long cavity enables the use of high injection current as the maximum current density is 25 to 30 KA/cm².[29] Two SOA converters taken from the same water but with cavity lengths of 450 μm and 1200 μm were tested with bias currents of 150 and 200mA, respectively. The signal and CW probe powers were –5 and –8dBm, and –3 and –13dBm for the short and long SOAs, respectively.[13] The long SOA had better conversion performance than the short one. The long SOA had no penality at 10Gb/s; the short SOA had a penality of 2.7 dB; however, the optical bandwidth of the long amplifier was 30 nm compared to 60 nm for the short one.

Joergensen et al. reported on experiments using identical SOAs with cavity lengths of 450, 800, and 1200 μm.[29] They were operated with the same injection current and gave modulation bandwidth around 17 GHz regardless of cavity length. When an injection current density of 25 kA/cm² was applied, however, modulation bandwidths of 7, 17, and 20 GHz were obtained for the 450, 800, and 1200 μm long SOAs, respectively. As can be seen, the modulation bandwidth increases with cavity length, and a bandwidth near 90 GHz is predicted for a 3600 μm long SOA; however, the optical bandwidth decreases from 50 nm for the 1200 μm to 25 nm for the 3600 μm long SOA. Though high bandwidth is obtained from the long SOA, it has three drawbacks:

1. It has large transit time which could limit the bandwidth, and causes jitter in the converted signal
2. It is difficult to fabricate a long active region with uniform crosssections
3. The optical bandwidth decreases as the cavity increases

The third parameter to be considered in achieving a high bandwidth is through the adjustment of the confinement factor. Increasing the confinement factor from 0.2 to 0.8 led to the increase of modulation bandwidth from 5 to 45 GHz, but at the expense of having smaller optical bandwidth. The last parameter that will lead to higher bandwidth is through the use of high differential gain.

The advantages of XGM WCs include simplicity, high conversion efficiency, insensitivity to input wavelength, and polarization independence (provided the SOA gain is polarization independent).[12] The drawbacks, however, include bit stream inversion, relatively high chirp, and extinction ratio degradation, which is about 7 dB when converting from shorter to longer wavelengths and around 9 dB when converting from longer to shorter wavelengths. This asymmetry limits the cascadability of such devices. Other drawbacks include low SNR and phase modulation of the target wave due to change in carrier density and index of refraction. This problem limits data rate transmission over fiber and leads to cross-phase modulation (XPM).

21.5.2 Cross-phase modulation in semiconductor optical amplifiers

In the SOA, the index of refraction in the active region depends on the carrier density. The input optical signal depletes the carrier concentration in the active region through stimulated emission, and thus modulates the carrier density and the index of refraction. The modulation of the index of refraction causes phase modulation of the probe signal with wavelength λc coupled into the converter. The phase-modulated signal can then be changed into intensity-modulated signal through the use of interferometers such as the MZI[13,29–32,36–39] or Michelson interferometer.[13, 34, 35]

In the XPM, a phase change of 180° needed to operate the interferometer is obtained for a gain variation of 4 dB, as compared with 10 dB for the XGM.[13] The phase required to achieve interferometric switching can be obtained by using different structures of MZI or MI. Figure 21.6a is an asymmetric MZI where the arms couplers have different splitting ratios. When an input information signal and CW probe signal are applied to the asymmetric configuration, power is split according to the arms-splitting ratio. Consequently, the power through each SOA leads to a phase difference between the two arms due to different saturations of SOA1 and SOA2.

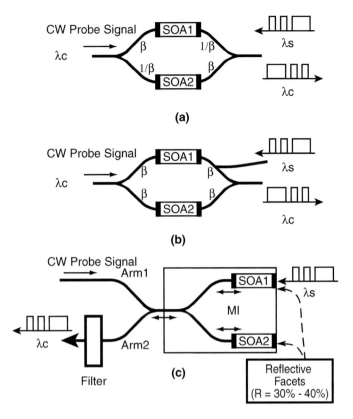

Figure 21.6 Configurations for interferometric SOA wavelength converter: (a) an asymmetric Mach–Zehnder interferometer; (b) a symmetric Mach–Zehnder interferometer; (c) Michelson interferometer.

In Figure 21.6b, a symmetric MZI is used. The input signal at wavelength λs is fed to SOA1, which only creates an asymmetric saturation because SOA2 is not affected by this signal. A CW probe signal at the target wavelength λc, which is injected as illustrated in Figure 21.6b, will be subject to phase modulation by the intensity modulated input signal at wavelength λs. In other words, for a logic 1 (high) of the applied signal, the probe signal undergoes constructive interference, thus getting 1 (high) probe signal; for logic zero (low) it undergoes destructive interference, thus getting 0 (low) probe signal. Thus, a converted output signal is obtained without inversion.

Another possible arrangement using MZI is to place one SOA in one of the arms to get the required phase shift; however, this arrangement is polarization-sensitive and gives less power.[13] Figure 21.6c is a Michelson interferometer with SOAs inserted into its arms. At phase difference of 180° between the two arms, the CW probe signal at λc is coupled into arm1 and will be reflected into this arm. If the information signal at wavelength λs is coupled to the SOA1, additional phase difference of 180° between arm1 and arm2 will occur due to change in the index of refraction caused by carrier depletion. As a result, the CW probe signal will be reflected into arm2. Then, the CW probe signal at λc will be modulated by the information of the input signal λs. An optical band-pass filter is used to isolate the target wavelength λc. It should be noted that the converted signal using an MI SOA is noninverted, and this configuration is insensitive to polarization of the incoming input signal if the gain of the SOA is polarization insensitive.[34]

Experiments at 40Gb/s using an MI wavelength converter demonstrated high-quality converted signals with greater than 10 dB extinction ratios and greater than 25 dB optical SNR (OSNR) in (1 nm). Theoretical studies have demonstrated that an MI converter is generally faster than an MZI converter due to the reflection facet in the MI configuration that allows the converted signal to transverse the SOA twice, and moreover allows for direct coupling of the input signal to the amplifier sections.[29]

Other interferometric WCs have been reported in the literature. Nakamura et al. reported on error-free, all-optical wavelength conversion at 168Gb/s using a Symmetric-Mach–Zehnder (SMZ)-type switch.[36] Wolfson et al. used an SOA-based, all-active Mach–Zehnder interferometer, which is an all-optical device with 2R and 3R regeneration.[37] When employed as a wavelength converter, it improved the optical signal-to-noise ratio (OSNR) by more than 20 dB at 40Gb/s. Simultaneous all-optical wavelength conversion and demultiplexing from 40 to 10Gb/s were demonstrated. Based on these features, it lends itself to be used in a combined optical time division multiplexing (OTDM) and WDM network.

An all-optical Mach–Zehnder wavelength converter with monolithically integrated DFB probe source was reported by Spiekman et al.[38] The device is inherently stable, provides low-loss coupling, and enables injection of large optical power into the SOA. Its drawback is in the sensitivity of the DFB laser to feedback. Input power at -5.4dBm was converted with penalty of 0.6 dB and at a bit rate of 2.5Gb/s. In 1998, Spiekman et al. presented a modified version of the device reported by Speikman et al. in 1997,[38] using preamplifiers to replace the DFB. Penalty-free conversion at 2.5Gb/s was obtained for power range of -6 to +4dBm.

In conclusion, XPM SOA converters have many advantages because of partial regeneration, relatively high conversion efficiency, extinction ratio enhancement, low chirp characteristics, high bit rate capabilities, quality of the converted signal, and availability of integrated versions of this WC. Their drawbacks lie in their restriction to amplitude modulation format, complex control of bias current due to sharp transfer characteristics, and limited transparency.

21.5.3 Semiconductor with saturable absorption wavelength converter

Wavelength converters using absorber saturation in FP SOA[43] and distributed feedback (DFB) SOA[44] were reported. In the FP SOA, the information signal at λs = 1.310 μm is applied and saturates an absorber section. It changes the gain experienced by a probe beam at $\lambda c = 1.55$ μm and data are thus converted from $\lambda s = 1.31$ μm to $\lambda c = 1.55$ μm. In the DFB SOA, a commercial DFB laser is driven below lasing threshold. Input information signal at $\lambda s = 1.31$ μm applied to the DFB SOA is absorbed and generates carriers in the active region, thus increasing its gain and decreasing its index of refraction. Now applying the probe signal, at $\lambda c = 1.55$ μm, data will be transferred from 1.31 to 1.55 μm due to optical gain pumping.

21.5.4 Nonlinear optical loop mirror (NOLM) wavelength converter

The NOLM WC operation principle was explained in Section 21.3.2. It is a fiber version of the classical Sagnac interferometer. Figure 21.7a presents this interferometer, where a probe signal at wavelength λc is coupled via a 3 dB directional coupler 1. Half of the signal will propagate in a clockwise direction and the other half will propagate in a counterclockwise direction. In the absence of nonlinear interaction, the output sees no probe signal. The information signal of wavelength λs is coupled via coupler 2 and propagates counterclockwise. This signal modulates the index of refraction of the fiber due to Kerr effect, which causes the phase of the probe signal propagating in

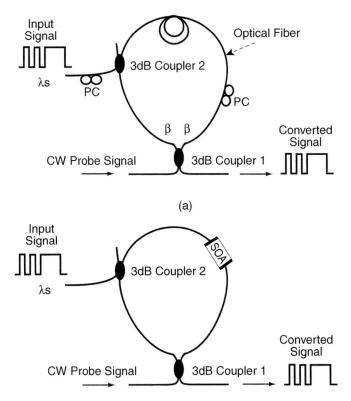

Figure 21.7 NOLM wavelength converters: (a) fiber loop; (b) SOA as the nonlinear medium.

the counterclockwise direction to increase relative to the probe signal propagating in the clockwise direction. This asymmetry causes the probe signal to appear at the output of coupler 1. Due to the weak nonlinearity, a long fiber of 1 km or more is required to get enough interaction. Due to finite propagation time, the probe signal has to be synchronized with the input signal.

Nonlinearity can effectively be increased through insertion of SOA in the loop as illustrated in Figure 21.7b. The SOA is placed asymmetrically with respect to the loop center. Coupler 1 splits a probe signal at λc into two equal signals; one propagates clockwise and the other propagates counterclockwise. An information signal at λs is coupled to the loop through coupler 2, saturates the SOA gain, and induces a change in the index of refraction. The two probe signals are subject to different phase shifts because one of them reaches the SOA before saturation and the other reaches the SOA after saturation. When the two signals arrive at coupler 1, they interfere and generate an output pulse at the output of coupler 1; thus conversion is achieved. This device is known as terahertz optical asymmetric demultiplexer (TOAD).[45] This device was operated as a 40Gb/s wavelength converter and a simultaneous 4:1 demultiplexer.[12]

Optical nonlinearity can also be enhanced through increased concentration of germanium in the core of a dispersion shifted fiber (DSF), thus getting highly nonlinear dispersion shifted fiber (HNL-DSF).[20] The high nonlinear coefficient of the HNL-DSF makes the fiber much shorter and thus more compact than the other designs. A 50m long HNL-DSF was constructed and demonstration of a 26 nm wavelength conversion of 500fs pulse trains with peak power of 4W was reported.[19] NOLM WCs have less noise because the wavelength conversion efficiency is a nonlinear function of the input signal power. The NOLM requires short pulses compared with the bit period, and suffers from stability problems.

21.6 Wavelength mixing wavelength converters

Certain materials change their properties when subjected to electric fields. The parameter that is influenced is the index of refraction. The index of refraction change due to the applied electric field is called Pockels effect, and the index of refraction change due to the square of the applied electric field is called Kerr effect. Thus, applying light with certain intensity will cause nonlinearity in the optical medium due to Kerr effect. Four-wave mixing (FWM) is a type of Kerr effect that occurs when light of two or more different wavelengths is injected into the nonlinear optical material. It generates a new wave with an intensity that is proportional to the product of the interacting waves' intensities. The phase and frequency of the generated wave are a linear combination of the interacting waves. Thus amplitude, phase, and frequency of information are preserved, which means that strict transparency is offered by this type of wavelength converter. It also offers conversion of a set of multiple wavelengths to another set of output wavelengths and can accommodate bit rates exceeding 100Gb/s.[11] Two kinds of wave mixing are used: three-wave mixing (two input

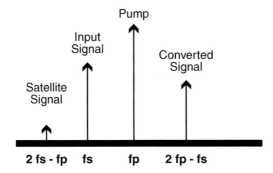

Figure 21.8 Schematic of four-wave mixing in the frequency domain.

waves and one output wave) due to the second-order electric susceptibility, and four-wave-mixing (three input waves and one output wave) due to the third-order electric susceptibility (Kerr effect).

Figure 21.8 is a representation of wave mixing in the frequency domain. Two optical signals are the pump wave, which comes from a high power CW source, and one that contains the information to be converted. If the pump signal at frequency f_p and the information signal at frequency f_s are present at the nonlinear medium, they cause grating to be formed. The formation of the grating causes the pump wave to be scattered and the frequency of the resulting signal is $2f_p$-f_s, which is called the converted signal. Scattering of the signal wave by the grating gives a signal wave with frequency $2f_s$-f_p, which is called the satellite signal. These new frequency components are formed on each side of the injected signals and spaced by the beat frequency as given in the figure. The intensity ratio of the converted signal to the satellite signal is 20 dB.[11] Filtering of the satellite signal is necessary to avoid cross talk. Different techniques of wave mixing are explored in the following sections.

21.6.1 Four-wave mixing in semiconductor optical amplifiers

Four-wave mixing in semiconductor optical amplifiers (FWM SOA) has received much attention by many research groups.[12,46–60] Different parameters of this technique are explored theoretically and experimentally. FWM SOA WCs offer strict format transparency, arbitrary wavelength mapping, high bit rate operation, and simple implementation; but these advantages are offset by polarization dependence, filtering of spurious wavelengths, and not very high conversion efficiency. Four-wave mixing in an active medium, such as an SOA, allows relatively efficient conversion due to increased optical intensities and enhanced nonlinearities. When two beams are injected, a grating is formed due to the following mechanisms:[11]

1. Carrier density modulation
2. Dynamic carrier heating
3. Spectral hole burning

Contribution from all these mechanisms could lead to FWM; however, the most dominant cause is due to carrier density modulation. When two signals are injected into an SOA, they beat and cause carrier density modulation with an envelope that is the same as that of the optical intensity.[12]

FWM efficiency using SOA increases as the square of the device length and the SNR of the converted signal increases in proportion to the device length. A limit to increasing SOA length is due to thermal effects at high injection current or to phase matching conditions no longer satisfied although phase matching condition for short SOA is not a problem.

To overcome some of the drawbacks such as polarization dependence, different polarization schemes are utilized. Lin et al. reported on polarization-insensitive wavelength conversion of up to 10Gb/s, for a conversion range from 6.4 nm wavelength downshift to 4.8 nm upshift. Mak et al. reported on constant conversion efficiency (less than 3 dB variation) over 36 nm interval with less than 0.34 dB polarization sensitivity. The power penalty at BER of 10^{-9} for 10Gb/s is less than 0.9. Results of bit error rate (BER) measurements show that there is tradeoff between maximizing OSNR and minimizing intersymbol interference (ISI).[60]

21.6.2 Four-wave mixing in optical fibers

Nonlinearity in optical fiber, due to Kerr effect, is utilized in wavelength conversion. Optical fibers exhibit FWM, but due to the weak nonlinearity in the fiber, lengths from 1 to 10 km are required, which make the FWM WC bulky. Enhanced nonlinearity is achieved through doping the core of the fiber with large amounts of germanium, achieving effective linearity approximately 50 times larger than that of conventional dispersion shifted fiber.[20] A 500m HNL-DSF is used to build a converter with relatively high conversion efficiency with conversion bandwidth of 20 nm. Aso et al. reported on using a 200m HNL-DSF reaching a conversion bandwidth of 23.3 nm. By FWM in polarization maintaining (PM) HNL-DSF, wavelength conversion width over 70 nm was achieved, and a simultaneous wavelength conversion of 32 x 10Gb/s WDM signals between C-band and L-band was demonstrated. Signal waves start at $\lambda s_1 = 1535.8$ nm and end at $\lambda s_{32} = 1560.7$ nm with a spacing of 0.8 nm. The converted signals start at $\lambda c_1 = 1599.4$ nm and end at $\lambda c_{32} = 1573.4$ nm.[61]

21.6.3 Difference-frequency generation

Nonlinear interaction of two optical waves, a pump wave, and a signal wave are used to get difference-frequency generation in periodically domain-inverted waveguides, such as a lithium niobate ($LiNbO_3$) channel waveguide. PFG conversion efficiency of 40% w^{-1} cm^{-2} was achieved and found to be almost consistant over a wavelength range of 30 nm.[62] DFG is also obtained using AlGaAs,[63] which was used to simultaneously convert eight input wavelengths in the 1546-to-1560 nm region to a set of eight outputs wavelengths in the 1524 to 1538 nm region. DFG offers bandwidth exceeding 90 nm, strict transparency, polarization-independent conversion efficiency, and spectrum inversion capabilities. Its drawback resides in the difficulty of fabricating low-loss waveguide to achieve high conversion efficiency.

21.7 Conclusion

Wavelength converters are necessary components for future networks, particularly large ones with dynamic traffic as they convert data from wavelength channel to another which resolves wavelength contention, increases flexibility and capacity for a fixed set of wavelengths, and allows decentralized network management with maximum degrees of transparency and interoperability. Efforts to achieve those attributes were intensified among various research groups. Different types of wavelength conversion techniques were investigated.

The optoelectronic WCs are readily available and their technology is mature, but they have limited transparency. Information in the form of phase, frequency, and amplitude is lost during the conversion process and such WCs need high plug power, especially at high data rate. The rest of the WCs discussed are considered all-optical, which enables translation of information from one wavelength to another without entering the electrical domain. XGM WCs are attractive because of their simplicity, high conversion efficiency, polarization independence, and insensitivity to input wavelength. Their drawbacks are in the bitstream inversion, limitation to amplitude modulated signals, and degraded extinction ratio. XPM converters can be used to get high extinction ratio, low chirp, partial regeneration, an inverted or noninverted bitstream, and polarization and wavelength independence, but require good control of bias current. Integrated all-optical MZI and MI converters have demonstrated good performance, which makes them good candidates for wavelength conversion in photonic switch fabrics.

The NOLM WCs also promise to be good candidates for all-optical WCs, particularly when HNL-DSF is used. Wave mixing has received much attention because of strict transparency and high bit rate. FWM in SOA is attractive because of its transparency to modulation format and high bit rate capabilities, but it has the problem of low conversion efficiency that decreases with increasing conversion span. FWM WCs were made using PM-HNL-DSF which allowed simultaneous conversion of 32 wavelengths; conversion of 8 wavelengths was also achieved using DFG WC. A pertinent question remains unanswered: which WC technology is best? The answer depends on many factors such as system requirements, reliability, and cost. In order to have a seamless network, however, it is essential to have WCs with strict transparency that leaves the door open for the wave-mixing techniques discussed in this chapter.

Glossary

ASE	amplified spontaneous emission
BER	bit error rate
CW	continuous wave
DFB	distributed feedback
DFG	difference frequency generation
DSF	dispersion shifted fiber
DWDM	dense wavelength division multiplexing
EDFA	Erbium doped fiber amplifier
FPA	Fabry–Perot amplifier
FWM	four-wave mixing
HNL-DSF	highly nonlinear dispersion shifted fiber
ISI	inter-symbol interference
MI	Michelson interferometer
MZI	Mach–Zehnder interferometer
NF	noise figure
NOLM	nonlinear-optical loop mirror
OSNR	optical signal-to-noise ratio
RF	radio frequency
SMZ	symmetric Mach–Zehnder
SNR	signal-to-noise ratio
SOA	semiconductor optical amplifier
TOAD	terahertz optical asymmetric demultiplexer
TWA	traveling wave amplifier
WC	wavelength converter
XGM	cross-gain modulation
XPM	cross-phase modulation

References

1. Alferness, R.C. et al., A practical vision for optical transport networking, *Bell Labs Tech. J.*, 4, 3, 1999.
2. Yano, Y. et al., 2.6Tbit/s WDM transmission experiment using optical duobinary coding, *ECOC '96*, paper ThB3.1.
3. Kawanishi, S. et al., 3Tbit/s (160Gbit/s x 19 channel) transmission by optical TDM and WDM, *IEEE LEOS News*, Oct. 1999, pp. 7, 8.
4. Charplyvy, A.R., High capacity lightwave transmission experiments, *Bell Labs Tech. J.*, 4, 230, 1999.
5. Bigo, S. and Idler, W., Multi-terabit/s transmission over Alcatel teralight™ fiber, *Alcated Telecom. Rev.*, 4, 288, 2000.
6. Ito, T. et al., 6.4Tbit/s (160 x 40Gbit/s) WDM transmission experiment with 0.8 bit/s/Hz spectral efficiency, *ECOC 2000*, Postdeadline paper 1.1.
7. Kartalopoulous, S.V., Introduction to DWDM technology-data in a rainbow, IEEE Press, NJ, 2000, ch. 14.
8. Bracket, C.A., Dense wavelength division multiplexing networks: Principles and applications, *IEEE, J. Select Areas Comm.*, 8, 948, 1990.

9. Elmirghani, J.M.H. and Mouftah, H.T., All-optical wavelength conversion: technologies and applications in DWDM networks, *IEEE Commun. Mag.,* 38, 86, March 2000.

10. Saleh, A.A.M., Transport optical networks for the next generation information infrastructure, *OFC '95,* paper THE1.

11. Yoo, S.J.B., Wavelength conversion technologies for WDM network applications, *IEEE/OSA J. Lightwave Technol.,* 14, 955, 1996.

12. Nesset, D., Kelly, T., and Marcenac, D., All-optical wavelength conversion using SOA nonlinearity, *IEEE Commun. Mag.,* 36, 56, Dec. 1998.

13. Durhuus, T. et al., All-optical wavelength conversion by semiconductor optical amplifiers, *IEEE/OSA J. Ligthwave Technol.,* 14, 942, 1996.

14. Keiser, G., *Optical Fiber Communication,* 3rd ed., McGraw-Hill International Edition, Singapore, 2000, ch. 11.

15. Simon, J.C., GaInAsP semiconductor laser amplifiers for single mode fiber communications, *IEEE/OSA J. Lightwave Technol.,* 5, 1287, 1987.

16. Bernard, J.J. and Renaud, M., Semiconductor optical amplifiers, *SPIE OE Mag.,* 1, 36, Sept. 2001.

17. Rauschenbach, K.A. et al., All-optical pulse width and wavelength conversion at 10Gb/s using a nonlinear optical loop mirror, *IEEE Photon. Technol. Lett.,* 6, 1130, 1994.

18. Aso, O., Tadakuma, M., and Namiki, S., Four-wave mixing in optical fibers and its applications, *Furukawa Rev.,* 19, 63, 2000.

19. Sakamoto, T. et al., All-optical wavelength conversion of 500-fs pulse trains by using a nonlinear-optical loop mirror composed of a highly nonlinear DSF, *IEEE Photon. Technol. Lett.,* 13, 502, 2001.

20. Okuno, T. et al., Silica-based functional fibers with enhanced nonlinearity and their applications, *IEEE J. Selected Topics Quantum Electr.,* 5, 1385, 1999.

21. Wiesenfeld, J.M. et al., Wavelength conversion at 10Gb/s using a semiconductor optical amplifier, *IEEE Photon. Technol. Lett.,* 5, 1300, 1993.

22. Freeman, P.N., Dutta, N.K., and Lopata, J., Semiconductor optical amplifier for wavelength conversion in subcarrier multiplexed systems, *IEEE Photon. Technol. Lett.,* 9, 46, 1997.

23. Vaughn, M.D. and Blumenthal, D.J., All-optical updating of subscriber encoded packet headers with simultaneous wavelength conversion of baseband payload in semiconductor optical amplifiers, *IEEE Photon. Technol. Lett.,* 9, 827, 1997.

24. Norte, D. and Willner, A.E., Experimental demonstration of a multiple-λ wavelength shifter for dynamically reconfigurable WDM networks, *IEEE Photon. Technol. Lett.,* 9, 922, 1997.

25. Deming, L., Hong, N.J., and Chao, L., Wavelength conversion based on cross-gain modulation of ASE spectrum of SOA, *IEEE Photon. Technol. Lett.,* 12, 1222, 2000.

26. Jeon, M.Y. et al., All optical wavelength conversion for 20-Gb/s RZ format data, *IEEE Photon. Technol. Lett.,* 12, 1528, 2000.

27. Stephens, M.F.C. et al., Demonstration of ultra-fast all-optical wavelength conversion utilizing birefringence in semiconductor optical amplifier, *IEEE Photon. Technol. Lett.,* 9, 449, 1997.

28. Asghari, M., White, I.H., and Penty, R.V., Wavelength conversion using semiconductor optical amplifiers, *IEEE/OSA J. Lightwave Technol.,* 15, 1181, 1997.

29. Joergensen, C. et al., All-optical wavelength conversion at bit rates above 10Gb/s using semiconductor optical amplifiers, *IEEE Selected. Topics Quantum Electr.*, 3, 1168, 1997.

30. Durhuus, T. et al., All optical wavelength conversion by SOAs in a Mach-Zehnder configuration, *IEEE Photon. Technol. Lett.*, 6, 53, 1994.

31. Idler, W. et al., 10Gb/s wavelength conversion with integrated multiquantum-well-based 3-port Mach–Zehnder interferometer, *IEEE Photon. Technol. Lett.*, 8, 1163, 1996.

32. Cho, P.S. et al., RZ wavelength conversion with reduced power penality using a semiconductor optical amplifier/fiber-grating hybrid device, *IEEE Photon. Technol. Lett.*, 10, 66, 1998.

33. Dagens, B. et al., Design optimization of all-active Mach–Zehnder wavelength converters, *IEEE Photon. Tech. Lett.*, 11, 424, 1999.

34. Mikkelsen, T. et al., Polarization-insensitive wavelength conversion of 10Gb/s signals with SOAs in Michelson interferometer, *Electron. Lett.*, 30, 261, 1994.

35. Sato, R. et al., Polarization-insensitive SOA–PLC hybrid integrated Michelson interferometric wavelength converter and its application to DWDM networks, *IEICE Trans. Commun.*, E84-B, 1197, 2001.

36. Nakamura, S., Ueno, Y., and Tajimak, K., 168Gb/s all-optical wavelength conversion with a Symmetric-Mach–Zehnder-type switch, *IEEE Photon. Technol. Lett.*, 13, 1091, 2001.

37. Wolfson, D. et al., 40-GHz all-optical wavelength conversion regeneration and demultiplexing in an SOA based all-active Mach–Zehnder interferometer, *IEEE Photon. Technol. Lett.*, 12, 332, 2000.

38. Spiekman, L.H. et al., All-optical Mach–Zehnder wavelength converter with monothically integrated PFB probe source, *IEEE Photon. Technol. Lett.*, 9, 1349, 1997.

39. Spiekman, L.H. et al., All-optical Mach–Zehnder wavelength converter with monothically integrated preamplifier, *IEEE Photon. Technol. Lett.*, 10, 1115, 1998.

40. Mikkelsen, B. et al., Wavelength conversion of high-speed data signals, *Electron. Lett.*, 29, 1716, 1993.

41. Lu, C.C. et al., Wavelength conversion using a T-gate laser, *IEEE Photon. Tech. Lett.*, 8, 52, 1996.

42. Schilling, M. et al., 6THz range tunable 2.5Gb/s frequency conversion with a MQW Y-laser, *IEEE J. Quantum Electron.*, 29, 1835,1993.

43. Barnsley, P.E. and Fiddyment, P.J., Wavelength conversion from 1.3 to 1.55 μm using split contact optical amplifiers, *IEEE Photon. Technol. Lett.*, 3, 256, 1991.

44. Maywar, D.N., Nakano, Y., and Agarwal G.P., 1.31- to 1.55μm wavelength conversion by optically pumping a distributed feedback amplifier, *IEEE Photon. Technol. Lett.*, 12, 858, 2000.

45. Sokoloff, J.P. et al., A terahertz optical asymmetric demultiplexer (TOAD), *IEEE Photon. Tech. Lett.*, 5, 787, 1993.

46. Zhou, J. et al., Four-wave mixing wavelength conversion efficiency in semiconductor traveling wave amplifiers measured to 65 nm of wavelength shift, *IEEE Photon. Tech. Lett.*, 5, 984, 1994.

47. Summerfield, M.A. and Tucker, R.S. Optimization of pump and signal powers for wavelength converters based on FWM in semiconductor optical amplifiers, *IEEE Photon. Technol. Lett.*, 8, 1316, 1996.

48. Geraghty, D.F. et al., Wavelength conversion up to 18 nm at 10Gb/s by four-wave mixing in a semiconductor optical amplifier, *IEEE Photon. Technol. Lett.*, 9, 452, 1997.

49. Hunziker, G. et al., Folded-path self-pumped wavelength converter based on four-wave mixing in a semiconductor optical amplifier, *IEEE Photon. Technol. Lett.*, 9, 1352, 1997.

50. Diez, S. et al., Four-wave mixing in semiconductor optical amplifiers for frequency conversion and fast optical switching, *IEEE J. Selected. Topics Quantum Electr.*, 3, 1131, 1997.

51. Geraghty, D.F. et al., Wavelength conversion for WDM communication systems using four-wave mixing in semiconductor optical amplifiers, *IEEE J,. Selected. Topics Quantum Electr.*, 3, 1146, 1997.

52. Scotti, S. et al., Effects of ultrafast processes on frequency converters based on four-wave mixing in semiconductor optical amplifiers, *IEEE J. Selected Topics Quantum Electr.*, 3, 1156, 1997.

53. Lin, L.Y. et al., Polarization-insensitive wavelength conversion up to 10Gb/s based on four-wave mixing in a semiconductor optical amplifier, *IEEE Photon. Technol. Lett.*, 10, 955, 1998.

54. D'Ottavi, A. et al., Wavelength conversion at 10Gb/s by four-wave mixing over a 30 nm interval, *IEEE Photon. Technol. Lett.*, 10, 952, 1998.

55. Morgan, T.J. et al., Widely tunable four-wave mixing in semiconductor optical amplifiers with constant conversion efficiency, *IEEE Photon. Technol. Lett.*, 10, 1401, 1998.

56. Tomkos, I. et al., Performance of reconfigurable wavelength converter based on dual-pump-wave mixing in a semiconductor optical amplifier, *IEEE Photon. Technol. Lett.*, 10, 1404, 1998.

57. Lacey, J.P.R., Summerfield, M.A., and Modden, S.J., Tunability of polarization-insensitive wavelength converters based on four-wave mixing in semiconductor optical amplifiers, *IEEE/OSA J. Lightwave Technol.*, 16, 2419, 1998.

58. Sotobayashi, H. and Kitayama, K., Observation of phase conservation in a pulse sequence at 10Gb/s in a semiconductor optical amplifier wavelength converter by four-wave mixing, *IEEE Photon. Technol. Lett.*, 11, 45, 1999.

59. Mak, W.K.M., Tsang, H.K., and Chan, K., Widely tunable polarization independent all-optical wavelength converter using a semiconductor optical amplifier, *IEEE Photon. Technol. Lett.*, 12, 525, 2000.

60. Lu, L. et al., Bit-error rate performance dependence on pump and signal powers of the wavelength converter based on FWM in semiconductor optical amplifiers, *IEEE Photon. Tech. Lett.*, 12, 855, 2000.

61. Watanabe, S. and Futami, F., All-optical signal processing using highly nonlinear optical fibers, *IEICE Trans. Commun.*, E84-B, 1179, 2001.

62. Xu, C.Q., Okayama, H., and Kawahara, M., 1.5 µm band efficient broadband wavelength conversion by difference frequency generation in a periodically domain-inverted $LiNbO_3$ channel waveguide, *Appl. Phys. Lett.*, 63, 3559, 1993.

63. Yoo, S.J.B. et al., Wavelength conversion by difference frequency generation in AlGaAs waveguide with periodic domain inversion achieved by wafer bonding, *Appl. Phys. Lett.*, 68, 2611, 1996.

Index

A

U

V

W

9780367395261

T - #0076 - 240420 - C0 - 234/156/22 - PB - 9780367395261